NUMERICAL AND ANALYTICAL METHODS WITH MATLAB®

CRC Series in
COMPUTATIONAL MECHANICS
and APPLIED ANALYSIS

Series Editor: J.N. Reddy
Texas A&M University

Published Titles

ADVANCED THERMODYNAMICS ENGINEERING
Kalyan Annamalai and Ishwar K. Puri

APPLIED FUNCTIONAL ANALYSIS
J. Tinsley Oden and Leszek F. Demkowicz

COMBUSTION SCIENCE AND ENGINEERING
Kalyan Annamalai and Ishwar K. Puri

EXACT SOLUTIONS FOR BUCKLING OF STRUCTURAL MEMBERS
C.M. Wang, C.Y. Wang, and J.N. Reddy

**THE FINITE ELEMENT METHOD IN HEAT TRANSFER AND FLUID DYNAMICS,
Second Edition**
J.N. Reddy and D.K. Gartling

**MECHANICS OF LAMINATED COMPOSITE PLATES AND SHELLS: THEORY
AND ANALYSIS, Second Edition**
J.N. Reddy

NUMERICAL AND ANALYTICAL METHODS WITH MATLAB®
William Bober, Chi-Tay Tsai, and Oren Masory

PRACTICAL ANALYSIS OF COMPOSITE LAMINATES
J.N. Reddy and Antonio Miravete

**SOLVING ORDINARY and PARTIAL BOUNDARY VALUE PROBLEMS in
SCIENCE and ENGINEERING**
Karel Rektorys

CRC Series in
COMPUTATIONAL MECHANICS and APPLIED ANALYSIS

NUMERICAL AND ANALYTICAL METHODS WITH MATLAB®

WILLIAM BOBER
CHI-TAY TSAI
OREN MASORY

CRC Press
Taylor & Francis Group
Boca Raton   London   New York

CRC Press is an imprint of the
Taylor & Francis Group, an **informa** business

CRC Press
Taylor & Francis Group
6000 Broken Sound Parkway NW, Suite 300
Boca Raton, FL 33487-2742

© 2009 by Taylor and Francis Group, LLC
CRC Press is an imprint of Taylor & Francis Group, an Informa business

No claim to original U.S. Government works

Printed in the United States of America on acid-free paper
10 9 8 7 6 5 4 3 2 1

International Standard Book Number: 978-1-4200-9356-8 (Hardback)

Library of Congress Cataloging-in-Publication Data

Bober, William.
    Numerical and analytical methods with MATLAB / authors, William Bober, Chi-Tay Tsai, Oren Masory.
        p. cm. -- (CRC series in computational mechanics and applied analysis)
    Includes bibliographical references and index.
    ISBN 978-1-4200-9356-8 (hardcover : alk. paper)
    1. Numerical analysis--Data processing. 2. MATLAB. I. Tsai, Chi-Tay. II. Masory, Oren. III. Title. IV. Series.

QA297.B565 2010
518.0285--dc22

2009018493

Visit the Taylor & Francis Web site at
http://www.taylorandfrancis.com

and the CRC Press Web site at
http://www.crcpress.com

Contents

Preface

I have been teaching computer applications in mechanical engineering (ME) at Florida Atlantic University (FAU) for many years. The Department of Mechanical Engineering at FAU offers two courses in computer applications in ME. The first course is usually taken in the student's sophomore year, while the second course is usually taken in the student's junior or senior year. Students entering FAU from the community colleges are given credit for the first course if they have completed a course in C or C++. Both computer classes are taught as a lecture–computer lab course. The MATLAB® software program is used in both courses. To familiarize students with engineering-type problems, approximately six to seven engineering-type projects are assigned during the semester. Students have, depending on the difficulty of the project, either one or two weeks to complete the project. Since I have *not* found a satisfactory MATLAB text for the type of course that I teach, I have written supplementary material manuals for these courses, which students are required to purchase from our department office. I believe that the best source for students to complete my assigned projects is my Supplementary Material Manual. As a result, I have converted and expanded this material into a textbook, which other schools may use. The textbook includes many of the projects that I have assigned to my classes over several years. In addition I have asked two of my colleagues to contribute to the textbook. Dr. T. C. Tsai has contributed a chapter on the finite element method using MATLAB ("MATLAB's Partial Differential Equation Toolbox") and Dr. Oren Masory has contributed a chapter on control systems using MATLAB ("MATLAB's Control System Toolbox").

The advantage of using the MATLAB software program over other software programs is that it contains built-in functions that numerically solve systems of linear equations, systems of ordinary differential equations, roots of transcendental equations, integrals, statistical problems, optimization problems, control systems problems, stress analysis problems using finite elements, and many other types of problems encountered in engineering. A student version of the MATLAB program is available at a reasonable cost. However, to students, these built-in functions are essentially black boxes. By combining a textbook on MATLAB with basic numerical and analytical analyses (although I am sure that MATLAB uses more sophisticated numerical techniques than are described in this text), the mystery of what these black boxes might contain is somewhat alleviated. The text contains many sample MATLAB programs (scripts) that should provide guidance to the student on completing the assigned projects. *I believe that the projects in this textbook are more like what a graduate engineer might see in the industry rather than the usual small problems that are contained in many textbooks on the subject.*

Furthermore, we believe that there is enough material in this textbook for two courses, especially if the courses are run as lecture–computer laboratory courses. The advantage of running these courses (especially the first course) as a lecture–laboratory course is that the instructor is in the computer laboratory to help the student debug his or her program. This includes the sample programs as well as the projects. The first course should be at a lower level, perhaps during the first or second semester of the sophomore year. Chapters 1 through 6 would be appropriate for this first course. These chapters include:

Chapter 1: Numerical Modeling for Engineering
Chapter 2: MATLAB Fundamentals
Chapter 3: Matrices
Chapter 4: Roots of Algebraic and Transcendental Equations
Chapter 5: Numerical Integration
Chapter 6: Numerical Integration of Ordinary Differential Equations

Chapters 7 through 14 are appropriate for a course at the senior or first year graduate levels. These chapters include:

Chapter 7: Simulink
Chapter 8: Curve Fitting
Chapter 9: Optimization
Chapter 10: Partial Differential Equations
Chapter 11: Iteration Method
Chapter 12: Laplace Transforms
Chapter 13: An Introduction to the Finite Element Method (requires the Partial Differential toolbox)
Chapter 14: Control Systems (requires the Control System toolbox)

All chapters, except for Chapters 1 and 12, contain projects. Chapters 3, 5, and 12 also contain several exercises. Projects differ from exercises in that the student is required to write a computer program in MATLAB that will produce tables or graphs, or both. Exercises only require the student to use pencil and paper and produce a single answer.

Schools offering a course in computer applications similar to FAU's Computer Applications in ME II, but not having offered an earlier course using

MATLAB, would have to include Chapter 2 as part of the course, since it covers MATLAB fundamentals.

The governing equations for most projects are derived either in the main body or in the project description itself, or in the Appendices. A chapter on Laplace Transforms has been included because the control systems chapter (Chapter 14) utilizes Laplace Transforms; this chapter also utilizes Simulink. An introduction to Simulink is covered in Chapter 7.

MATLAB is a registered trademark of The MathWorks, Inc. For product information, please contact:

The MathWorks, Inc.
3 Apple Hill Drive
Natick, MA 01760-2098 USA
Tel: 508 647 7000
Fax: 508-647-7001
E-mail: info@mathworks.com
Web: www.mathworks.com

Acknowledgments

The authors wish to thank Jonathan Plant of Taylor & Francis Group/CRC Press for his confidence and encouragement in writing this textbook. We also wish to thank Jennifer Ahringer and Linda Leggio for guiding us through the textbook submission process. Finally, we wish to express our deep gratitude to our wives for tolerating the many hours we spent on preparation of this manuscript, time which otherwise would have been devoted to our families.

Authors

William Bober, Ph.D., received his B.S. degree in civil engineering from City College of New York (CCNY), his M.S. degree in engineering science from Pratt Institute, and his Ph.D. in engineering science and aerospace engineering from Purdue University. At Purdue University he was on a Ford Foundation Fellowship and was assigned to teach one engineering course during each semester. After receiving his Ph.D., Dr. Bober went to work as an associate engineering physicist in the Applied Mechanics Department at Cornell Aeronautical Laboratory, in Buffalo, New York. After leaving Cornell Labs, he was employed as an associate professor in the Department of Mechanical Engineering at the Rochester Institute of Technology (RIT) for the following 12 years. After leaving RIT he was, and still is, employed at Florida Atlantic University in the Department of Mechanical Engineering. While at RIT, Dr. Bober was the principal author of a textbook on fluid mechanics (*Fluid Mechanics*, 1980), published by John Wiley & Sons. He has written several papers for the *International Journal of Mechanical Engineering Education* (IJMEE).

C. T. Tsai, Ph.D., is a professor of mechanical engineering at Florida Atlantic University. Before joining the FAU faculty in August 1990, he was a research scientist at the Air Force Institute of Technology, Wright Patterson Air Force Base, Ohio. He is the author and coauthor of more than 80 articles in national and international journals and conferences. Dr. Tsai is one of the first pioneers for modeling dislocation multiplication in semiconductor crystals and has made significant contributions for developing a numerical model to predict the quantity of dislocations generated in the semiconductor crystals grown from the melt. His computational model has been cited by the *Handbook of Crystal Growth* (Elsevier Science B.V., 1994) as the first numerical model for predicting the dislocation densities generated in the growing of semiconductor crystals.

Dr. Tsai received his Ph.D. in engineering mechanics from the University of Kentucky in 1995. He has guided five Ph.D. students and nine master's students to graduation. He was the recipient of the Presidential Research Development Award in 2001, NASA summer faculty fellowship in 1991, and Air Force summer faculty research fellowship in 1994, 1995, and 1997. He is listed in *Who's Who in the World*, *Who's Who in American Education*, and *Who's Who in Science and Engineering*.

Oren Masory, Ph.D., is currently serving as the Chairman of the Department of Mechanical Engineering at Florida Atlantic University. He earned his B.Sc., M.Sc., and Ph.D. from the Technion, Israel Institute of Technology, Haifa, Israel, in 1974, 1977, and 1980, respectively. During his academic career he has taught a variety of courses including mechanics, vibrations, controls, manufacturing methods, automatic assembly, industrial automation, computer control of manufacturing systems, robotics application, controls, mechatronics, and microelectromechanical systems (MEMS). Dr. Masory has been employed by Florida Atlantic University since 1988. He worked for Gould Inc. (1980–1983) on the development of a robotic assembly line for the assembly of nonstandard electrical components. He has also consulted and performed research for General Dynamics, Pratt & Whitney, Sensormatic Electronics Corporation, Motorola, and others. Dr. Masory's interests include robotics, automation, vehicle dynamics, and product liability.

Chapter 1

Numerical Modeling for Engineering

1.1 Computer Usage in Engineering

1.1.1 Importance of the Computer

Nearly all engineering firms today utilize the computer in one way or another. Therefore, if a student has a programming capability, he will be more valuable to the firm that hires him than someone without such a capability.

1.1.2 Computer Usage

In engineering, the computer is mainly used in

   a. Solving mathematical models of physical phenomena.
   b. Storing and reducing experimental data.
   c. Controlling machine operations.

The engineer's interest lies in

   a. Design of equipment.
   b. Evaluation of performance of equipment.
   c. Research and development.

These can be accomplished by

a. Full-scale experiments. May be prohibitively expensive.
b. Small-scale model experiments. Still very expensive and extrapolation is frequently questionable.
c. A mathematical model, which is the least expensive and fastest. It can give more detailed answers and different cases under different conditions can be run quickly. If there is confidence in a mathematical model it will be used in preference to an experiment.

1.2 The Mathematical Model

Physical phenomena are described by a set of governing equations. Numerical methods are frequently used to solve the set of governing equations. *Reason:* We don't have methods for obtaining closed-form solutions for many types of problems involving general geometric conditions.

Numerical methods invariably involve the computer. The computer performs arithmetic operations upon discrete numbers in a defined sequence of steps. The sequence of steps is defined in the program. A useful solution is obtained if

a. The mathematical model accurately represents the physical phenomena; that is, the model has the correct governing equations.
b. The numerical method is accurate. *Note:* If the governing equations aren't correct, the solution will be worthless regardless of the accuracy.
c. The numerical method is programmed correctly.

This book is mainly concerned with items (b) and (c).

1.3 Computer Programming

The advantages of using the computer include:

a. It can carry out many, many calculations in a fraction of a second.
b. To get the computer to carry out the calculations, one has to write a set of instructions.

The modern electronic digital computer consists of the following:

a. *Input unit*—provides data and instructions to the computer.
b. *Memory/storage unit*—in which data and instructions are stored.

c. *Arithmetic-logic unit*—which performs the arithmetic operations and provides the decision-making ability (or logic) to the computer.

d. *Control unit*—takes instructions from memory; interprets and executes the instructions.

e. *Output unit*—prints out results of the program or displays results on a screen.

The control unit + the arithmetic logic unit are considered the central processing unit (CPU).

1.4 Preparing a Computer Program

For problems of interest in this book, the digital computer is only capable of performing arithmetic, logical, and graphical operations. Therefore, arithmetic procedures must be developed for solving differential equations, evaluating integrals, determining roots of an equation, solving a system of linear equations, etc. The arithmetic procedure usually involves a set of algebraic equations. A computer solution for such problems involves developing a computer program that defines a step-by-step procedure for obtaining an answer to the problem of interest. The method of solution is called an *algorithm*.

1.5 Recommended Procedures for Writing a Program

a. Study the problem to be programmed.
b. List the algebraic equations to be used in the program.
c. Create a flow chart or outline.
d. Carry out a sample calculation by hand.
e. Write the program using the outline and the list of algebraic equations.

1.6 Building Blocks in Writing a Program

a. Arithmetic Statements
b. Input/Output Statements
c. Loop Statements (*for* loop and *while* loop)
d. Alternative Path Statements (*if* and *elseif*)
e. Functions (built in and self-written)

Chapter 2

MATLAB Fundamentals

2.1 Introduction

MATLAB® is a software program for numeric computation, data analysis, and graphics. One advantage that MATLAB has for engineers over software packages such as C/C++ or Fortran is that the MATLAB program includes functions that numerically solve large systems of linear equations, a system of ordinary differential equations, roots of transcendental equations, integrals, statistical problems, optimization problems, control system problems, and many other types of problems encountered in engineering. MATLAB also contains toolboxes, which must be purchased separately, that are designed to solve problems in very specialized areas.

2.1.1 The MATLAB Windows

By clicking on the MATLAB icon on the desktop, several MATLAB windows open (the default is the MATLAB desktop; see Figure 2.1).

- *Command window* (in the center; in older versions, the command window is on the right)—In the command window you can enter commands and data, make calculations, and print results. You can write a program (or script) in the command window and execute the program. However, it will not be ✕ saved. Thus, if an error is made, the entire program needs to be retyped. By clicking the up (↑) arrow key on your keyboard the previous command can be reentered.
- *Command history window* (bottom right)—This window lists commands that the user has used in the command window.

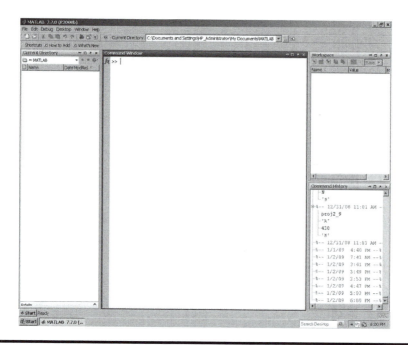

Figure 2.1   MATLAB windows: command, current directory, command history, and workspace windows. (From MATLAB. With permission.)

- *Current directory window* (on the top in the center of the task bar area)—This small window lists the current working directory. To run a MATLAB program, the program needs to be in the directory listed in this window. The directory listed in this window can be changed by clicking on the little box just to the right of the window. This produces a drop-down menu allowing one to select the directory in which the program resides (see Figure 2.2). If the working directory is not listed in the drop-down menu, one can click on the little box containing the three dots, which allows one to browse for the directory containing the program of interest (see Figure 2.3).
- *Second window* also called the current directory (on the left)—This window lists all the directories that are available.
- *Workspace window* (upper right)—This window lists all the files in the current workspace.
- *Editor window*—Programs are best written in the editor window. Programs are then saved as *m* files. These files have the extension *m*, such as *heat.m*. To execute the program, return to the command window and type in the name of the program without the extension (*.m*). The editor window can be obtained from the command window by clicking on *File* in the task bar section and selecting *New-M-file*.

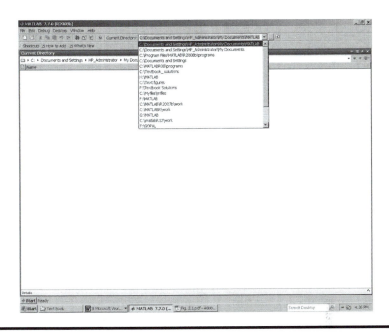

Figure 2.2 Drop-down menu from current directory window. (From MATLAB. With permission.)

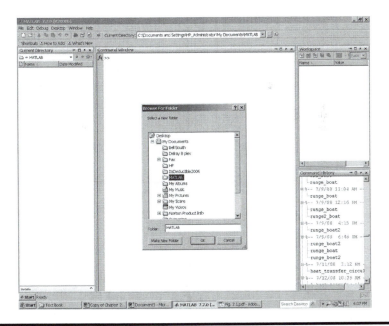

Figure 2.3 Browse for folder menu from current directory window. (From MATLAB. With permission.)

2.2 Constructing a Program in MATLAB

1. Start the MATLAB program by double clicking on the *MATLAB icon* on the desktop, bringing up the Command Window.
2. Click on *File-New-M-file*. This brings up the editor window.
3. Type program in the editor window.
4. Save the program in a directory of your own choosing. It is best that this directory contain only your own MATLAB programs.
5. To run the program, return to the Command Window and in the *current directory slot* (upper center) select the directory in which the program has been saved. Then in the Command Window type in the program name without the *.m* extension. Example: Suppose the program has been saved as *heat.m*, then after the (>>) characters type in *heat*.

2.3 The MATLAB Program

- Variable names
  - Must start with a letter.
  - Can contain letters, digits, and the underscore character.
  - Can be of any length, but must be unique within the first 19 characters.
- MATLAB is case sensitive in variable names, as well as in commands.
- Semicolon is usually placed after variable definition and program statements; otherwise, the defined variable appears on the screen; example:

$$A = [3 \quad 4 \quad 7 \quad 6]$$

- In the command window, you would see

$$A =$$
$$3 \quad 4 \quad 7 \quad 6$$

- % sign is used for a comment line.
- There is a graphics window to display plots and graphs.
- There are several commands for clearing windows.
  clc—clears the command window.
  clf—clears the graphics window.
  clear—removes all variables and data from memory.
  ctrl-c—aborts a program that may be running in an infinite loop.
- Commands are case sensitive; use lowercase letters for commands.
- *quit* or exit terminates a program.
- *save*—saves variables or data in workspace.
- *functions*—are also saved as M. files.

- Scripts (programs) and functions are saved as ASCII text files. Thus, they may be written in either the built-in editor, in notepad, or in a word processor (saved as a text file).
- The basic component in MATLAB is a matrix.
- A matrix is designated by brackets; example:

$$A = \begin{bmatrix} 1 & 3 \\ 6 & 5 \end{bmatrix}; \text{ or } A = [1 \ 3 ; 6 \ 5]$$

- A matrix of 1 row and 1 column is a scalar; example:

$$A = [3.5]$$

- However, MATLAB accepts A = 3.5 (without brackets) as a scalar.
- A matrix consisting of 1 row and several columns, or 1 column and several rows is considered a vector; example:

$$A = [2 \ 3 \ 6 \ 5] ; \text{ (row vector)} \qquad A = \begin{bmatrix} 2 \\ 3 \\ 6 \\ 5 \end{bmatrix}; \qquad \text{(column vector)}$$

- A matrix can be defined by including a second matrix as one of the elements; example:

$$B = [1.5 \quad 3.1];$$

$$C = [4.0 \quad B] = [4.0 \quad 1.5 \quad 3.1]$$

- A specific element of matrix *C* can be selected by writing

$$a = C(2); \quad \text{then,} \quad a = 1.5$$

- The Colon Operator may be used to
  1. Create a new matrix from an existing matrix; examples:

$$A = \begin{bmatrix} 5 & 7 & 10 \\ 2 & 5 & 2 \\ 1 & 3 & 1 \end{bmatrix}$$

$$x = A(:,1); \text{ gives } x = \begin{bmatrix} 5 \\ 2 \\ 1 \end{bmatrix}$$

The semicolon in the expression (:,1) implies all the rows, and the 1 implies column 1.

$$x = A(:, 2:3); \quad gives \ x = \begin{bmatrix} 7 & 10 \\ 5 & 2 \\ 3 & 1 \end{bmatrix}$$

2. Generate a series of numbers; for example:

$$n = 1:8; \quad gives \ n = [1\ 2\ 3\ 4\ 5\ 6\ 7\ 8]$$

To increment in steps of 2 use

$$n = 2:2:8; \quad gives \ n = [2\ 4\ 6\ 8].$$

■ Elementary math functions. (A complete set can be obtained by typing "help elfun" in the Command Window.)
Trigonometric functions:

sin	Sine
sinh	Hyperbolic sine
asin	Inverse sine
asinh	Inverse hyperbolic sine
cos	Cosine
cosh	Hyperbolic cosine
acos	Inverse cosine
acosh	Inverse hyperbolic cosine
tan	Tangent
tanh	Hyperbolic tangent
atan	Inverse tangent
atan2	Four quadrant inverse tangent
atanh	Inverse hyperbolic tangent
sec	Secant
sech	Hyperbolic secant
asec	Inverse secant
asech	Inverse hyperbolic secant
csc	Cosecant
csch	Hyperbolic cosecant
acsc	Inverse cosecant
acsch	Inverse hyperbolic cosecant
cot	Cotangent
coth	Hyperbolic cotangent
acot	Inverse cotangent
acoth	Inverse hyperbolic cotangent

Table 2.1   Special Values

Symbol	Value
pi	π
i or j	$\sqrt{-1}$
Inf	∞
Ans	A value computed by an expression in the Command window.

■ Exponential, logarithm, square root, and error function

exp	Exponential
log	Natural logarithm
log10	Common (base 10) logarithm
sqrt	Square root
erf	Error function

■ Complex numbers

abs	Absolute value
conj	Complex conjugate
imag	Complex imaginary part
real	Complex real part

■ Other useful functions

size(X)—Gives the size of matrix X (or vector X).

sum(X)—For vectors, sum(X) gives the sum of the elements in X. For matrices, sum(X) gives a row vector containing the sum of the elements in each column of the matrix.

max(X)—For vectors, max(X) gives the maximum element in X. For matrices, max(X) gives a row vector containing the maximum in each column of the matrix. If X is a column vector, it gives the maximum absolute value of X.

min(X)—Same as max(X) except it gives the minimum element in X.

sort(X)—For vectors, sort(X) sorts the elements of X in ascending order. For matrices, sort(X) sorts each column in the matrix in ascending order.

factorial(n)—n!.

■ See Table 2.1 for special values and special matrices.

■ Sometimes one needs to define the size of a matrix. This can be done by establishing a matrix of all zeros or all ones; examples:

$$A = \text{zeros}(3) = \begin{bmatrix} 0 & 0 & 0 \\ 0 & 0 & 0 \\ 0 & 0 & 0 \end{bmatrix}$$

$$B = \text{zeros}(3,2) = \begin{bmatrix} 0 & 0 \\ 0 & 0 \\ 0 & 0 \end{bmatrix}$$

$$A = \text{ones}\,(3) = \begin{bmatrix} 1 & 1 & 1 \\ 1 & 1 & 1 \\ 1 & 1 & 1 \end{bmatrix}$$

$$A = \text{ones}\,(2,3) = \begin{bmatrix} 1 & 1 & 1 \\ 1 & 1 & 1 \end{bmatrix}$$

The identity matrix symbol is eye; example:

$$\text{eye}\,(3) = \begin{bmatrix} 1 & 0 & 0 \\ 0 & 1 & 0 \\ 0 & 0 & 1 \end{bmatrix}$$

■ If you wish to have your program pause to have an input from the keyboard, use the input function; for example, suppose we wish to enter a 2×3 matrix, use

Z = input ('Enter values for Z in brackets \n');
type in:
[5.1   6.3   2.5;   3.1   4.2   1.3]

Note: \n means move the cursor to the next line.
To display a matrix X just type X and matrix X will appear on the screen.

■ The display command—prints contents of a matrix or alpha-numeric information; example:

p = 14.7;
disp (p); disp ('psia');
The following will be displayed on the screen:
14.7000
psia

■ The fprintf command; example:

p = 14.7;
fprintf ('the press is %*f* psia \n', p);

The following will appear on the screen:
the press is 14.7000 psia

■ The default for *f* format is 4 decimal places.
■ One can specify the number of spaces for the variable as well as the number of decimal places by using

% 8.2 *f*

■ This will allow 8 spaces for the variable to 2 decimal places.
■ Other formats:

% i—used for integers.
% e—exponential notation (default is 6 decimal places).

% g—computer decides whether to use f format or e format.

% s—used for a string of characters.

% c—used for a single character.

Strings and a character must be enclosed by apostrophe marks.

■ To print to a file use

p = 14.7

fid = fopen ('output.dat','w');

fprintf (fid, 'press = % 5.2 f psia', p)

■ To access the file output.dat after the program has run, the program had to include a statement fclose(fid) after all the output statements in the program or at the end of the program.

■ To read from a file use

A = zeros (n,m);

fid = fopen ('filename.dat', 'r');

[A] = fscanf (fid, '%f', [n,m]);

where n×m is the number of elements in the data file.

The n×m matrix is filled in column order. To use the data in its original order transpose the read-in matrix.

■ To transpose a matrix *B*, type in B′. This changes columns to rows and rows to columns.

■ A data file can also be entered into a program by the load command; example:

load filename.dat

x = filename (: , 1);

y = filename (: , 2);

The input file must have the same number of rows in each column.

■ Arithmetic Operators

+	addition
−	subtraction
*	multiplication
/	division
^	exponentiation

■ The *for loop* provides the means to carry out a series of statements with just a few lines of code.

Syntax:

for m = 1 : 20

      statement;

      statement;

      -----;

end

The computer sets the index *m* to 1, carries out the statements between the *for* and *end* statements, then returns to the top of the loop, changes *m* to 2, and repeats the process. This continues until *m* is set to 21, in which case the computer leaves the loop.

■ The *while* loop

Syntax:
n = 0;
while n < 10
    n = *n* +1;
    y (*n*) = *n*^2;
end

In the *while loop*, the computer will carry out the statements between the *while* and *end* statements as long as the condition in the *while* statement is satisfied.

■ *if* statement

Syntax:
if logical expression
    statement;
    statement;
else
    statement;
    statement;
end

If the logical expression is true, the upper set of statements is executed.
If the logical expression is false, the bottom set of statements is executed.

■ Logical expressions are of the form

a = = b;   a < = b;
a < b;   a > = b;
a > b;   a ~= b;

■ Compound logical expressions

a > b   &&   a ~= c   (and)
a > b || a < c   (or)

■ The *if-elseif* ladder

Syntax:
if (*logical expression* 1)
    statements;
elseif (*logical expression* 2)
    statements;
else
    statements;
end

The *if-elseif* ladder works from top down. If the top logical expression is true, the statements related to that logical expression are executed and the program will leave the ladder. If the top logical expression is not true, the program moves to the next logical expression. If that logical expression is true, the program will execute the group of statements associated with that logical expression and leave the ladder. If that logical expression is not true, the program moves to the next logical expression and continues the process. If none of the logical expressions are true the program will execute the statements associated with the *else* statement. The *else* statement is not required. In that case, if none of the logical expressions are true, no statements within the ladder will be executed.

■ The break command may be used with an *if* statement to end a loop; example:

for m = 1 : 20
    -----;
    -----;
    if m > 10
      break;
    end
    -----;
end

In this example, the *for* loop is ended when *m* = 11.

■ The *switch* group

Syntax:
    switch(*var*)
    case var1
        statement; ...; statement;
    case var2
        statement; ...; statement;
    case var3
        statement; ...; statement;
    otherwise
        statement; ...; statement;
    end

where *var* takes on values of *var1*, *var2*, *var3*, etc.

If *var* equals *var1*, those statements associated with *var1* are executed and the program leaves the switch group. If *var* does not equal *var1*, the program tests if *var* equals *var2*; if yes, the program executes those statements associated with *var2* and leaves the switch group. If *var* does not equal any of *var1*, *var2*, etc., the program executes the statements associated with the *otherwise* statement. If *var1*, *var2*, etc., are strings, they need to be enclosed by apostrophe mark. It should be noted that *var* cannot be a logical expression, such as *var1* $>= 80$.

■ For plot commands see Table 2.2.
■ Multiple plots

plot(x,y,w,z), gives 2 plots, y vs. x and z vs. w.

■ Example, suppose

$$A = \begin{bmatrix} t_1 & y_1 & z_1 & w_1 \\ t_2 & y_2 & z_2 & w_2 \\ t_3 & . & . & . \\ \vdots & \vdots & \vdots & \vdots \\ t_n & y_n & z_n & w_n \end{bmatrix}, T = \begin{bmatrix} t_1 \\ t_2 \\ t_3 \\ . \\ \vdots \\ t_A \end{bmatrix}, Y = \begin{bmatrix} y_1 \\ y_2 \\ . \\ . \\ . \\ y_n \end{bmatrix}, Z = \begin{bmatrix} z_1 \\ z_2 \\ . \\ . \\ . \\ z_n \end{bmatrix}, W = \begin{bmatrix} w_1 \\ w_2 \\ . \\ . \\ . \\ w_n \end{bmatrix}$$

Then to plot *Y* vs. *T*, *Z* vs. *T*, and *W* vs. *T* all on the same graph, use
   plot (*A*(: , 1), *A*(: , 2), *A*(: , 1), *A*(: , 3), *A*(: , 1), *A*(: , 4))
   or *plot* (*T,Y, T,Z, T,W*);
To label the x and y axes as well as adding a title to the plot use:
   xlabel('*T*'), *ylabel*('*Y, Z, W*'), and *title*('*Y, Z, W* vs. *T*').
You can add a grid to the plot with the command
   grid.
You can also add text to the plot with the command
   text(*x position, y position,* '*text*').
Greek letters and mathematical symbols can be used in xlabel, ylabel, title, and text by writing "\Greek symbol name". Example:
   ylabel(' \omega'), *title*(' \omega vs. t'), *text*(10,5,'\omega1');
To obtain a list of Greek symbols, in the command window, select from the upper task bar "help—product help—index—G," then scroll down until you reach "Greek letters and mathematical symbols." This will provide a list of Greek symbols that can be used with the plot commands.

Table 2.2   Plot Commands

Command	Type
plot(x,y)	linear plot of y vs. x
semilogx(x,y)	semi-log plot; log scale for x axis, linear scale for y axis
semilogy(x,y)	semi-log plot; linear scale for x axis, log scale for y axis
loglog(x,y)	log-log plot; log scale for both x and y axes

■ Subplot command

Suppose one chooses to plot each of the curves in the previous example as a separate plot, but all on the same page. The subplot command provides the means to do so. The command *subplot(m,n,p)* breaks the page into an $m \times n$ matrix of small plots; p selects the matrix position of the plot. The following code would create the 3 plots in the previous example all on the same page.

```
for i=1:3
    subplot(2,2,i),
    if i==1
        plot(T,Y),grid,title('Y VS. T'),xlabel('T'),ylabel('Y');
    end
    if i==2
        plot(T,Z),grid,title('Z VS. T'),xlabel('T'),ylabel('Z');
    end
    if i==3
        plot(T,W),grid,title('W' VS. T'),xlabel('T'),ylabel('W');
    end
end
```

■ Function Files (saved as m files)

Functions are useful if one has a complicated program and wishes to break it down into smaller parts. Also, if a series of statements is to be used many times, it is convenient to place them in a function.

Syntax:

function [output variables] = function_name (input variables)

The function file name is to be saved as function_name.m (see Table 2.3).

The first executable statement in the function must be "function." If there is more than one output value, one needs to put the output variables in brackets. If there is only one output value, no brackets are necessary. If there are no output values, use brackets that are blank.

A function differs from a script in that the arguments may be passed to another function or to the command window. Variables defined and manipulated inside the function are local to the function.

For an example of the calculation of e^x see Table 2.4.

To determine e^x by series, one approach is to start with sum = 1, evaluate each term using exponentiation and MATLAB's factorial function, and add the

Table 2.3   Examples of "Function" Usage

Function Definition Line	File Name
function [q, Q, tf] = heat (x, y, tinf)	heat.m
function ex = exf (x)	exf.m
function [ ] = output (x, y)	output.m

Table 2.4 e^x Series

Term Index	1	2	3	4
$e^x = 1 + x + \dfrac{x^2}{2!} + \dfrac{x^3}{3!} + \dfrac{x^4}{4!}$				

obtained term to the sum. For some series expressions, MATLAB's factorial function will not be applicable. As a result, another approach will be needed. It can be seen from this example that term 3 can be obtained from term 2 by multiplying term 2 by x and dividing term 2 by the index; i.e.,

(term3 = term2 × x/3), similarly (term4 = term3 × x/4), etc.

This example is used in several of the following sample programs. Example *programs* demonstrating the use of *for loops, while loops, if* statement, *if-elseif* statement, *input* statement, *fprintf* statement, *fscanf* statement, *save* command, *load* command, *plot* command, and the use of a *self-written function* follow.

2.4 Program Examples

Example 2.1

```
% exa.m
% This program calculates e^x by series and by MATLAB's exp( ) function.
% e^x = 1+x+x^2/2!+x^3/3!+x^4/4! +...+
% The "for loop" ends when the term only affects the seventh significant
% figure.
clear; clc;
x=5.0;
sum=1.0;
for n=1:100
    term=x^n/factorial(n);
    sum=sum+term;
    if(abs(term) <= sum*1.0e-7)
        break;
    end
end
ex1=sum;
ex2=exp(x);
fprintf('x=%3.1f ex1=%8.5f ex2=%8.5f \n',x,ex1,ex2);
% Note: A variable name cannot be the same as the name of a file.
% Therefore, use exa as the file name and not ex.
```

Example 2.2

```
% exB.m
% The program calculates of e^x by both an arithmetic statement (ex2)
% and by a Taylor series (ex1), where -0.5< x <0.5. A 'for loop' is
% used to determine the x values. In the series part of the program,
% the 'sum' function is used to sum all the terms calculated in the
% inner 'for loop'. Fifty terms are used in the series.
    clear; clc; xmin=-0.5; dx=0.1;
% Table headings
```

```
    fprintf(' x            ex1          ex2 \n');
    for i=1:11
        x=xmin+(i-1)*dx;
        ex2=exp(x);
        for n=1:50
          term(n)=x^n/factorial(n);
        end
        ex1=1.0+sum(term);
        fprintf('%5.2f   %10.5f   %10.5f \n',x,ex1,ex2);
    end
```

Example 2.3

```
% while1.m
% Calculation of e^x by a Taylor series using a while loop. The 'input
% statement' is used to establish the exponent, x. A 'while loop' is
% used in determining the series solution. In this example term(n) is
% obtained by multiplying term(n-1) by x and dividing by the index n.
clear; clc;
x=input('enter a value for the exponent x \n');
n=1; term=1.0; ex=1.0;
while abs(term) > ex*1.0e-6
    term=term*x/n; ex=ex+term; n=n+1;
    if n > 50
        break;
    end
end
disp(x); disp(ex);
```

Example 2.4

```
% matrix1.m
% INPUTTING A 3 BY 3 MATRIX FROM THE KEYBOARD
% NESTED LOOPS ARE USED
clear; clc;
for n=1:3
    for m=1:3
        fprintf('n=%i m=%i ',n,m);
        a(n,m)=input('Enter a(n,m) ');
        fprintf('\n');
    end
end
a
```

Example 2.5

```
% exf.m
% ESTABLISHING A FUNCTION TO EVALUATE e^x
% In this example term(n) is obtained from term(n-1) by multiplying
% term(n-1) by x and dividing by index n.
function ex=exf(x)
sum=1.0; term=1.0;
for n=1:100
    term=term*x/n;
    sum=sum+term;
    if abs(term) <= sum*1.0e-5
        break;
    end
```

```
end
ex=sum;
% In Command Window type in
% exf(5.0) or y=exf(5.0)
```

Example 2.6

```
% table_plot.m
% This program creates a simple table and a simple plot.
% First a table of y1=n^2/10 and y2=n^3/100 is created.
% Then y1 and y2 are plotted.
clear; clc;
for n=1:10
    t(n)=n;
    y1(n)=n^2/10;
    y2(n)=n^3/100;
end
fo=fopen('output.dat','w');
% By printing the output to a file, one can edit the output
% such as lining up column headings, etc. This cannot be
% done if one prints to the screen.
% Column headings
fprintf(fo,'t    y1        y2      \n');
fprintf(fo,'-------------------------------------\n');
for n=1:10
    fprintf(fo,'%8.1f  %10.2f  %10.2f \n',t(n),y1(n),y2(n));
end
% Creating the plot, y1 is red, y2 is in green.
% Plot identification is also established by adding text to the plot;
plot(t,y1,'r',t,y2,'g'),grid, title('y1 & y2 vs. t');

% After viewing the plot, text can be entered to identify which
% curve is y1 and which curve is y2.
text(6.5,2.5,'y1'), text(4.2,2.4,'y2');
% The data t, y1 and y2 can be saved in a data file and loaded later
% to be used in another program. By the use of the 'clear' command
% the data file is removed from the workspace. The load command
% brings it back into the workspace and can be used in another program.
data1=[t y1 y2];
% data1 is saved as a mat file in the working directory.
save data1;
clear;
% the clear command removes data1 from the workspace.
% Color types
% blue      b
% green     g
% red       r
% cyan      c
% yellow    y
% black     k
```

Example 2.7

```
% plotc.m
% This script is a sample of a plot program loading data obtained
% in another program
clear; clc;
```

```
load data1
x=data1(:,1);
y=data1(:,2);
z=data1(:,3);
plot(x,y,'--',x,z,'-.');
xlabel('x'), ylabel('y,z'), title('y & z vs. x'), grid,
text(5.4,3.4,'y'),text(7.2,3.5,'z'),
% print;
% curve types:
% solid              default
% dashed       --    (minus sign)
% dashed-dot   -.
% doted        :
% point        .
% plus         +
% star         *
% circle       0
% x-mark       x
% The point and below are points on the plot that are not connected.
```

A neat table can also be created by copying the output data to EXCEL. To do this, follow these steps:

1. First create the data either in a file or on the screen.
2. Highlight the data to be copied and from the menu bar, select "edit" and "copy" from the drop-down menu list (the copy option can also be obtained by right-clicking the mouse).
3. Open the EXCEL program and click on "edit" in the menu bar. Select "paste" from the drop-down menu list giving a single block of data.
4. Then enlarge column "A" to include the entire data set.
5. Now select "data" from the menu bar, and select "text to columns" from the drop-down menu.
6. Click on "finish," then highlight the table area and select the "center" option in the menu bar to center the data in each column.
7. To add table headings, place the cursor in the top left margin and select "insert" from the menu bar and "rows" from the drop-down menu.
8. Type in table headings in each column.
9. To add lines separating rows and columns, first highlight the area that is to have the separating lines, then select "format" from the menu bar and "cells" from the drop-down menu list. Then select "border" from the "Format Cells" page. Click on the separating lines that you wish to appear in your table.
10. To print the table, click on the print icon in the menu bar.

Example 2.8

```
% subplot1.m
% This program is an example of the use of the subplot command.
% Values of y1 and y2 are taken from the saved data file 'data1'
% Separate plots of y1 vs. t & y2 vs. t are plotted on the same page.
clear; clc;
```

```
load data1;
for i=1:2
    subplot(1,2,i);
    if i==1
        plot(t,y1), xlabel('t'), ylabel('y1'), title('y1 vs. t'), grid;
    end
    if i==2
        plot(t,y2), xlabel('t'), ylabel('y2'), title('y2 vs. t'), grid;
    end
end
```

Example 2.9

```
% plotg.m
% This program is a demonstration of creating a semi-log plot
clc; clear;
x=1:0.5:5;
for i=1:9
    y(i)=3.0*exp(x(i));
end
semilogy(x,y), title('semi-log plot of y & x'), xlabel('x'),
    ylabel('y'),grid;
```

Before running Example 2.10 the data file "GAUSS6.DAT" needs to be created. To do this, from the COMMAND window click on file-Blank M-file. Type in the following data (numbers only):

8.77	2.40	5.66	1.55	1.0
4.93	1.21	4.48	1.10	1.0
3.53	1.46	2.92	1.21	1.0
5.05	4.04	2.51	2.01	1.0
3.54	1.04	3.47	1.02	1.0
-32.04	-20.07	-8.53	-6.30	-12.04

Save the file as GAUSS6.DAT in the working directory.

Example 2.10

```
% amatrix7.m
% MATRIX A(I,J) IS READ FROM A FILE
% amatrix7.m
% MATRIX A(I,J) IS READ FROM A FILE NAMED "GAUSS6.DAT"
% NOTE: BEFORE RUNNING THIS PROGRAM THE DATA FILE GAUSS6.DAT MUST
% FIRST BE CREATED AND SAVED IN THE WORKING DIRECTORY.
clear; clc;
n=5;
A=zeros(n); B=zeros(1,n);
fin=fopen('GAUSS6.DAT','r');
[A]=fscanf(fin,'%f',[n,n]);
[B]=fscanf(fin,'%f',[n]);
fclose(fin);
size(A)
fo=fopen('output.dat','w');
fprintf(fo,' output.dat \n');
fprintf(fo,' Matrix A elements \n');
for i=1:n
    for j=1:n
```

```
            fprintf(fo,'  %8.3f  ',A(i,j));
      end
      fprintf(fo,'\n');
end
% Note that rows become columns and columns become rows.
fprintf(fo,'\n\n Matrix B elements \n');
for i=1:n
      fprintf(fo, '  %8.3f \n',B(i));
% The A matrix is returned to its original state.
end
fclose(fo);
C=[A' B];
moutput(C);
```

Before running Example 2.10, the function moutput.m needs to be created.

```
% moutput.m
% This Function is to be created with "amatrix7.m"
% This function accepts matrix A & matrix B and prints them out to
% output2.dat
function D=moutput(C)
n=5;
A=C(:,1:n)
B=C(:,n+1)
fo=fopen('output2.dat','w');
fprintf(fo,'  output2.dat \n')
fprintf(fo,'  Matrix A elements \n')
for i=1:n
    for j=1:n
        fprintf(fo,'  %8.3f',A(i,j));
    end
    fprintf(fo,'\n');
end
fprintf(fo,'\n\nMatrix B elements \n');
for i=1:n
    fprintf(fo,'  %8.3f \n',B(i));
end
fclose(fo;
```

Example 2.11

```
% charmatrix.m
% Sometimes one might wish to print out a string of characters in a
% loop. This can be done by declaring a 2-D character string matrix as
% shown in this example. Note that all row character strings
% must have the same number of columns and that character strings must
% be enclosed by apostrophe marks.
clear; clc;
char=['Internal modem        '
      'Graphics circuit board'
      'CD drive              '
      'DVD drive             '
      'Floppy drive          '
      'Hard disk drive       '];
```

```
for i=1:5
    fprintf('%22s \n',char(i,1:22));
end
```

Example 2.12

```
% char1.m
% This example is a modification of Example 2.11. The program asks the
% user if he/she wishes to have the output go to the screen or to a
% file.
% This example illustrates the use of the switch statement.
clear; clc;
char=['Internal modem         '
      'External modem         '
      'Graphics circuit board'
      'CD drive               '
      'Hard disk drive        '];
fprintf('you have a choice, you can have the output go to the');
fprintf('screen or go to a file named output.dat \n');
var=input('enter s for screen or f for file ...
      each enclosed by apostrophe mark \n');
switch(var)
   case 's'
        for i=1:5
            fprintf('%22s \n',char(i,1:22));
        end
   case 'f'
        fo=fopen('output.dat','w');
        for i=1:5
            fprintf(fo,'%22s \n',char(i,1:22));
        end
   otherwise
        fprintf('you did not enter an s or an f, try again \n');
        exit;
        fclose(fo);
end
```

Example 2.13

```
% grades.m
% This example uses the if-elseif ladder.
% The program determines a letter grade depending on the score the user
% enters from the keyboard.
clear; clc;
gradearray=['A'; 'B'; 'C'; 'D'; 'F'];

score=input('Enter your test score \n');
fprintf('score is:%i \n',score);
    if score > 100,
        fprintf('error: score is out of range. Rerun program \n');
        break;
    elseif (score >= 90 && score <= 100)
        grade=gradearray(1);
    elseif (score >= 80 && score < 90)
        grade=gradearray(2);
    elseif (score >= 70 && score < 80)
        grade=gradearray(3);
```

```
      elseif (score >= 60 && score < 70)
         grade=gradearray(4);
      elseif (score < 60)
         grade=gradearray(5);
      end
fprintf('grade is:%c \n',grade);
```

Example 2.14

```
% grades3.m
% This example uses a loop to determine the correct interval of
% interest. For a large number of intervals, this method is more
% efficient (fewer statements) than the method in Example 2.13.
% The program determines a letter grade depending on the score the user
% enters from the keyboard.
clear; clc;
gradearray=['A'; 'B'; 'C'; 'D'; 'F'];
sarray=[100 90 80 70 60 0];
score=input('Enter your test score \n');
fprintf('score is:%i \n',score);
% The following 2 statements are needed for the case when score = 100.
if score == 100
    grade=gradearray(1);
else
    for i=1:5
        if (score >= sarray(i+1) && score < sarray(i))
            grade=gradearray(i);
        end
    end
end
fprintf('grade is:%c \n',grade);
```

MATLAB function interp1 (the last character is a one). The format for the function is

$$y_2 = \text{interp1}(x, y, x_2)$$

where *x,y* are the set of data points and x_2 is the set of *x* values at which the set of y_2 values is to be determined by linear interpolation. Arrays *x* and *y* have to be of the same length.

Example 2.15

```
% interp1f.m
% This program uses MATLAB's function interp1 to interpolate for
% specific volume of air. Air table values for specific volume (m^3/kg)
% vs. temperature (K) are given.
clear; clc;
Tt=[150 200 250 300 350 400 450 500];
vt=[0.04249 0.05665 0.07082 0.08498 0.09914 0.11330 0.12747 0.14163];
fprintf('This program interpolates for specific volume,v, at \n\n');
fprintf('a specified temp, T. Table temperature range ...
        is 150-500(K) \n\n');
T=input('Enter T at which v is to be determined \n\n');
v=interp1(Tt,vt,T);
fprintf(' T=%6.1f (K)    v=%9.5f (m^3/kg) \n',T,v);
```

2.5 Debugging a Program

It is not uncommon that in typing up a program, one makes a typographical error, such as omitting a parenthesis, a comma in a 2-D array, etc. This type of error is called a syntax error. When this occurs, MATLAB will provide an error message pointing out the line in which the error has occurred. However, there are cases where there are no syntax errors, but the program either doesn't run or gives an obvious incorrect answer. When this occurs, one should consider using the debug feature in MATLAB. The debug feature allows you to set break points in your program. The program will run up to the line containing the break point. To set a break point just left click the mouse in the narrow column next to the line that you wish to be a break point. A small *red circle* will appear next to the break-point line as shown in Figure 2.4. One can then click on the debug listing in the task bar and select several options from the drop-down menu (see Figure 2.5). One option is to have MATLAB execute one line at a time. This is done by successively selecting the *step* option in the debug drop-down menu (see Figure 2.6). If the program contains a self-written function, one can execute one line at a time in the function by selecting the *step in* option in the debug drop-down menu. Selecting the *step out* option returns the control to the main program. Selecting *Clear Breakpoints in All Files* will remove all break points from the program.

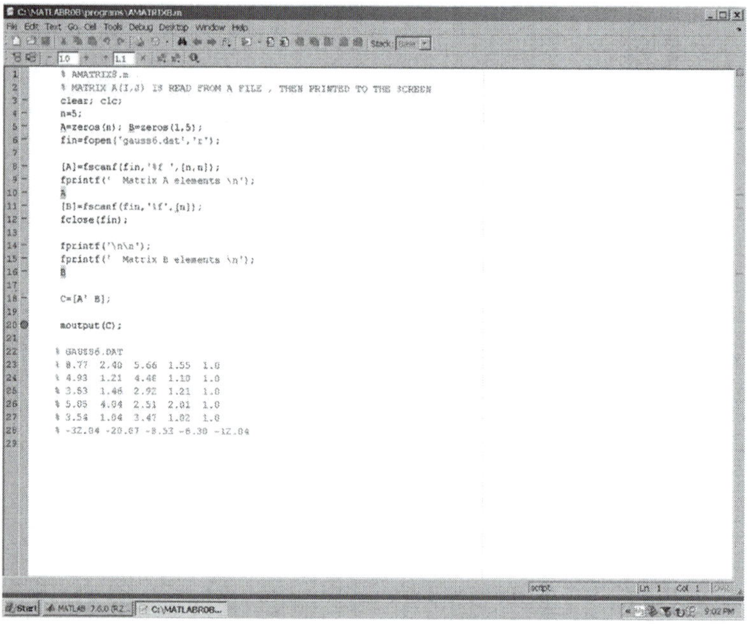

Figure 2.4   Selecting a break point. (From MATLAB. With permission.)

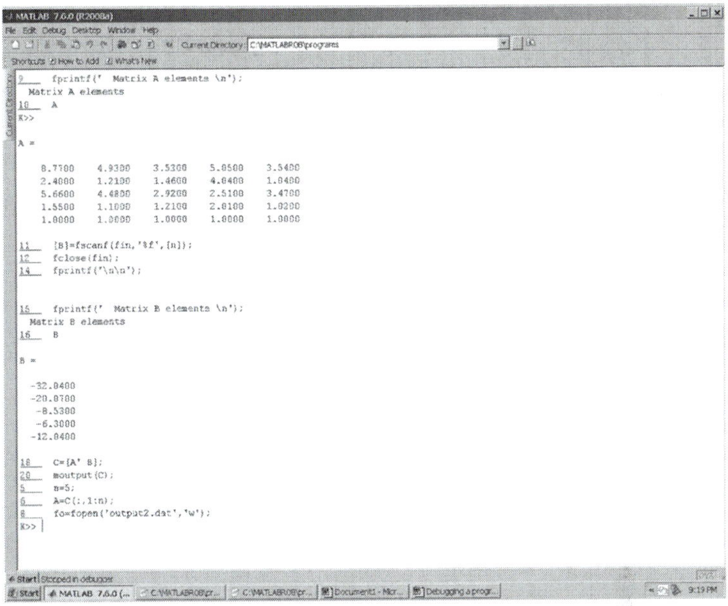

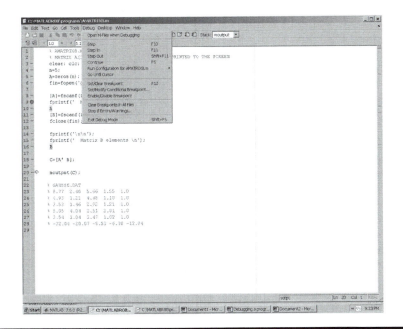

Figure 2.5 **Stepping through a program execution. (From MATLAB. With permission.)**

Figure 2.6 Debug drop-down menu. (From MATLAB. With permission.)

2.6 3-D and Contour Plots

Given $z = f(x,y)$, one can create a 3-D plot of z vs. x,y. To create the 3-D or contour plot, one first needs to create a meshgrid. The command is $[X,Y] = meshgrid(x,y)$. The number of elements in the grid will equal the number of elements in x times the number of elements in y. For each grid point there is an X element and a Y element. Suppose $-3 <= x <= 3$ in steps of 1 and $-2 <= y <= 2$ in steps of 1. Then, the X and Y matrices will appear as shown below.

$$X = \begin{bmatrix} -3 & -2 & -1 & 0 & 1 & 2 & 3 \\ -3 & -2 & -1 & 0 & 1 & 2 & 3 \\ -3 & -2 & -1 & 0 & 1 & 2 & 3 \\ -3 & -2 & -1 & 0 & 1 & 2 & 3 \\ -3 & -2 & -1 & 0 & 1 & 2 & 3 \end{bmatrix}$$

$$Y = \begin{bmatrix} -2 & -2 & -2 & -2 & -2 & -2 & -2 \\ -1 & -1 & -1 & -1 & -1 & -1 & -1 \\ 0 & 0 & 0 & 0 & 0 & 0 & 0 \\ 1 & 1 & 1 & 1 & 1 & 1 & 1 \\ 2 & 2 & 2 & 2 & 2 & 2 & 2 \end{bmatrix}$$

The meshgrid statement is used to produce several different types of 3-D plots. The example below uses the meshgrid statement to produce a 3-D surface plot and a contour plot. In expressing the surface equation, one needs to do element-by-element multiplication and division.

Example 2.16

```
% contour2.m
% This program develops a 3-D surface plot of an elliptic cone
% and a contour plot.
clear; clc;
x=-5:0.1:5;
y=-4:0.1:4;
a=5; b=4; c=3;
[X,Y]=meshgrid(x,y);
% Elliptic cone
Z=c*sqrt((X/5).^2+(Y/4).^2);
% the mesh (x, y, Z) statement creates a meshed surface plot.
mesh(x,y,Z)
figure;
% The surf(x, y, Z) statement creates a surface plot.
surf(X,Y,Z), xlabel('x'), ylabel('y'), zlabel('z'),grid;
figure;
% The contour(X, Y, Z) statement creates a contour plot.
ct=contour(X,Y,Z); clabel(ct); xlabel('x'); ylabel('y');
```

To produce plots of complete spheres and ellipsoids use ellipsoid and sphere functions. A description of these MATLAB functions can be obtained by typing

help ellipsoid
help sphere
help cylinder

in the command window.

Projects

Project 2.1

Determine the Taylor Series expansion of cos(x) about $x = 0$ and develop a computer program in MATLAB that will evaluate the function from $-\pi \le x \le \pi$ in steps of 0.1 π by

a. An arithmetic statement.
b. Series allowing for as many as 50 terms. However, stop adding terms when the last term only affects the 6th place in the answer.

Print out a table (to a file) in the format shown in Table P2.1; use 6 decimal places for cos(x).

Also create a plot of cos(x) for $-\pi \le x \le \pi$.

Table P2.1   Table Format

X	cos(x) (by arithmetic stm)	cos(x) (by series)	Terms in the series
$-1.0\,\pi$			
$-0.9\,\pi$			
$-0.8\,\pi$			
—			
—			
—			
$0.9\,\pi$			
$1.0\,\pi$			

Project 2.2

Determine the Taylor Series expansion of sin(x) about $x = 0$ and develop a computer program in MATLAB that will evaluate the function from $-\pi \leq x \leq \pi$ in steps of 0.1 π by

a. An arithmetic statement.
b. Series allowing for as many as 50 terms. However, end adding terms when the last term only affects the 6th place in the answer.

Create a table similar to the table shown in Project 2.1, except replace cos(x) with sin(x). Also create a plot of sin(x) for $-\pi \leq x \leq \pi$.

Project 2.3

Using the Taylor Series expansion for e^x, show that $e^{ix} = \cos(x) + i \sin(x)$ and that $e^{-ix} = \cos(x) - i \sin(x)$, where $i = \sqrt{-1}$.

Project 2.4

Develop a computer program in MATLAB that will evaluate the following function for $-0.9 \leq x \leq 0.9$ in steps of 0.1 by

a. An arithmetic statement.
b. Series allowing for as many as 50 terms. However, end adding terms when the last term only affects the 6th place in the answer.

The function and its series expansion is

$$f(x) = (1 + x^2)^{-1/2} = 1 - \frac{1}{2}x^2 + \frac{1 \cdot 3}{2 \cdot 4}x^4 - \frac{1 \cdot 3 \cdot 5}{2 \cdot 4 \cdot 6}x^6 + \frac{1 \cdot 3 \cdot 5 \cdot 7}{2 \cdot 4 \cdot 6 \cdot 8}x^8 - + \cdots$$

Print out a table (to a file) in the format shown in Table P2.4; use 6 decimal places for $f(x)$.

Table P2.4   Table Format for Project 2.4

X	f(x) (by arith stm)	f(x) (by series)	No. of terms used in the series
−0.9	—		
−0.8	—		
−0.7	—		
—	—		
—	—		
0.7	—		
0.8	—		
0.9	—		

Project 2.5

The binomial expansion for $(1 + x)^n$, where n is an integer, is as follows:

$$(1 + x)^n = 1 + nx + \frac{n(n-1)x^2}{2!} + \frac{n(n-1)(n-2)x^3}{3!}$$
$$+ \cdots + \frac{n(n-1)(n-2)\ldots(n-r+2)x^{r-1}}{(r-1)!} + \cdots + x^n$$

Construct a MATLAB program that will evaluate $(1 + x)^n$ by both the above series and by an arithmetic statement for $n = 10$ and $1.0 \le x \le 10.0$ in steps of 0.5. Print out the results in a table as shown in Table P2.5.

Table P2.5   Table Format for Project 2.5

x	$(1 + x)^n$ (by arith stm)	$(1 + x)^n$ (by series)
1.0	—	
1.5	—	
2.0	—	
2.5	—	
—	—	
—	—	
—	—	
10.0	—	

Project 2.6

Although atmospheric conditions vary from day to day, it is convenient for design purposes to have a model for atmospheric properties as a function of altitude. The U.S. Standard Atmosphere, modified in 1976, is such a model. The model consists of two types of regions, one in which the temperature varies linearly with altitude, and the other where the temperature is a constant. The temperature and approximate pressure relations are as follows:

a. For a region where the temperature varies linearly,

$$p = p_i \left(1 - \frac{\lambda(z - z_i)}{T_i} \right)^{\frac{g}{\lambda R}}$$

$$T = T_i - \lambda(z - z_i)$$

$$\rho = \frac{p}{RT}$$

where
  z = the altitude.
  z_i = the altitude at the beginning of the region of interest.
  (p_i, T_i) = the pressure and temperature at the beginning of the region of interest.
  λ = the lapse rate.
  R = the air gas constant.
  g = the gravitational constant, which varies slightly with altitude. The
     above expression for p assumes that within the region of interest, g is a
     constant; otherwise the expression for p would be a lot more compli-
     cated than the one shown above.
  ρ = air density.

b. For a region where the temperature is constant

$$p = p_i \exp\left(-\frac{g(z - z_i)}{RT_i} \right)$$

$$T = T_i$$

Table P2.6 gives the values of pressure, temperature, and the gravitational constant
at the beginning of each region, as well as the lapse rate of the region.
   Determine the property values of (T, p, ρ) for every 1,000 m from sea level
$(z = 0)$ to $z = 80,000$ m and

a. Save the results as a data file to be used in Project 2.7.
b. Print the results in a table format for every 1,000 m.

Table P2.6 Regional Properties of U.S. Standard Atmosphere

Region	z (km)	T (K)	p (Pa)	λ (K/m)	g (m/s²)
	0	288.15	101325		9.810
1				0.0065	
	11.0	216.65	22632.05		9.776
2				0.0000	
	20.0	216.65	5474.88		9.749
3				−0.001	
	32.0	228.65	868.02		9.712
4				−0.0028	
	47.0	270.65	110.91		9.666
5				0.0000	
	51.0	270.65	66.88		9.654
6				0.0028	
	71.0	214.65	3.956		9.594
7				0.0020	
	84.9	186.95	0.373		9.553

Project 2.7

Project 2.6 provided a table of pressure, p, temperature, T, and density, ρ, every 1,000 m of altitude. Atmospheric property values at altitudes between table values can be determined by linear interpolation. The linear interpolation formula is as follows:

$$y = y_1 + (z - z_1) \times (y_2 - y_1)/(z_2 - z_1)$$

where y represents an atmospheric property to be determined, z is the altitude at which the unknown property is to be determined, y_1 and y_2 are the nearest table property values, and z_1 and z_2 are the altitudes at y_1 and y_2.

Construct a MATLAB program to determine by linear interpolation the properties of (T, p, ρ) at the following altitudes: 5,170 m, 8,435 m, 13,320 m, 22,250 m, 34,370 m, 48,550 m, and 64,220 m. Make the program interactive; that is, the program is to ask the user to enter the above elevations one at a time. After each entry the program is to print the results to the screen and then ask the user if he/she wishes to enter another altitude. The program is to continue as long as the response to the question is 'Y.'

Project 2.8

Repeat Project 2.7, but this time create a function *interp*, with input arguments (z, z_1, z_2, y_1, y_2), that will do the interpolation and that will return the requested property to the main program. The variable y is to represent an atmospheric property of either pressure, temperature, or density at an altitude z entered from the keyboard; z_1 and z_2 are the nearest altitudes to z and y_1 and y_2 are the property values at z_1 and z_2, respectively.

Project 2.9

The properties of specific volume, v, and pressure, p, as a function of temperature, T, for three gasses based on the Redlich-Kwong equation of state are given in Tables P2.9a, P2.9b, and P2.9c. Table P2.9d specifies the gas and the temperature at which the gas properties are to be determined by interpolation using MATLAB's *interp1* function.

Write a computer program in MATLAB that will do the following:

a. Construct three different data files, each containing the gas properties tabulated in Tables P2.9a, P2.9b, and P2.9c. Name the first air.dat; the second O2.dat; and the third CO2.dat. *Note*: The data files do not contain column headings.

b. Request the user to enter from the keyboard the gas type and the temperature at which the gas properties are to be obtained. The user is to enter, one at a time, the gas type and temperature as shown in Table P2.9d.

c. The program is to use a *switch* statement to select the proper table for interpolation and MATLAB's built-in function *interp1*.

d. After the user selects the gas and the temperature, the program should print to the screen the desired properties identifying the gas, the temperature, T, the specific volume, v, and the pressure, p, including units.

e. After each entry from the keyboard, the program should ask the user if he/she wishes to make another entry at which the properties are to be determined. The user is to enter 'Y' for yes and 'N' for no. The program is to continue determining gas properties by interpolation until the user enters 'N.' At that point, the program should terminate.

Table P2.9a   Air Properties

T (K)	v (m³/kmol)	p (N/m²)
350	0.28	10430330
400	0.32	10565630
450	0.36	10638510
500	0.40	10677250
550	0.44	10696520
600	0.48	10704340
650	0.52	10705300
700	0.56	10702130
750	0.60	10696500

Table P2.9b   Oxygen Properties

T (K)	v (m³/kmol)	p (N/m²)
350	0.28	10188750
400	0.32	10371810
450	0.36	10477620
500	0.40	10540230
550	0.44	10577460
600	0.48	10599220
650	0.52	10611290
700	0.56	10617160
750	0.60	10619010

Table P2.9c   Carbon Dioxide Properties

T (K)	v (m³/kmol)	p (N/m²)
350	0.28	7649998
400	0.32	8573591
450	0.36	9159231
500	0.40	9547238
550	0.44	9813341
600	0.48	10000960
650	0.52	10136240
700	0.56	10235580
750	0.60	10309620

Table P2.9d   Gas Type and Temperature

Gas	Air	CO_2	Air	O_2	CO_2	O_2
Temp (K)	366	523	728	364	572	685

Project 2.10

We wish to plot the various motions that can occur with a mass-spring-dashpot system. A sketch of such a system is shown in Figure P2.10. Development of the governing equation [1] describing the motion of the mass follows:

Equilibrium state:

$$W - k\bar{y}_0 = 0$$

Nonequilibrium state:

$$W - k\bar{y} - c\bar{y}' = m\bar{y}'' \qquad\qquad \text{(P2.10a)}$$

where

k = the spring constant
c = the damping factor
m = the object mass
$\bar{y}$ = mass displacement from unstretched position
$\bar{y}', \bar{y}''$ = velocity and acceleration of the mass, respectively.

Let y be the mass displacement from the equilibrium position, then

$$\bar{y} = \bar{y}_0 + y, \ \bar{y}' = y' \text{ and } \bar{y}'' = y''$$

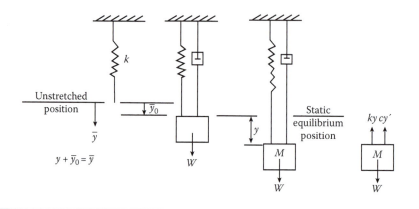

Figure P2.10   Spring-dashpot system.

Substituting these values into Equation (2.10a) gives

$$W - k(\bar{y}_0 + y) - c\,y' = m\,y''$$

Since $W - k\bar{y}_0 = 0$, the above equation reduces to

$$y'' + \frac{c}{m}\,y' + \frac{k}{m}\,y = 0 \qquad\qquad \text{(P2.10b)}$$

We seek a function that satisfies this differential equation. Such a function is one in which its derivatives reproduce the function multiplied by a constant. A function that satisfies this condition is e^{pt}. Assume that $y = a\,e^{pt}$, then

$$y' = pa\,e^{pt}, \quad \text{and} \quad y'' = p^2 a e^{pt}$$

Substituting these terms into the differential equation gives

$$\left(p^2 + \frac{c}{m}\,p + \frac{k}{m}\right)a e^{pt} = 0$$

Now $e^{pt} \neq 0$, so

$$\left(p^2 + \frac{c}{m}\,p + \frac{k}{m}\right) = 0$$

Thus,

$$p = -\frac{1}{2}\frac{c}{m} \pm \frac{1}{2}\sqrt{\left(\frac{c}{m}\right)^2 - 4\frac{k}{m}} = -\frac{1}{2}\frac{c}{m} \pm \sqrt{\left(\frac{c}{2m}\right)^2 - \frac{k}{m}} \qquad \text{(P2.10c)}$$

We see that there are two solutions that satisfy the differential equation. It can be shown that the sum of the two solutions is also a solution to the differential equation. The general solution is

$$y = A\exp\left(-\frac{c}{2m}t + \sqrt{\left(\frac{c}{2m}\right)^2 - \frac{k}{m}}\,t\right) + B\exp\left(-\frac{c}{2m}t - \sqrt{\left(\frac{c}{2m}\right)^2 - \frac{k}{m}}\,t\right)$$

or

$$y = \exp\left(-\frac{c}{2m}t\right)\left\{A\exp\left(\sqrt{\left(\frac{c}{2m}\right)^2 - \frac{k}{m}}\,t\right) + B\exp\left(-\sqrt{\left(\frac{c}{2m}\right)^2 - \frac{k}{m}}\,t\right)\right\} \quad \text{(P2.10d)}$$

where $\exp(x) = e^x$. The type of motion depends on the variables k, c, and m.

If $\left(\dfrac{c}{2m}\right)^2 > \dfrac{k}{m}$, then the above equation is the one to use. The system is said to be *overdamped*.

If $\left(\dfrac{c}{2m}\right)^2 < \dfrac{k}{m}$, then the square root term becomes imaginary and the system is said to be *underdamped*. Noting that $e^{ix} = \cos x + i\sin x$ and $e^{-ix} = \cos x - i\sin x$, Equation (P2.10d) reduces to

$$y = \exp\left(-\frac{c}{2m}t\right)\left\{A\cos\left(\sqrt{\frac{k}{m} - \left(\frac{c}{2m}\right)^2}\,t\right) + B\sin\left(\sqrt{\frac{k}{m} - \left(\frac{c}{2m}\right)^2}\,t\right)\right\} \quad \text{(P2.10e)}$$

If $\left(\dfrac{c}{2m}\right)^2 = \dfrac{k}{m}$, then the square root term is zero and the system is said to be *critically damped*. For this case, the solution is

$$y = A\exp\left(-\frac{c}{2m}t\right) \quad \text{(P2.10f)}$$

(a) Given the following parameters:

$$m = 75 \text{ kg}, \quad k = 87.5\frac{N}{m}, \quad c = 875\frac{N-s}{m}$$

develop a computer program to determine the coefficients A and B for the following cases:

1. $y(0) = 0.5m$, $y'(0) = 1.0\dfrac{m}{s}$.

2. $y(0) = 0.5m$, $y'(0) = -1.0\dfrac{m}{s}$.

3. $y(0) = 0.5m$, $y'(0) = 0\dfrac{m}{s}$.

For each case:

    a. Determine y (t) for $0 \leq t \leq 10$ seconds in steps of 0.1 seconds.
    b. Print out a table of y vs. t every 1 second.
    c. Plot y vs. t for all three cases on the same graph. Label each curve with the value of y'.

  (b) Given the following parameters:

$$m = 25 \ kg, \quad k = 200\frac{N}{m}, \quad c = 5\frac{N-s}{m}$$

$$y(0) = 5m, \quad y'(0) = 0\frac{m}{s}$$

The envelope of the solution graph for this case is given by

$$y = \pm A \exp\left(-\frac{c}{2m}t\right)$$

Determine y (t) for $0 \leq t \leq 40$ seconds in steps of 0.1 seconds.
    Plot y vs. t for both the oscillating function and its envelope on the same graph.

Project 2.11

We wish now to consider the system described in Project 2.10 to be subjected to a driving force F_0. The governing equation then becomes

$$y'' + \frac{c}{m}y' + \frac{k}{m}y = \frac{F_0}{m}\sin \omega t \qquad \text{(P2.11a)}$$

The solution can be obtained by assuming the solution is the sum of the complementary solution plus a particular solution. The complementary solution is the solution to the homogeneous equation, which was obtained in Project 2.10. To obtain the particular solution, y_p, assume

$$y_p = a \sin \omega t + b \cos \omega t \qquad \text{(P2.11b)}$$

then

$$y'_p = \omega(a \cos \omega t - b \sin \omega t)$$

and

$$y_p'' = \omega^2(-a\sin\omega t - b\cos\omega t)$$

Substituting these expressions into Equation (2.11a) gives

$$\left(-a\omega^2 - \frac{c}{m}\omega b + \frac{k}{m}a\right)\sin\omega t + \left(-b\omega^2 + \frac{c}{m}\omega a + \frac{k}{m}b\right)\cos\omega t = \frac{F_0}{m}\sin\omega t \quad (P2.11c)$$

Collecting coefficients of the *sin* and *cos* terms on the left side of Equation (P2.11c) and equating them to the *sin* and *cos* coefficients on the right side of that equation gives

$$\left(\frac{k}{m} - \omega^2\right)a - \frac{c\omega}{m}b = \frac{F_0}{m}$$

$$\frac{c\omega}{m}a + \left(\frac{k}{m} - \omega^2\right)b = 0$$

Solving the above two equations for *a* and *b* gives

$$a = \frac{F_0}{m} \times \frac{\frac{k}{m} - \omega^2}{\left(\frac{c\omega}{m}\right)^2 + \left(\frac{k}{m} - \omega^2\right)^2} \quad (P2.11d)$$

$$b = -\frac{F_0}{m} \times \frac{\frac{c\omega}{m}}{\left(\frac{c\omega}{m}\right)^2 + \left(\frac{k}{m} - \omega^2\right)} \quad (P2.11e)$$

Substituting Equations (P2.11d) and (P2.11e) into Equation (P2.11b) gives

$$y_p = \frac{F_0}{m} \times \frac{1}{\left(\frac{c\omega}{m}\right)^2 + \left(\frac{k}{m} - \omega^2\right)^2}\left\{\left(\frac{k}{m} - \omega^2\right)\sin\omega t - \frac{c\omega}{m}\cos\omega t\right\} \quad (P2.11f)$$

We can rewrite Equation (P2.11f) using the trig identity

$$\alpha\sin\omega t + \beta\cos\omega t = \gamma\sin(\omega t - \phi)$$

where

$$\gamma = \sqrt{\alpha^2 + \beta^2}$$

and

$$\phi = \tan^{-1}\frac{\beta}{\alpha}$$

Applying these relations to Equation (2.11f) gives

$$y_p = \frac{F_0}{m} \times \frac{1}{\sqrt{\left(\frac{k}{m}-\omega^2\right)^2 + \left(\frac{c\omega}{m}\right)^2}} \sin(\omega t - \phi) \qquad (P2.11g)$$

For a system with no driving force and no damping, the system would oscillate at a frequency, $\omega_n = \sqrt{\frac{k}{m}}$ (see Equation P2.10e). It is convenient to introduce the ratio $\zeta = \frac{c}{c_c}$ where $c_c = 2m\omega_n$. After some algebraic manipulation, Equation (P2.11g) can be put into the form

$$y_p = \frac{F_0}{k} \frac{1}{\sqrt{\left(1-\frac{\omega}{\omega_n}\right)^2 + \left(2\zeta\frac{\omega}{\omega_n}\right)^2}} \sin(\omega t - \phi) \qquad (P2.11h)$$

The term

$$\frac{F_0}{k} \frac{1}{\sqrt{\left(1-\frac{\omega}{\omega_n}\right)^2 + \left(2\zeta\frac{\omega}{\omega_n}\right)^2}}$$

is the amplitude of the oscillation. For a given $\frac{F_0}{k}$, the larger the term

$$\frac{1}{\sqrt{\left(1-\frac{\omega}{\omega_n}\right)^2 + \left(2\zeta\frac{\omega}{\omega_n}\right)^2}},$$

the larger is the amplitude.

$$\text{Let ampl} = \frac{1}{\sqrt{\left(1-\frac{\omega}{\omega_n}\right)^2 + \left(2\zeta\frac{\omega}{\omega_n}\right)^2}}.$$

Construct a MATLAB program to create a plot of ampl vs. ω/ω_n for values of $\zeta = 1.0$, 0.5, 0.25, 0.10, 0.05, and $0 < \omega/\omega_n < 2$ in steps of 0.01. What happens as $\omega \to \omega_n$?

Project 2.12

The motion of a piston in an internal combustion engine is shown in Figure P2.12a,b. The piston's position, s, as seen from the crank shaft center can be

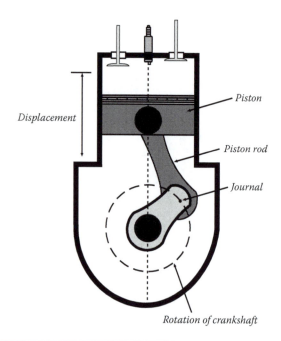

Figure P2.12a   Sketch of a typical piston-cylinder engine.

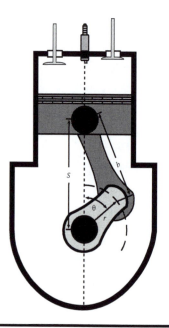

Figure P2.12b   Geometry of piston motion.

determined by the Law of Cosines; that is,

$$b^2 = s^2 + r^2 - 2sr\cos\theta$$

or

$$s^2 - (2r\cos\theta)s + (r^2 - b^2) = 0 \qquad \text{(P2.12a)}$$

where
  b = the length of the piston rod.
  r = the radius of the crankshaft.

Equation (P2.12a) is a quadratic equation in s and therefore

$$s = \frac{1}{2}(2r\cos\theta + \sqrt{4r^2\cos^2\theta - 4(r^2 - b^2)}) = r\cos\theta + \sqrt{r^2(\cos^2\theta - 1) + b^2}$$

or

$$s = r\cos\theta + \sqrt{b^2 - r^2\sin^2\theta} \qquad \text{(P2.12b)}$$

The piston is constrained to move in the vertical direction and its position, s, varies as the crankshaft rotates. The angle, θ, varies with time, t, and can be expressed in terms of the rotational speed, ω, of the crankshaft. The angle θ is thus given by

$$\theta = 2\pi\omega t \qquad \text{(P2.12c)}$$

where ω is in revolutions per second. Substituting Equation (P2.12c) into Equation (P2.12b) gives

$$s(t) = r\cos(2\pi\omega t) + \sqrt{b^2 - r^2\sin^2(2\pi\omega t)} \qquad \text{(P2.12d)}$$

The piston velocity, v, can be obtained by taking the derivative of Equation (P2.12d) with respect to time, giving

$$v(t) = -2\pi\omega r\sin(2\pi\omega t) - \frac{2\pi\omega r^2\sin(2\pi\omega t)\cos(2\pi\omega t)}{\sqrt{b^2 - r^2\sin^2(2\pi\omega t)}} \qquad \text{(P2.12e)}$$

The piston acceleration, a, can be obtained by taking the derivative of Equation (P2.12e) with respect to time, giving

$$a(t) = -4\pi^2\omega^2 r\cos(2\pi\omega t) - \frac{4\pi^2\omega^2 r^4\sin^2(2\pi\omega t)\cos^2(2\pi\omega t)}{[b^2 - r^2\sin^2(2\pi\omega t)]^{3/2}}$$

$$- \frac{4\pi^2\omega^2 r^2\cos^2(2\pi\omega t)}{\sqrt{b^2 - r^2\sin^2(2\pi\omega t)}} + \frac{4\pi^2\omega^2 r^2\sin^2(2\pi\omega t)}{\sqrt{b^2 - r^2\sin^2(2\pi\omega t)}} \qquad \text{(P2.12f)}$$

(a) In MATLAB, create a matrix consisting of *s* vs. *t*, v vs. *t*, and *a* vs. *t* for $0 \leq t \leq 0.01$ second. Use 50 subdivisions on the *t* domain. Take $r = 9$ cm, $\omega = 100$ revolutions per second, and $b = 14$ cm. Plot *s* vs. *t*, v vs. *t*, and *a* vs. *t* on three separate graphs.

(b) Using MATLAB's *max* function and the matrix obtained in part (a), determine the approximate maximum velocity and maximum acceleration in one revolution, and print out those values.

(c) Plot on a single graph *s* vs. *t* for ω = [50 100 150 200] revolutions per second.

References

1. Thomson, W. T., *Theory of Vibration with Applications*, Prentice-Hall, Englewood Cliffs, NJ, 1972.

Chapter 3

Matrices

3.1 Matrix Operations

■ A rectangle array of numbers of the form shown below is called a matrix.

$$
\begin{bmatrix}
a_{11} & a_{12} & .. & a_{1m} \\
a_{21} & a_{22} & .. & a_{2m} \\
. & . & .. & \\
. & . & .. & \\
a_{m1} & a_{m2} & .. & a_{mm}
\end{bmatrix}
$$

The numbers a_{ij} in the array are called the elements of the matrix.

■ A matrix of M rows and one column is called a column vector.
■ A matrix of one row and n columns is called a row vector.
■ Matrices obey certain rules of addition, subtraction, and multiplication.
■ If matrices A and B have the same number of rows and columns, then

$$
C = A + B = \begin{bmatrix}
(a_{11} + b_{11}) & (a_{12} + b_{12}) & (a_{13} + b_{13}).. \\
(a_{21} + b_{21}) & (a_{22} + b_{22}) & (a_{23} + b_{23}).. \\
.. & .. & .. \\
(a_{m1} + b_{m1}) & (a_{m2} + b_{m2})... &
\end{bmatrix}
$$

$$C = A - B = \begin{bmatrix} (a_{11} - b_{11}) & (a_{12} - b_{12}) & (a_{13} - b_{13}).. \\ (a_{21} - b_{21}) & .. & .. \\ .. & .. & .. \\ (a_{m1} - b_{m1}) & (a_{m2} - b_{m2})... & \end{bmatrix}$$

- Addition and subtraction of matrices A and B are only defined if A and B have the same number of rows and columns.
- The product AB is only defined if the number of columns in A equals the number of rows in B.
- If C = AB, where

$$A = \begin{bmatrix} a_{11} & a_{12} & a_{13} \\ a_{21} & a_{22} & a_{23} \end{bmatrix}, \quad B = \begin{bmatrix} b_{11} & b_{12} \\ b_{21} & b_{22} \\ b_{31} & b_{32} \end{bmatrix}$$

then

$$C = \begin{bmatrix} (a_{11}b_{11} + a_{12}b_{21} + a_{13}b_{31}) & (a_{11}b_{12} + a_{12}b_{22} + a_{13}b_{32}) \\ (a_{21}b_{11} + a_{22}b_{21} + a_{23}b_{31}) & (a_{21}b_{12} + a_{22}b_{22} + a_{23}b_{32}) \end{bmatrix}$$

- If A has m rows and B has k columns, then C = AB will have m rows and k columns.

$$c_{ij} = \sum_{k=1}^{K} a_{ik}b_{kj}$$

In MATLAB®, the multiplication of matrices A and B is A*B.

- Transpose of a matrix (rows become columns and columns become rows). If

$$A = \begin{bmatrix} 2 & 5 & 1 \\ 7 & 3 & 8 \\ 4 & 5 & 21 \\ 16 & 3 & 10 \end{bmatrix}$$

then

$$A^T = \begin{bmatrix} 2 & 7 & 4 & 16 \\ 5 & 3 & 5 & 3 \\ 1 & 8 & 21 & 10 \end{bmatrix}$$

In MATLAB, $A^T = A'$

■ Dot product of two vectors

$$A \cdot B = \sum a_i b_i$$

If $A = [4 \quad -1 \quad 3]$, $B = [-2 \quad 5 \quad 2]$,
then $A \cdot B = -8 - 5 + 6 = -7$
In MATLAB

$$A \cdot B = dot(A, B)$$

■ Matrix Inverse

The inverse of a matrix, denoted A^{-1}, is a matrix such that

$$A^{-1}A = AA^{-1} = I = \begin{bmatrix} 1 & 0 & 0 \\ 0 & 1 & 0 \\ 0 & 0 & 1 \end{bmatrix} \quad \text{(for a } 3 \times 3 \text{ matrix)}$$

In MATLAB, $A^{-1} = \text{inv}(A)$

■ The Identity Matrix, I

I is a matrix where the main diagonal elements are all 1 and all other elements are 0.

$$I A = A I = A$$

$$\begin{bmatrix} a_{11} & a_{12} & a_{13} \\ a_{21} & a_{22} & a_{23} \\ a_{31} & a_{32} & a_{33} \end{bmatrix} \begin{bmatrix} 1 & 0 & 0 \\ 0 & 1 & 0 \\ 0 & 0 & 1 \end{bmatrix} = \begin{bmatrix} (a_{11} + 0 + 0) & (0 + a_{12} + 0) & (0 + 0 + a_{13}) \\ (a_{21} + 0 + 0) & (0 + a_{22} + 0) & (0 + 0 + a_{23}) \\ (a_{31} + 0 + 0) & (0 + a_{22} + 0) & (0 + 0 + a_{33}) \end{bmatrix}$$

■ The Determinant of a Matrix

In MATLAB the determinant of matrix A is written

$$det\,(A)$$

■ Element-by-Element Operations

Given: $A = [a_1 \ a_2 \ a_3]$, $B = [b_1 \ b_2 \ b_3]$,

$$C = A \,.* B = [a_1 b_1 \quad a_2 b_2 \quad a_3 b_3]$$

$$C = A \,./B = \begin{bmatrix} \dfrac{a_1}{b_1} & \dfrac{a_2}{b_2} & \dfrac{a_3}{b_3} \end{bmatrix}$$

$$\text{sum (A)} = a_1 + a_2 + a_3$$

$$\text{Given:} \quad A = \begin{bmatrix} a_{11} & a_{12} & a_{13} \\ a_{21} & a_{22} & a_{23} \\ a_{31} & a_{32} & a_{33} \end{bmatrix}$$

$$sum(A) = [(a_{11} + a_{21} + a_{31}) \quad (a_{12} + a_{22} + a_{32}) \quad (a_{13} + a_{23} + a_{33})]$$

■ *Given:* $A = [a_1 \; a_2 \; a_3]$, $B = [b_1 \; b_2 \; b_3]$

$$A \cdot B = sum(A \,.* B) = sum([a_1 b_1 \; a_2 b_2 \; a_3 b_3]) = a_1 b_1 + a_2 b_2 + a_3 b$$

The following example illustrates several matrix operations.

Example 3.1

```
% matrixalg.m
% This program demonstrates matrix algebra in MATLAB
clear; clc;
a=[1 5 9]
b=[2 6 12]
c=a+b
d=dot(a,b)
e=a.*b
f=a./b
g=sum(a.*b)
h=a*b'
```

3.2 System of Linear Equations

Given the set of equations

$$a_{11}x_1 + a_{12}x_2 + a_{13}x_3 + \cdots + a_{1n}x_n = c_1$$

$$a_{21}x_1 + a_{22}x_2 + a_{23}x_3 + \cdots + a_{2n}x_n = c_2$$

$$\vdots \qquad\qquad\qquad\qquad\qquad\qquad (3.1)$$

$$a_{n1}x_1 + a_{n2}x_2 + a_{n3}x_3 + \cdots + a_{nn}x_n = c_n$$

This set can be represented by the matrix equation

$$AX = C \tag{3.2}$$

where

$$A = \begin{bmatrix} a_{11} & a_{12} & a_{13} & \cdots & a_{1n} \\ a_{21} & a_{22} & a_{23} & \cdots & a_{2n} \\ \vdots & & & & \\ a_{n1} & a_{n2} & a_{n3} & \cdots & a_{nn} \end{bmatrix}, \quad X = \begin{bmatrix} x_1 \\ x_2 \\ \vdots \\ x_n \end{bmatrix}, \quad C = \begin{bmatrix} c_1 \\ c_2 \\ \vdots \\ c_n \end{bmatrix}$$

Matrix A has n rows and n columns
Matrix X has n rows and one column
Matrix C has n rows and one column
In matrix algebra, X can be obtained by multiplying both sides by A^{-1}; that is,

$$\underbrace{A^{-1}A}_{I} X = A^{-1}C \quad \Rightarrow IX = X = A^{-1}C$$

3.2.1 The inv Function

To solve a system of linear equations in MATLAB, one can use

$$X = inv\,(A)*C \tag{3.3}$$

The method of solving a system of linear equations by using the *inv* function is more complicated than a method called Gauss elimination, which is discussed below.

3.2.2 The Gauss Elimination Function

MATLAB represents the Gauss Elimination method by use of the backslash; that is, if $AX = B$, then

$$X = A\backslash B \;(X \text{ will be solved by Gauss elimination})$$

The following example solves the three-linear-equation system shown below:

$$3x_1 + 2x_2 - x_3 = 10$$
$$-x_1 + 3x_2 + 2x_3 = 5$$
$$x_1 - x_2 - x_3 = -1$$

3.2.3 Examples

Example 3.2

```
% matrix2.m
% This program solves a simple linear system of equations
clc; clear;
A=[3 2 -1; -1 3 2; 1 -1 -1]
C=[10 5 -1]'
x=inv(A)*C
y=A\C
A*x
% To print the A matrix & x to a file use:
fid=fopen('output.dat','w');
fprintf(fid,'The A matrix is:\n');
for i=1:3
    for j=1:3
        fprintf(fid,' %3.1f',A(i,j));
    end
    fprintf(fid,'\n');
end
fprintf(fid,'\n');
% An alternative way of printing out the matrix.
fprintf(fid,'The A matrix is:\n');
for i=1:3
    fprintf(fid,'%5.1f %5.1f %5.1f \n',A(i,:));
end
fprintf(fid,'\n');
fprintf(fid,' x= %5.2f %5.2f %5.2f',x);
fclose(fid);
```

Example 3.3 An Example in Statics

An engineering example involving a large system of linear equations can be found in the field of statics. The problem is to solve for the internal forces in a truss. For illustration purposes a truss consisting of five structural members is considered (see Figure 3.1).

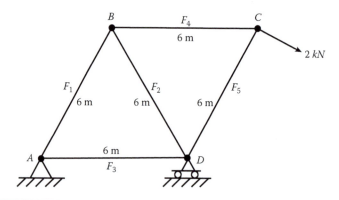

Figure 3.1   Simple truss.

As can be seen in Figure 3.1, there are five structural members; thus, there are five unknowns (reactions can be obtained from the equilibrium equations involving only the external forces and are thus considered to be known). Also note that there are four joints. We can write two scalar equations at each joint; that is,

$$\sum_n (F_x)_n = 0 \quad \text{and} \quad \sum_n (F_y)_n = 0 \tag{3.4}$$

However, we can only use five independent equations. We will write two scalar equations at Joints A and B and one scalar equation (in the x direction) at Joint C. The system of equations can then be represented by the matrix equation:

$$AF = P \tag{3.5}$$

where
    A = coefficient matrix
    F = column matrix of the unknown internal forces
    P = column matrix involving given external forces

The coefficient matrix A will be made up of elements a_{ij} where the first index is the equation number and the second index represents the force member associated with that element. In writing the equations, take all unknown internal forces to be in tension. If F_i comes out negative, then the internal force is in compression. First we solve for the reactions at A and D (see Figure 3.2).

$$\sum M_A = 0 = 6D_y - 2\cos(30°) \times 6\sin(60°) - 2\sin(30°) \times 6(1 + \cos(60°))$$

or

$$D_y = \frac{18}{6} = 3 \text{ kN}$$

$$\sum F_x = 0 = A_x + 2\cos(30°) \qquad \Rightarrow A_x = -2\cos(30°) \text{ kN}$$

$$\sum F_y = 0 = A_y + D_y - 2\sin(30°) \quad \Rightarrow A_y = -D_y + 2\sin(30°) \text{ kN}$$

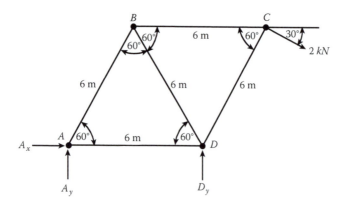

Figure 3.2   Reactions of simple truss.

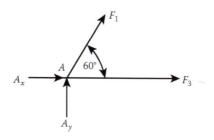

Figure 3.3   Internal forces at Joint A.

■ Internal forces at Joint A (see Figure 3.3):

$$\sum F_x = 0 = F_1 \cos(60°) + F_3 + A_x \tag{1}$$

$$a_{11} = \cos(60°), \ a_{13} = 1, \ P_1 = -A_x$$

$$\sum F_y = 0 = F_1 \sin(60°) + A_y \tag{2}$$

$$a_{21} = \sin(60°), \ P_2 = -A_y$$

■ Internal forces at Joint B (see Figure 3.4):

$$\sum F_x = 0 = F_4 + F_2 \cos(60°) - F_1 \cos(60°) \tag{3}$$

$$a_{32} = \cos(60°), \ a_{31} = -\cos(60°), \ a_{34} = 1, \ P_3 = 0$$

$$\sum F_y = 0 = -F_1 \sin(60°) - F_2 \sin(60°) \tag{4}$$

$$a_{41} = -\sin(60°), \ a_{42} = -\sin(60°), \ P_4 = 0$$

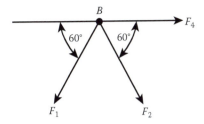

Figure 3.4   Internal forces at Joint B.

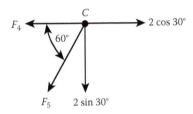

Figure 3.5   Internal forces at Joint C.

■ Internal forces at Joint C (see Figure 3.5):

$$\sum F_x = 0 = -F_4 - F_5 \cos(60°) + 2 \cos(30°)$$ (5)

$$a_{54} = -1, \; a_{55} = -\cos(60°), \; P_5 = -2 \cos(30°)$$

Since Joint D has not been used in establishing the coefficient matrix, it can be used as a check on the obtained solution; that is, the sum of all the forces acting at Joint D, both in the x and y directions, must be zero.

■ Forces at Joint D (see Figure 3.6):

$$CKDx = -F_3 - F_2 \cos(60°) + F_5 \cos(60°)$$ (3.6)

$$CKDy = F_2 \sin(60°) + F_5 \sin(60°) + D_y$$ (3.7)

The following program solves for the internal forces in the truss.

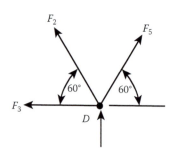

Figure 3.6   Internal forces at Joint D.

Example 3.4

```
% STATICS_TEST_F07.m
% This program solves for the internal forces in a truss by the matrix
% method
clear; clc;
th60=pi/3; th30=pi/6; ie=5; je=5;
a=zeros(5); p=zeros(5,1);
sth60=sin(th60); cth60=cos(th60); sth30=sin(th30); cth30=cos(th30);
% Calculation of reactions:
% Sum of moments about Joint A gives the vertical component of
% the reaction at joint
Dy=(2*cth30*6*sth60+2*sth30*6*(1+cth60))/6;
% Sum of external forces in the x direction gives the horizontal component
% of the reaction at Joint A.
Ax=-2*cth30;
% the sum of external forces in vertical direction gives the vertical
%component of the reaction at Joint A.
Ay=2*sth30-Dy;
fid=fopen('output.dat','w');
fprintf(fid,' statics_problem.m \n');
fprintf(fid,' Program solves for the internal forces of a truss \n\n');
fprintf(fid,' Reactions in N \n');
fprintf(fid,' Ax=%8.3f \t\t Ay=%8.3f \t\t Dy=%8.1f \n\n',Ax,Ay,Dy);
% Overwrite the non-zero elements of matrix a and matrix p.
a(1,1)=cth60; a(1,3)=1.0; p(1)=-Ax;
a(2,1)=sth60; p(2)=-Ay;
a(3,1)=-cth60; a(3,2)=cth60; a(3,4)=1.0;
a(4,1)=-sth60; a(4,2)=-sth60;
a(5,4)=-1.0; a(5,5)=-cth60; p(5)=-2*cth30;
fprintf(fid,' A matrix \n\n');
jindex=1:je;
fprintf(fid,'  ');
for i=1:ie
    fprintf(fid,'%6i',jindex(i));
end
fprintf(fid,'\n');
fprintf(fid,' ---------------------------------------------------------------\n');
```

```
for i=1:ie
    fprintf(fid,'%4i ',i);
    for j=1:je
        fprintf(fid,'%8.4f',a(i,j));
    end
fprintf(fid,'\n');
end
F=a\p;
fprintf('f1=%5.2f \n,f2=%5.2f \n,f3=%5.2f \n,f4=%5.2f \n,f5=%5.2f \n',F);
fprintf(fid,'\n\n');
fprintf(fid,'Internal forces,F(1)-F(5)& external forces p(i),in N \n\n');
fprintf(fid,' Member # F(i) Equation # p(i) \n');
fprintf(fid,'    ===================================================\n');
for i=1:ie
    fprintf(fid,' %3.0f %9.4f %3.0f %5.3f \n',i,F(i),i,p(i));
end
% check if the sum of the horizontal components at Joint D is zero:
fprintf(fid,'\n check if sum of the y components at Joint D is zero \n');
ckDx=-F(2)*cth60+F(5)*cth60-F(3);
fprintf(fid,' ckDx=%12.4e \n',ckDx);
% check if the sum of the vertical components at Joint D is zero:
fprintf(fid,'\n check if sum of the y components at Joint D is zero \n');
ckDy=Dy+F(2)*sth60+F(5)*sth60;
fprintf(fid,' ckDy=%12.4e \n',ckDy);
fclose(fid);
```

3.3 Gauss Elimination

In engineering we are frequently confronted with the problem of solving a set of linear equations. For n equations in n unknowns, it is convenient to express the set in the following form:

$$a_{11}x_1 + a_{12}x_2 + a_{13}x_3 + \cdots + a_{1n}x_n = c_1$$

$$a_{21}x_1 + a_{22}x_2 + a_{23}x_3 + \cdots + a_{2n}x_n = c_2$$

$$\cdots$$

$$\cdots$$

$$a_{n1}x_1 + a_{n2}x_2 + a_{n3}x_3 + \cdots + a_{nn}x_n = c_n \qquad (3.8)$$

Cramer's rule gives a theoretical method for obtaining a solution. Use for $n \le 3$. However, for larger n, the calculations become excessive. For example, for $n = 10$, the number of calculations required by the use of determinants is approximately 3×10^8. Using the Gauss–Jordan method, this is reduced to approximately 600 calculations. The system of equations, Equation (3.8), can be expressed in matrix form; that is,

$$AX = C \qquad (3.9)$$

where

$$
A = \begin{bmatrix}
a_{11} & a_{12} & a_{13} & \cdots & a_{1n} \\
a_{21} & a_{22} & a_{23} & \cdots & a_{2n} \\
\vdots & & & & \\
a_{n1} & a_{n2} & a_{n3} & \cdots & a_{nn}
\end{bmatrix}
$$

$$
X = \begin{bmatrix} x_1 \\ x_2 \\ x_3 \\ \vdots \\ x_n \end{bmatrix}, \qquad
C = \begin{bmatrix} c_1 \\ c_2 \\ c_3 \\ \vdots \\ c_n \end{bmatrix}
$$

Note: The number of columns in A has to equal the number of rows in X, otherwise matrix multiplication is not defined. In the Gauss Elimination method, the original system is reduced to an equivalent triangular set that can readily be solved by back substitution.

The reduced equivalent set would appear like the following set of equations:

$$
\tilde{a}_{11}x_1 + \tilde{a}_{12}x_2 + \tilde{a}_{13}x_3 + \cdots + \tilde{a}_{1n}x_n = \tilde{c}_1
$$

$$
\tilde{a}_{22}x_2 + \tilde{a}_{23}x_3 + \cdots + \tilde{a}_{2n}x_n = \tilde{c}_2
$$

$$
\tilde{a}_{33}x_3 + \cdots + \tilde{a}_{3n}x_n = \tilde{c}_3
$$

$$
\cdots
$$

$$
\cdots
$$

$$
\tilde{a}_{n-1,n-1}x_{n-1} + \tilde{a}_{n-1,n}x_n = \tilde{c}_{n-1}
$$

$$
\tilde{a}_{n,n}x_n = \tilde{c}_n
$$

Then

$$
x_n = \tilde{c}_n / \tilde{a}_{n,n}
$$

$$
x_{n-1} = \frac{1}{\tilde{a}_{n-1,n-1}} (\tilde{c}_{n-1} - \tilde{a}_{n-1,n}x_n)
$$

etc.

To accomplish the reduced equivalent set it is convenient to augment the coefficient matrix with the C matrix; that is,

$$A = \begin{bmatrix} a_{11} & a_{12} & a_{13} & \cdots & a_{1n} & | & c_1 \\ a_{21} & a_{22} & a_{23} & \cdots & a_{2n} & | & c_2 \\ a_{31} & a_{32} & a_{33} & & & | & c_3 \\ \cdots & & & & & | & \\ \cdots & & & & & | & \\ a_{n1} & a_{n2} & a_{n3} & \cdots & a_{nn} & | & c_n \end{bmatrix} \qquad (3.10)$$

The following procedure is used to obtain the reduced equivalent set:

■ Multiply the first row of Equation (3.10) by a_{21}/a_{11} and subtract from the second row, giving

$$a'_{21} = a_{21} - \frac{a_{21}}{a_{11}} \times a_{11}, \quad a'_{22} = a_{22} - \frac{a_{21}}{a_{11}} \times a_{12}$$

$$a'_{23} = a_{23} - \frac{a_{21}}{a_{11}} \times a_{13}, \quad \text{etc.}$$

$$c'_2 = c_2 - \frac{a_{21}}{a_{11}} \times c_1$$

This gives $a'_{21} = 0$

■ For the third row: Multiply the first row of Equation (3.10) by a_{31}/a_{11} and subtract from row 3, giving

$$a'_{31} = a_{31} - \frac{a_{31}}{a_{11}} \times a_{11}, \quad a'_{32} = a_{32} - \frac{a_{31}}{a_{11}} \times a_{12}$$

$$a'_{33} = a_{33} - \frac{a_{31}}{a_{11}} \times a_{13}, \quad \text{etc.}$$

$$c'_3 = c_3 - \frac{a_{31}}{a_{11}} \times c_1$$

This gives $a'_{31} = 0$

This process is carried out for rows 2, 3, 4,..., n. The original row 1 is kept in its original form. All other rows have been modified and the new coefficients are designated by a′. Except for the first row in Equation (3.10), the resulting set does not contain x_1.

For this step, the first row of Equation (3.10) was used as the pivot row and a_{11} as the pivot element. We now use the new row 2 as the pivot row.

Multiply the new row 2 by $\dfrac{a'_{32}}{a'_{22}}$ and subtract from row 3, giving

$$a''_{32} = a'_{32} - \frac{a'_{32}}{a'_{22}} \times a'_{22}, \quad a''_{33} = a'_{33} - \frac{a'_{32}}{a'_{22}} \times a'_{23}$$

$$a''_{34} = a'_{34} - \frac{a'_{32}}{a'_{22}} \times a'_{24}, \quad \text{etc.}$$

$$c''_3 = c'_3 - \frac{a'_{32}}{a'_{22}} \times c'_2$$

This gives $a''_{32} = 0$

Multiply the new row 2 by $\dfrac{a'_{42}}{a'_{22}}$ and subtract from row 4, giving

$$a''_{42} = a'_{42} - \frac{a'_{42}}{a'_{22}} \times a'_{22}, \quad a''_{43} = a'_{43} - \frac{a'_{42}}{a'_{22}} \times a'_{23}$$

$$c''_4 = c'_4 - \frac{a'_{42}}{a'_{22}} \times c'_2$$

This gives $a''_{42} = 0$

This process is continued for rows 3, 4, 5,..., n.

Except for the second row 2 the set does not contain x_2.

The next row is used as a pivot row and the process is continued until the $(n-1)$ row is used as the pivot row. When this is complete the new system is triangular and the system can be solved by back substitution.

The general expression for the new coefficients is

$$a'_{ij} = a_{ij} - \frac{a_{ik}}{a_{kk}} \times a_{kj} \qquad \textit{for } i = k+1, k+2,...,n$$

$$j = k+1, k+2,...,n$$

where k is the pivot row.

These operations only affect the a_{ij} and c_j terms. Thus, we need only operate on the A and C matrices.

Example 3.5

Given:

$$x_1 - 3x_2 + x_3 = -4$$

$$-3x_1 + 4x_2 + 3x_3 = -10$$

$$2x_1 + 3x_2 - 2x_3 = 18$$

The augmented coefficient matrix is

$$A_{aug} = \begin{bmatrix} 1 & -3 & 1 & | & -4 \\ -3 & 4 & 3 & | & -10 \\ 2 & 3 & -2 & | & 18 \end{bmatrix} \quad \text{multiply row 1 by } \frac{a_{21}}{a_{11}} = \frac{-3}{1}$$

and subtract from row 2

row 2 becomes: $(-3+3), (4-9), (3+3), (-10-12) = (0, -5, 6, -22)$

The new matrix is

$$Equiv \ A_{aug} = \begin{bmatrix} 1 & -3 & 1 & | & -4 \\ 0 & -5 & 6 & | & -22 \\ 2 & 3 & -2 & | & 18 \end{bmatrix} \quad \text{multiply row 1 by } \frac{a_{31}}{a_{11}} = \frac{2}{1}$$

and subtract from row 3

row 3 becomes: $(2-2), (3+6), (-2-2), (18+8) = (0, 9, -4, 26)$

The new matrix is

$$Equiv \ A_{aug} = \begin{bmatrix} 1 & -3 & 1 & | & -4 \\ 0 & -5 & 6 & | & -22 \\ 0 & 9 & -4 & | & 26 \end{bmatrix}$$

We now use row 2 as the pivot row.

$$Equiv\ A_{aug} = \begin{bmatrix} 1 & -3 & 1 & | & -4 \\ 0 & -5 & 6 & | & -22 \\ 0 & 9 & -4 & | & 26 \end{bmatrix} \quad \text{multiply row 2 by } \frac{a_{32}}{a_{22}} = -\frac{9}{5}$$

and subtract from row 3

row 3 becomes:

$$(0-0), \left(9 - \frac{9}{5} \times 5\right), \left(-4 + \frac{9}{5} \times 6\right), \left(26 - \frac{9}{5} \times 22\right) = \left(0, 0, \frac{34}{5}, -\frac{68}{5}\right)$$

The new matrix is

$$Equiv\ A_{aug} = \begin{bmatrix} 1 & -3 & 1 & | & -4 \\ 0 & -5 & 6 & | & -22 \\ 0 & 0 & \frac{34}{5} & | & -\frac{68}{5} \end{bmatrix}$$

The system is now triangular.

$$Equiv\ A\ X = C$$

gives

$$x_1 - 3x_2 + x_3 = -4$$

$$-5x_2 + 6x_3 = -22$$

$$\frac{34}{5}x_3 = -\frac{68}{5}$$

$$\frac{34}{5}x_3 = -\frac{68}{5} \quad \Rightarrow x_3 = -2$$

$$x_2 = \frac{1}{5}[22 + 6(-2)] = 2$$

$$x_1 = -4 + 3(2) + 2 = 4$$

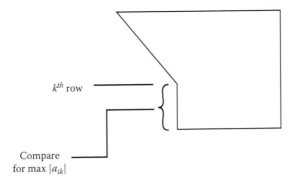

k^{th} row

Compare
for max $|a_{ik}|$

Figure 3.7   Row interchange.

Two important considerations:

1. If a_{kk} is zero, where k is the pivot row, then the process cannot be carried out.
2. Greater accuracy in the solution is obtained if the pivot element is the absolute maximum available from the set. That is, if the pivot row is k, one compares the a_{ik}'s for $i = k + 1, k + 2,..., n$ (see Figure 3.7). If $|a_{ik}|_{max} \neq |a_{kk}|$, then the row containing the $|a_{ik}|_{max}$ is interchanged with the kth row. This only affects the ordering of the equations and does not affect the solution.

If after row interchange is carried out and one of the a_{kk}'s remains zero, then the system is singular and no solution can be obtained.

One last consideration: It can be shown that if the magnitude of the pivot element is much smaller than other elements in the matrix, the use of the small pivot element will cause a decrease in the accuracy of the solution. To check if this is the case, one can first scale the equations; that is, divide each equation by the absolute maximum coefficient in that equation. This makes the absolute maximum coefficient in that equation equal to 1.0. If $|a_{kk}|$ in the pivot row $\ll 1$, then the solution may be inaccurate.

3.4 The Gauss–Jordan Method

The Gauss–Jordan method is a modification of the Gauss Elimination method. It also treats the problem of solving a system of linear equations of the form AX = C. In this method the objective is to obtain an equivalent coefficient matrix that, except for the main diagonal, all elements are zero. The method starts out, as in the Gauss Elimination method, by finding an equivalent matrix that is triangular. It then continues, assuming that A is an $n \times n$ matrix, using the n row as the pivot row, multiplying the nth row by $a_{n-1,n}/a_{n,n}$, and subtracting the result from row $n - 1$, thus making the new $a_{n-1,n} = 0$. The process is repeated for rows $n - 2, n - 3,..., 1$.

Example 3.6

As an example, this method is applied to Example 3.5. Starting from the triangular equivalent matrix

$$Equiv\ A_{aug} = \begin{bmatrix} 1 & -3 & 1 & | & -4 \\ 0 & -5 & 6 & | & -22 \\ 0 & 0 & \dfrac{34}{5} & | & -\dfrac{68}{5} \end{bmatrix}$$

multiply row 3 by $a_{23}/a_{33} = 6 \times 5/34$ and subtract from row 2. Row 2 becomes

$$(0 - 30/34 \times 0),\ (-5 - 30/34 \times 0),\ (6 - 30/34 \times 34/5),$$
$$(-22 + 30/34 \times 68/5) = (0\ \ -5\ \ \ 0\ \ -10)$$

Multiply row 3 by $a_{13}/a_{33} = 5/34$ and subtract from row 1. Row 1 becomes

$$(1 - 5/34 \times 0),\ (-3 - 5/34 \times 0),\ (1 - 5/34 \times 34/5),\ (-4 + 5/34 \times 68/5) = (1\ -3\ \ \ 0\ -2)$$

The new *Equiv A*$_{aug}$ becomes

$$Equiv\ A_{aug} = \begin{bmatrix} 1 & -3 & 0 & | & -2 \\ 0 & -5 & 0 & | & -10 \\ 0 & 0 & \dfrac{34}{5} & | & -\dfrac{68}{5} \end{bmatrix}$$

Now row 2 is used as the pivot row. Multiply row 2 by $a_{12}/a_{22} = 3/5$ and subtract from row 1. Row 1 becomes

$$(1 - 3/5 \times 0),\ (-3 + 3/5 \times 5),\ (0 - 3/5 \times 0),\ (-2 + 3/5 \times 10) = (1\ \ \ 0\ \ \ 0\ \ \ 4)$$

The new *Equiv A*$_{aug}$ becomes

$$Equiv\ A_{aug} = \begin{bmatrix} 1 & 0 & 0 & | & 4 \\ 0 & -5 & 0 & | & -10 \\ 0 & 0 & \dfrac{34}{5} & | & -\dfrac{68}{5} \end{bmatrix}$$

Thus, the equivalent set of equations becomes

$$x_1 = 4, -5\, x_2 = -10,\ 34/5\ x_3 = -68/5$$
$$\text{or } x_1 = 4,\ x_2 = 2,\ x_3 = -2$$

which is the same answer that was obtained earlier by the Gauss Elimination method.

3.5 Number of Solutions

Suppose a Gauss Elimination program is carried out and the following results are obtained:

$$a_{11}\, x_1 + a_{12}\, x_2 + a_{13}\, x_3 + \cdots + a_{1n}\, x_n = c_1$$
$$a_{22}\, x_2 + a_{23}\, x_3 + \cdots + a_{2n}\, x_n = c_2$$
$$a_{33}\, x_3 + \cdots + a_{3n}\, x_n = c_3$$
$$+ \cdots\cdots\cdots\cdots\cdots$$
$$+ \cdots\cdots\cdots\cdots\cdots$$
$$a_{rr}\, x_r + \cdots = c_r$$
$$0 = c_{r+1}$$
$$0 = c_{r+2}$$
$$\cdots$$
$$\cdots$$
$$0 = c_n$$

where $r < n$ and $a_{11}, a_{22}, \ldots, a_{rr}$ are not zero. There are two possible cases:

(1) No solution if any one of the c_{r+1} through c_n is not zero.
(2) Infinitely many solutions if c_{r+1} through c_n are all zero.

If $r = n$ and $a_{11}, a_{22}, \ldots a_{nn}$ are not zero, then the system would appear as follows:

$$a_{11}\, x_1 + a_{12}\, x_2 + a_{13}\, x_3 + \cdots + a_{1n}\, x_n = c_1$$
$$a_{22}\, x_2 + a_{23}\, x_3 + \cdots + a_{2n}\, x_n = c_2$$
$$a_{33}\, x_3 + \cdots + a_{3n}\, x_n = c_3$$
$$+ \cdots\cdots\cdots\cdots\cdots$$
$$+ \cdots\cdots\cdots\cdots\cdots$$
$$a_{nn}\, x_n = c_n$$

For this case there is only one solution.

3.6 Inverse Matrix

Given:

$$AX = B$$

$$A^{-1}AX = IX = X = A^{-1}B$$

MATLAB's method of solution

$$X = inv(A) \ *B \text{ (solves by } A^{-1})$$

or

$$X = A/B \text{ (solves by Gauss Elimination)}$$

Let us see what is involved by determining A^{-1}:

$$A^{-1}A = I$$

Let $B = A^{-1}$, then $B*A = I$
It will be demonstrated for a 3×3 matrix.

$$\begin{bmatrix} b_{11} & b_{12} & b_{13} \\ b_{21} & b_{22} & b_{23} \\ b_{31} & b_{32} & b_{33} \end{bmatrix} \begin{bmatrix} a_{11} & a_{12} & a_{13} \\ a_{21} & a_{22} & a_{23} \\ a_{31} & a_{32} & a_{33} \end{bmatrix} = \begin{bmatrix} 1 & 0 & 0 \\ 0 & 1 & 0 \\ 0 & 0 & 1 \end{bmatrix}$$

First row of B*A

Element (1,1): $b_{11}a_{11} + b_{12}a_{21} + b_{13}a_{31} = 1$

Element (1,2): $b_{11}a_{12} + b_{12}a_{22} + b_{13}a_{32} = 0$

Element (1,3): $b_{11}a_{13} + b_{12}a_{23} + b_{13}a_{33} = 0$

$$A^T = \begin{bmatrix} a_{11} & a_{21} & a_{31} \\ a_{12} & a_{22} & a_{32} \\ a_{13} & a_{23} & a_{33} \end{bmatrix}$$

Here b_{11}, b_{12}, and b_{13} are the unknowns.

$$\text{Let } B_1 = \begin{bmatrix} b_{11} \\ b_{12} \\ b_{13} \end{bmatrix}, \quad \text{then} \quad A^T B_1 = \begin{bmatrix} 1 \\ 0 \\ 0 \end{bmatrix}$$

Solve for b_{11}, b_{12}, b_{13} by Gauss Elimination.

Second row of B*A

Element (2,1): $b_{21}a_{11} + b_{22}a_{21} + b_{23}a_{31} = 0$

Element (2,2): $b_{21}a_{12} + b_{22}a_{22} + b_{23}a_{32} = 1$

Element (2,3): $b_{21}a_{13} + b_{22}a_{23} + b_{23}a_{33} = 0$

Here b_{21}, b_{22}, and b_{23} are the unknowns.

$$\text{Let } B_2 = \begin{bmatrix} b_{21} \\ b_{22} \\ b_{23} \end{bmatrix}, \quad \text{then} \quad A^T B_2 = \begin{bmatrix} 0 \\ 1 \\ 0 \end{bmatrix}$$

Solve for b_{21}, b_{22}, and b_{23} by Gauss Elimination.

   Third row of B*A

Element (3,1): $b_{31}a_{11} + b_{32}a_{21} + b_{33}a_{31} = 0$

Element (3,2): $b_{31}a_{12} + b_{32}a_{22} + b_{33}a_{32} = 0$

Element (3,3): $b_{31}a_{13} + b_{32}a_{23} + b_{33}a_{33} = 1$

Here b_{31}, b_{32}, and b_{33} are the unknowns.

$$\text{Let } B_3 = \begin{bmatrix} b_{31} \\ b_{32} \\ b_{33} \end{bmatrix}, \quad \text{then} \quad A^T B_3 = \begin{bmatrix} 0 \\ 0 \\ 1 \end{bmatrix}$$

Solve for b_{31}, b_{32}, and b_{33} by Gauss Elimination.

An alternative [1] to the method described above is to augment the coefficient matrix with the identity matrix, then apply the Gauss–Jordan method making the coefficient matrix the identity matrix. The original identity matrix then becomes A⁻¹. The starting augmented matrix for a 3 × 3 coefficient matrix is shown below:

$$\begin{bmatrix} a_{11} & a_{12} & a_{13} & | & 1 & 0 & 0 \\ a_{21} & a_{22} & a_{23} & | & 0 & 1 & 0 \\ a_{31} & a_{32} & a_{33} & | & 0 & 0 & 1 \end{bmatrix}$$

Example 3.7

This method is illustrated by the following example:

$$x_1 - 3x_2 + x_3 = -4$$

$$-3x_1 + 4x_2 + 3x_3 = -10$$

$$2x_1 + 3x_2 - 2x_3 = 18$$

$$A_{aug} = \begin{bmatrix} 1 & -3 & 1 & | & 1 & 0 & 0 \\ -3 & 4 & 3 & | & 0 & 1 & 0 \\ 2 & 3 & -2 & | & 0 & 0 & 1 \end{bmatrix}$$

Multiply row 1 by –3/1 and subtract from row 2, giving:

$$(-3 + 3 \times 1), (4 - 3 \times 3), (3 + 3 \times 1), (0 + 3 \times 1), (1 + 3 \times 0),$$
$$(0 + 3 \times 0) = (0 \quad -5 \quad 6 \quad 3 \quad 1 \quad 0)$$

Multiply row 1 by 2/1 = 2 and subtract from row 3, giving:

$$(2 - 2 \times 1), (3 - 2 \times (-6)), (-2 - 2 \times 1), (0 - 2 \times 1), (0 - 2 \times 0),$$
$$(1 - 2 \times 0) = (0 \quad 9 \quad -4 \quad -2 \quad 0 \quad 1)$$

$$equiv\ A_{aug} = \begin{bmatrix} 1 & -3 & 1 & | & 1 & 0 & 0 \\ 0 & -5 & 6 & | & 3 & 1 & 0 \\ 0 & 9 & -4 & | & -2 & 0 & 1 \end{bmatrix}$$

Multiply row 2 by –9/5 and subtract from row 3, giving:

$$(0 + 9/5 \times 0), (9 + 9/5 \times (-5)), (-4 + 9/5 \times 6), (-2 + 9/5 \times 3),$$
$$(0 + 9/5 \times 1), (1 + 9/5 \times 0) = (0 \quad 0 \quad 34/5 \quad 17/5 \quad 9/5 \quad 1)$$

$$equiv\ A_{aug} = \begin{bmatrix} 1 & -3 & 1 & | & 1 & 0 & 0 \\ 0 & -5 & 6 & | & 3 & 1 & 0 \\ 0 & 0 & 34/5 & | & 17/5 & 9/5 & 1 \end{bmatrix}$$

Multiply row 3 by 30/34 and subtract from row 2, giving:

$(0 - 30/34 \times 0)$, $(-5 - 30/34 \times 0)$, $(6 - 30/34 \times 34/5)$, $(3 - 30/34 \times 17/5)$,

$(1 - 30/34 \times 9/5)$, $(0 - 30/34 \times 1) = (0 \quad -5 \quad 0 \quad 0 \quad -20/34 \quad -30/34)$

Multiply row 3 by 5/34 and subtract from row 1, giving:

$(1 - 5/34 \times 0)$, $(-3 - 5/34 \times 0)$, $(1 - 5/34 \times 34/5)$, $(1 - 5/34 \times 17/5)$,

$(0 - 5/34 \times 9/5)$, $(0 - 5/34 \times 1) = (1 \quad -3 \quad 0 \quad 1/2 \quad -9/34 \quad -5/34)$

$$equiv\ A_{aug} = \begin{bmatrix} 1 & -3 & 0 & | & 1/2 & -9/34 & -5/34 \\ 0 & -5 & 0 & | & 0 & -20/34 & -30/34 \\ 0 & 0 & 34/5 & | & 17/5 & 9/5 & 1 \end{bmatrix}$$

Multiply row 2 by 3/5 and subtract from row 1, giving:

$(1 - 3/5 \times 0)$, $(-3 - 3/5 \times (-5))$, $(0 - 3/5 \times 0)$, $(1/2 - 3/5 \times 0)$,

$(-9/34 + 3/5 \times 20/34)$, $(-5/34 + 3/5 \times 30/34) = (1 \quad 0 \quad 0 \quad 1/2 \quad 3/34 \quad 13/34)$

$$equiv\ A_{aug} = \begin{bmatrix} 1 & 0 & 0 & | & 1/2 & 3/34 & 13/34 \\ 0 & -5 & 0 & | & 0 & -20/34 & -30/34 \\ 0 & 0 & 34/5 & | & 17/5 & 9/5 & 1 \end{bmatrix}$$

Divide row 2 by −5 and row 3 by 34/5, giving:

$$equiv\ A_{aug} = \begin{bmatrix} 1 & 0 & 0 & | & 1/2 & 3/34 & 13/34 \\ 0 & 1 & 0 & | & 0 & 4/34 & 6/34 \\ 0 & 0 & 1 & | & 1/2 & 9/34 & 5/34 \end{bmatrix}$$

Thus,

$$A^{-1} = \begin{bmatrix} 1/2 & 3/34 & 13/34 \\ 0 & 4/34 & 6/34 \\ 1/2 & 9/34 & 5/34 \end{bmatrix}$$

It is left as a student exercise to show that $AA^{-1} = I$.

3.7 The Eigenvalue Problem

One very important application of the eigenvalue problem is in the theory of vibrations. Consider the two-degrees-of-freedom problem shown in Figure 3.8.

The governing differential equations describing the motion of the two masses are

$$m_1\ddot{x}_1 = k_2(x_2 - x_1) - k_1 x_1 \tag{3.11}$$

$$m_2\ddot{x}_2 = -k_2(x_2 - x_1) - k_3 x_2 \tag{3.12}$$

We wish to determine the modes of oscillation such that each mass undergoes harmonic motion at the same frequency. To obtain such a solution, set

$$x_1 = A_1 \exp(i\omega t) \tag{3.13}$$

$$x_2 = A_2 \exp(i\omega t) \tag{3.14}$$

Substituting Equations (3.13) and (3.14) into Equations (3.11) and (3.12) gives

$$\left(\frac{k_1 + k_2}{m_1} - \omega^2\right)A_1 - \frac{k_2}{m_1}A_2 = 0 \tag{3.15}$$

$$-\frac{k_2}{m_2}A_1 + \left(\frac{k_2 + k_3}{m_2} - \omega^2\right)A_2 = 0 \tag{3.16}$$

Equations (3.15) and (3.16) are two homogeneous linear algebraic equations in two unknowns. There is a theorem in linear algebra that says that the only way for two

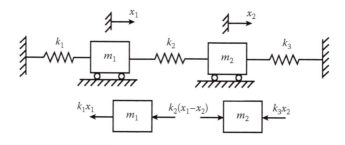

Figure 3.8 Two-degrees-of-freedom vibration system.

homogeneous linear algebraic equations in two unknowns to have a nontrivial solution is for the determinant of the coefficient matrix to be zero.

$$
\begin{vmatrix}
\left(\dfrac{k_1 + k_2}{m} - \omega^2 \right) & -\dfrac{k_2}{m_1} \\[3mm]
-\dfrac{k_2}{m_2} & \left(\dfrac{k_2 + k_3}{m_2} - \omega^2 \right)
\end{vmatrix}
$$

Letting

$$
\frac{k_1 + k_2}{m_1} = a_{11}, \quad -\frac{k_2}{m_1} = a_{12}, \quad -\frac{k_2}{m_2} = a_{21}, \quad \frac{k_2 + k_3}{m_2} = a_{22} \quad and \quad \omega^2 = \lambda
$$

The equation for λ becomes

$$
\lambda^2 - (a_{11} + a_{22})\lambda + (a_{11}a_{22} - a_{12}a_{21}) = 0 \tag{3.17}
$$

The solution of Equation (3.17) gives the eigenvalues, λ_1 *and* λ_2, which are the square of the two natural frequency of oscillations for this system. The ratio of the amplitudes of the oscillation of the two masses can be obtained by substituting the values of λ into Equation (3.15) or Equation (3.16); that is,

$$
\frac{A_2}{A_1} = -(a_{11} - \lambda_1)/a_{12} = \frac{m_1}{k_2}\left(\frac{k_1 + k_2}{m_1} - \lambda_1 \right) \quad \text{for the first mode}
$$

and

$$
\frac{A_2}{A_1} = -(a_{11} - \lambda_2)/a_{12} = \frac{m_1}{k_2}\left(\frac{k_1 + k_2}{m_1} - \lambda_2 \right) \quad \text{for the second mode}
$$

The eigenvector, V_1, associated with λ_1 is

$$
\begin{bmatrix}
A_1 \\
-A_1(a_{11} - \lambda_1)/a_{12}
\end{bmatrix}
$$

and the eigenvector, V_2, associated with λ_2 is

$$
\begin{bmatrix}
A_1 \\
-A_1(a_{11} - \lambda_2)/a_{12}
\end{bmatrix}
$$

Since A_1 is arbitrary, we can select $A_1 = 1$, then

$$V_1 = \begin{bmatrix} 1 \\ -(a_{11} - \lambda_1)/a_{12} \end{bmatrix} \tag{3.18}$$

and

$$V_2 = \begin{bmatrix} 1 \\ -(a_{11} - \lambda_2)/a_{12} \end{bmatrix} \tag{3.19}$$

If [1] V is an eigenvector of a matrix a corresponding to an eigenvalue λ, so is bV with any $b \neq 0$.

MATLAB's eig *Function*

MATLAB has a built-in function that gives the eigenvalues of a square matrix X. MATLAB's description of the function follows:

E = *eig*(X) is a vector containing the eigenvalues of a square matrix X.
[V,D] = *eig*(X) produces a diagonal matrix D of eigenvalues and a full matrix V whose columns are the corresponding eigenvectors so that X*V = V*D.

For the problem under discussion, matrix a replaces matrix X. Thus, the statement [V, D] = *eig*(a) gives the eigenvectors associated with λ_1 and λ_2. V(:,1) is associated with λ_1 and V(:,2) is associated λ_2.

Example 3.8

Suppose in Figure 3.8 the following parameters were given:

$$m_1 = m_2 = 1500 \ kg, k_1 = 3250 \ N/m, k_2 = 3500 \ N/m, k_3 = 3000 \ N/m$$

A program that will determine

1. The eigenvalues of the system by both Equation (3.17) and by MATLAB's eig function
2. The eigenfunctions by both Equations (3.18) and (3.19) and MATLAB's [V,D] function

follows.

Example 3.9

```
% eigen2.m
% Eigenvalues and eigenvectors.
% E = eig(a) is a vector containing the eigenvalues of a square matrix a.
% [V,D] = eig(a) produces a diagonal matrix D of eigenvalues and a
% full matrix V whose columns are the corresponding eigenvectors so
% that a*V = V*D.
% Units are in SI units.
    clear; clc;
    k1=3250; k2=3500; k3=3000; m1=1500; m2=1500;
    a(1,1)=(k1+k2)/m1;
    a(1,2)=-k2/m1;
    a(2,1)=-k2/m2;
    a(2,2)=(k2+k3)/m2;
% Lamda^2-(a(1,1)+a(2,2))*Lamda+(a(1,1)*a(2,2)-a(1,2)*a(2,1))=0
    b=-(a(1,1)+a(2,2)); c=a(1,1)*a(2,2)-a(1,2)*a(2,1);
    Lamda1=(-b-sqrt(b^2-4*c))/2;
    Lamda2=(-b+sqrt(b^2-4*c))/2;
    E=eig(a);
    fprintf('Lamda1=%7.5f Lamda2=%7.5f \n', Lamda1,Lamda2)
    fprintf('E(1)= %7.5f E(2)=%7.5f \n',E(1),E(2));
    v1=[1;-(a(1,1)-Lamda1)/a(1,2)]
    v2=[1;-(a(1,1)-Lamda2)/a(1,2)]
    [V,D] = eig(a);
    V1=V(:,1)
    V2=V(:,2)
    D
```

The following results were obtained:

```
Lamda1=2.08185   Lamda2=6.75149
E(1)=2.08185     E(2)=6.75149
v1 =
        1.0000
        1.0364
v2 =
        1.0000
       -0.9649
V1 =
       -0.6944
       -0.7196
V2 =
       -0.7196
        0.6944
D =
        2.0818       0

             0       6.7515
```

Examining the results, it can be seen that Lamda1 = E(1) and Lamda2 = E(2). Also, V1 is a scalar multiple of v1 and V2 is a scalar multiple of v2.

Exercises

Exercise 3.1

Use pencil and paper and the Gauss Elimination method to solve the following system of equations:

a. $2x_1 + 3x_2 - x_3 = 20$
$4x_1 - x_2 + 3x_3 = -14$

$x_1 + 5x_2 + x_3 = 21$

b. $4x_1 + 8x_2 + x_3 = 8$
$-2x_1 - 3x_2 + 2x_3 = 14$

$x_1 + 3x_2 + 4x_3 = 30$

c. $2x_1 + x_2 + x_3 - 11x_4 = 1$
$5x_1 - 2x_2 + 5x_3 - 4x_4 = 5$

$x_1 - x_2 + 3x_3 - 3x_4 = 3$

$3x_1 + 4x_2 - 7x_3 + 2x_4 = -7$

Projects

Project 3.1

For the following truss structure (see Figure P3.1) write a MATLAB program that will determine the internal forces in the structural members by the method described in Example 3.3. Print out the reactions, the coefficient matrix, the members' internal forces, and a check on the solution.

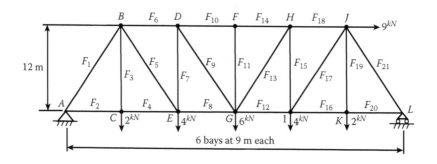

Figure P3.1   Truss structure for Project 3.1.

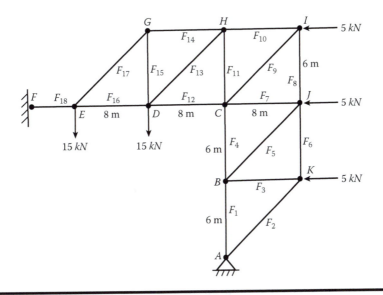

Figure P3.2   Truss structure for Project 3.2.

Project 3.2

For the truss structure shown in Figure P3.2 write a MATLAB program that will determine the internal forces in the structural members by the method described in Example 3.3. Print out the reactions, the coefficient matrix, the members' internal forces, and a check on the solution.

Project 3.3

An automobile suspension system is simulated by two springs connected by a bar supporting the automobile's weight as shown in Figure P3.3a. We shall assume that the ends of the spring are at the same elevation and X is measured downward from this position (Figure P3.2b). The automobile weight is applied on the bar at point G. The equilibrium position of the bar is shown in Figure P3.3c. The governing equations at the equilibrium position follow:

$$\sum_i F_{x,i} = 0 = -k_1(X_0 - L_1\vartheta_0) - k_2(X_0 + L_2\vartheta_0) + W \qquad (P3.3a)$$

$$\sum_i M_{G,i} = 0 = k_1(X_0 - L_1\vartheta_0)L_1 - k_2(X_0 + L_2\vartheta_0)L_2 \qquad (P3.3b)$$

If the system is disturbed and released, it will vibrate at its natural frequencies. The system has two degrees of freedom, resulting in a vertical and a rotational vibration.

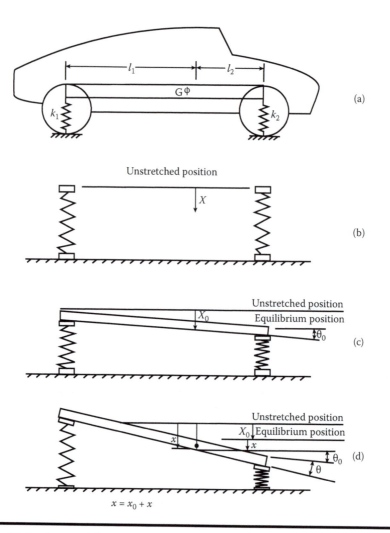

Figure P3.3 **Automobile suspension system.**

Let x be measured from the equilibrium position, then

$$X = X_0 + x \quad \text{and} \quad \ddot{X} = \ddot{x}$$

The governing equations describing the vibrating system follow:

$$M\ddot{x} = -k_1[X_0 + x - L_1(\vartheta_0 + \vartheta)] - k_2[X_0 + x + L_2(\vartheta_0 + \vartheta)] + W \quad \text{(P3.3c)}$$

$$I_{zz}\ddot{\vartheta} = k_1[X_0 + x - L_1(\vartheta_0 + \vartheta)]L_1 - k_2[X_0 + x + L_2(\vartheta_0 + \vartheta)]L_2 \quad \text{(P3.3d)}$$

$$M\ddot{x} = -k_1(X_0 - L_1\vartheta_0) - k_2(X_0 - L_2\vartheta_0) + W - k_1(x - L_1\vartheta) - k_2(x + L_2\vartheta) \quad \text{(P3.3e)}$$

By Equation (P3.3a), the sum of the first three terms on the right-hand side of Equation (P3.3e) is zero. Thus,

$$M\ddot{x} = -k_1(x - L_1\vartheta) - k_2(x + L_2\vartheta) \tag{P3.3f}$$

Similarly, Equation (P3.3d) can be rewritten as

$$I_{zz}\ddot{\vartheta} = k_1(X_0 - L_1\vartheta_0)L_1 - k_2(X_0 + L_2\vartheta_0)L_2 + k_1(x - L_1\vartheta)L_1 - k_2(x + L_2\vartheta)L_2 \tag{P3.3g}$$

By Equation (P3.3b), the sum of the first two terms on the right-hand side of Equation (P3.3g) is zero. Thus,

$$I_{zz}\ddot{\vartheta} = k_1(x - L_1\vartheta)L_1 - k_2(x + L_2\vartheta)L_2 \tag{P3.3h}$$

We wish to determine the modes of oscillation such that the vertical and rotational vibrations are at the same frequency. To obtain such a solution, set

$$x = A\exp(i\omega t) \tag{P3.3i}$$

$$\vartheta = B\exp(i\omega t) \tag{P3.3j}$$

Substituting Equations (P3.3i) and (P3.3j) into Equations (P3.3f) and (P3.3h), respectively, gives

$$\left(\frac{k_1 + k_2}{M} - \omega^2\right)A - \left(\frac{k_2 L_2}{M} - \frac{k_1 L_1}{M}\right)B = 0 \tag{P3.3k}$$

$$\left(\frac{k_2 L_2 - k_1 L_1}{I_{zz}}\right)A + \left(\frac{k_2 L_2^2 + k_1 L_1^2}{I_{zz}} - \omega^2\right)B = 0 \tag{P3.3l}$$

Using MATLAB's eig function, determine the natural frequencies of oscillation for the system. Use the following variable values:

$k_1 = 35$ kN/m, $k_2 = 38$ kN/m, $L_1 = 1.4$ m, $L_2 = 1.7$ m, $M = 1500$ kg, $I_{zz} = 2170$ kg-m².

Project 3.4

Suppose a manufacturer wishes to purchase a piece of equipment that costs $40,000. He plans to borrow the money from a bank and pay off the loan in 10 years in 120 equal payments. The annual interest rate is 6%. Each month the interest charged will be on the unpaid balance of the loan. He wishes to determine what his monthly payment will be. This problem can be solved by a system of linear equations.

Let x_j = the amount in the jth payment that goes toward paying off the principal. Then the equation describing the jth payment is

$$jth\ payment = M = x_j + \left(P - \sum_{n=1}^{n=j-1} x_n \right) I \tag{P3.4a}$$

where
  M = the monthly payment.
  P = the amount borrowed.
  I = the monthly interest rate = annual interest rate/12.

The total number of unknowns is 121 (120 x values and M).

Applying Equation (P3.4a) to each month gives 120 equations. One additional equation is

$$P = \sum_{n=1}^{n=120} x_n \tag{P3.4b}$$

Develop a computer program that will

1. Ask the user to enter from the keyboard the amount of the loan (P), the annual interest rate, I, and the time period, Y, in years.
2. Set up the system of linear equations, using $A_{n,m}$ as the coefficient matrix of the system of linear equations. The n represents the equation number and m represents the coefficient of x_m in that equation. Set $x_{121} = M$.
3. Solve the system of linear equations in MATLAB.
4. Print out a table consisting of four columns. The first column should be the month number, the second column the monthly payment, the third column the amount of the monthly payment that goes toward paying off the principal, and the fourth column the interest payment for that month.

Reference

1. Kreyszig, E., *Advanced Engineering Mathematics*, 8th Ed., John Wiley & Sons, 1999.

Chapter 4

Roots of Algebraic and Transcendental Equations

4.1 The Search Method

The equation whose roots are to be determined should be put into the form of Equation (4.1), as shown below:

$$f(x) = 0 \qquad (4.1)$$

First a search is made to obtain intervals in which real roots lie. This is accomplished by subdividing the x domain into N equal subdivisions, giving

$$x_1, x_2, x_3, \ldots, x_{N+1} \text{ and } x_{i+1} = x_i + \Delta x$$

Then locate where f(x) changes sign (see Figure 4.1). This occurs when

$$f(x_i)\, f(x_{i+1}) < 0$$

The sign change usually indicates that a real root has been passed. However, it may also indicate a discontinuity in the function. (Example: tan x is discontinuous at $x = \pi/2$.) Once the intervals in which the roots lie have been established, one can use several methods for obtaining the real roots.

4.2 Bisection Method

Suppose it has been established that a root lies between x_j and x_{j+1}. Cut the interval in half (see Figure 4.2), then

$$x_{j+1/2} = x_j + \frac{\Delta x}{2}$$

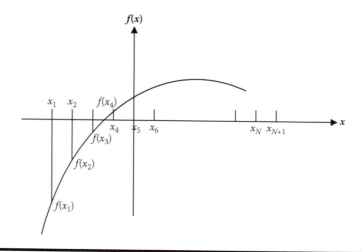

Figure 4.1 Selecting the interval in which a root lies.

Now compute $f(x_j)\, f(x_{j+1/2})$.

Case 1: If $f(x_j)\, f(x_{j+1/2}) < 0$, then the root lies between x_j and $x_{j+1/2}$.

Case 2: If $f(x_j)\, f(x_{j+1/2}) > 0$, then the root lies between $x_{j+1/2} + x_{j+1}$.

Case 3: If $f(x_j)\, f(x_{j+1/2}) = 0$, then x_j or $x_{j+1/2}$ is a real root.

For cases 1 and 2, select the interval containing the root and repeat the process. Continue repeating the process until $(\Delta x)_f = \dfrac{\Delta x}{2^r}$ is sufficiently small, where r is the number of bisections. Then the real root lies within the last interval.

Note: For 20 bisections,

$$(\Delta x)_f = \frac{\Delta x}{2^{20}} \approx \Delta x \times 1.05 \times 10^{-6}$$

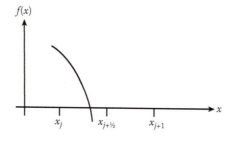

Figure 4.2 Selecting the interval containing the root in the Bisection Method.

Program method:

Set
$$x_A = x_j$$

$$x_B = x_{j+1}$$

$$x_C = \frac{1}{2}(x_A + x_B)$$

If $f(x_A)f(x_C) < 0$ (root lies between x_A & x_C)
Set $x_B = x_C$ and

$$x_C = \frac{1}{2}(x_A + x_B)$$

and repeat the process.

If $f(x_A)f(x_C) > 0$ (root lies between x_B & x_C)
Set $x_A = x_C$ and

$$x_C = \frac{1}{2}(x_A + x_B)$$

and repeat the process.

If $f(x_A)f(x_C) = 0$
Either x_A or x_C is a root.

4.3 Newton–Raphson Method

This method uses the tangent to the curve $f(x) = 0$ to estimate the root. One needs to obtain an expression for $f'(x)$. One also needs to make an initial guess for the root, say, x_1 (see Figure 4.3). $f'(x)$ gives the slope of the tangent to the curve at x.

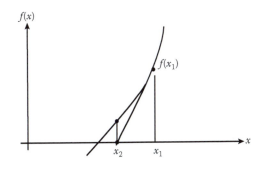

Figure 4.3   Predicting root in the Newton–Raphson Method.

On the tangent to the curve

$$\frac{f(x_1) - f(x_2)}{x_1 - x_2} = f'(x_1)$$

Set $f(x_2) = 0$ and solve for x_2; that is,

$$x_2 = x_1 - \frac{f(x_1)}{f'(x_2)}$$

Check if $|f(x_2)| < \in$. If yes, quit. x_2 is the root—print out x_2. If no, set $x_1 = x_2$ and repeat the process.

Continue repeating the process until $|f(x_2)| < \in$. An alternate condition for convergence is

$$|\div f(x_1)/f'(x_1) \div| < \in$$

This method is widely used for its rapid convergence. However, there are cases where convergence will not occur. This can happen if

a. $f'(x)$ changes sign near the root.
b. The initial guess for the root is too far from the true root.

If one combines the Newton–Raphson method with the search method for obtaining a small interval in which the real root lies, convergence will not be a problem.

4.4 The fzero Function

MATLAB® has a built-in function to obtain the real roots of a transcendental equation. It is the fzero function. To get started click on the Command window and type in:

```
>> help fzero
```

This gives several options on using the fzero function, some of which are shown below. The first is appropriate if one has some idea where the root lies. The second is more appropriate when there is more than one root and all roots need to be obtained. The second should be used in combination with the search method described earlier.

```
X = FZERO(FUN,X0) tries to find a zero of the function FUN near
X0. FUN accepts real scalar input X and returns a real scalar function
value F evaluated at X. The value X returned by FZERO is near a point
where FUN changes sign (if FUN is continuous), or NaN if the search fails.

X = FZERO(FUN,X0), where X is a vector of length 2, assumes X0 is an
interval where the sign of FUN(X0(1)) differs from the sign of FUN(X0(2)).
An error occurs if this is not true. Calling FZERO with an interval
guarantees FZERO will return a value near a point where FUN changes sign.

X = FZERO(FUN,X0,OPTIONS) minimizes with the default optimization
```

parameters replaced by values in the structure OPTIONS, an
argument created with the OPTIMSET function. See OPTIMSET for
details. Used options are Display and TolX. Use OPTIONS = [] as a
place holder if no options are set.

X = FZERO(FUN,X0,OPTIONS,P1,P2,...) allows for additional arguments which
are passed to the function

[X,FVAL]= FZERO(FUN,...) returns the value of the objective
function, described in FUN, at X.
Examples
FUN can be specified using @:
X = fzero(@sin,3)
returns pi.
FUN can also be an inline object:
X = fzero(inline('sin(3*x)'),2);

Several examples using both the search method and the fzero function are shown
below.

4.4.1 Example Programs

Example 4.1

```
% search.m
% This program determines the real roots of a third degree polynomial.
% The program produces a plot of the function. Next the program searches
% for intervals where there are sign changes, then it calls the fzero
% function to obtain the real roots.
clear; clc;
xmin=-10.0; xmax=10.0; N=50; nr=0;
dx=(xmax-xmin)/N;
fprintf('Searching for roots, root is obtained by fzero function \n');
fprintf(' x              f\n');
for n=1:N+1
    x(n)=xmin+(n-1)*dx;
    fx(n)=func(x(n));
    fprintf('      %6.2f      %10.3f \n',x(n),fx(n));
end
plot(x,fx), title('f(x) vs. x'), xlabel('x'), ylabel('f(x)'), grid;
for n=1:N
    p=fx(n)*fx(n+1);
    if p < 0.0
        nr=nr+1;
        xr(1)=x(n);
        xr(2)=x(n+1);
        rt(nr)=fzero('func',xr);
    end
    if p==0.0
        nr=nr+1;
        rt(nr)=x(n+1);
        n=n+1;
    end
end

if nr ~= 0
    fprintf('\n\n Roots of function f(x)=0 \n\n');
```

```
    fprintf('root no. root \n');

    for n=1:nr
        fprintf(' %i %10.4f \n',n,rt(n));
    end
else
        fprintf('\n\n No roots lie within xmin <= x <= xmax');
end
```

```
% func.m
function f=func(x)
f=x^3-4.7*x^2-35.1*x+85.176;
```

There are two ways to enter parameters into the function func.m. These are

a. The use of the global statement: Add the following two statements to Example 4.1 just after the clear; clc; statements:

```
global a0 a1 a2 a3;
a3=1.0; a2=-4.7; a1=-35.1; a0=85.176;
Then add the exact global statement to func.m just after the function
statement as shown below:
function f=func(x)
global a0 a1 a2 a3;
f=a3*x^3+a2*x^2+a1*x+a0;
```

b. By modifying the fzero function to include parameters a3, a2, a1, and a0. For this case

```
a3=1.0; a2=-4.7; a1=-35.1; a0=85.176; would still be added
after the clear; clc; statements.
The statement rt(nr)=fzero('func',xr); would be replaced by:
rt(nr)= fzero('func',xr, [ ], a3,a2,a1,a0);
and the corresponding function statements would be replaced by
function f=func(x,a3,a2,a1,a0)
 f=a3*x^3+a2*x^2+a1*x+a0;
```

The coefficients a3, a2, a1, and a0 can be replaced by a row vector, then for case (a), the added statements would be

```
global a;
a(4)=1.0; a(3)=-4.7; a(2)=-35.1; a(1)=85.176;
function f=func(x)
global a;
f=a(4)*x^3+a(3)*x^2+a(2)*x+a(1);
```

Note: The index of a vector starts with 1 and not 0.

For case (b), the coefficients a3, a2, a1 and a0 can also be replaced by a row vector, then the added statements would be

```
a(4)=1.0; a(3)=-4.7; a(2)=-35.1; a(1)=85.176;
rt(nr)=fzero('func',xr,[],a);
```

```
function f=func(x,a)
f=a(4)*x^3+a(3)*x^2+a(2)*x+a(1);
```

MATLAB's Roots Function

MATLAB has a function to obtain the roots of a polynomial. The function is *roots*. To obtain the use of the function, in the COMMAND window type in

```
>> help roots (this gives)
ROOTS(C) computes the roots of the polynomial whose coefficients
are the elements of the vector C. If C has N+1 components,
the polynomial is C(1)*X^N + ··· + C(N)*X + C(N+1).
```

MATLAB gives the coefficients of the polynomial whose roots are V. To obtain the use of the function, in the COMMAND window type in:

```
>> help poly (this gives)
POLY(V),where V is a vector whose elements are
the coefficients of the polynomial whose roots are the
elements of V. For vectors, ROOTS and POLY are inverse
functions of each other, up to ordering, scaling, and
roundoff error.
real(x) gives the real part of x
imag(x) gives the imaginary part of x
```

Example 4.2

```
% roots_poly.m
% This program determines the roots of a polynomial using
% the built in function 'roots'.
% The first polynomial is: f=x^3-4.7*x^2-35.1*x+85.176. The
% roots of this polynomial are all real.
% The second polynomial is: f=x^3-9*x^2+23*x-65. The roots of
% this polynomial are both real and complex. Complex roots must
% be complex conjugates.
% To obtain more info on complex numbers do the following.
% In the command window type in "help complex numbers".
clear; clc;
c(1)=1.0; c(2)= -4.7; c(3)= -35.1; c(4)=85.176;
v=roots(c);
fprintf('The roots are: \n');
v
fprintf('The coefs. of the polynomial whose roots are v are:\n');
coef=poly(v)
fprintf('----------------------------------------\n');
a(1)=1.0; a(2)= -9.0; a(3)=23.0; a(4)= -65.0;
w=roots(a);
fprintf('The roots are: \n');
w
fprintf('The coefs. of the polynomial whose roots are v are:\n');
coef=poly(w)
re=real(w)
im=imag(w)
```

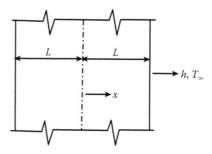

Figure P4.1a A thick plate undergoing convective heat transfer.

Projects

Project 4.1

The temperature distribution of a thick flat plate[*], initially at a uniform temperature, T_0, and which is suddenly immersed in a huge bath at a temperature T_∞, is given by (see Figure P4.1a)

$$T(x,t) = T_\infty + 2(T_0 - T_\infty) \sum_{n=1}^{\infty} \frac{\sin(\delta_n)\cos\left(\delta_n \frac{x}{L}\right)e^{-a\delta_n^2 t/L^2}}{\cos(\delta_n)\sin(\delta_n) + \delta_n} \qquad (P4.1a)$$

where

$L = 1/2$ of the plate thickness.

a = the thermal diffusivity of the plate material.

δ_n are the roots of the equation:

$$F(\delta) = \tan \delta - \frac{hL}{k\delta} = 0 \qquad (P4.1b)$$

where

h = the convective heat transfer coefficient for the bath.

k = the thermal conductivity of the plate material.

There are an infinite number of roots that satisfy Equation (P4.1b), these being δ_1, δ_2, δ_3,...,δ_n. Figure P4.1b shows that the intersection of the curve $hL/k\delta$ with the curves tan δ gives the roots of Equation (P4.1b). Note that δ_1 lies between 0 and $\pi/2$, δ_2 lies between π and $3\pi/2$, δ_3 lies between 2π and $5\pi/2$, etc. Subtracting T_∞ from Equation (P4.1a) and dividing by $T_0 - T_\infty$, we obtain Equation (P4.1c).

$$TRATIO = \frac{T\left(\frac{x}{L},t\right) - T_\infty}{T_0 - T_\infty} = 2\sum_{n=1}^{\infty} \frac{\sin(\delta_n)\cos\left(\delta_n \frac{x}{L}\right)e^{-a\delta_n^2 t/L^2}}{\cos(\delta_n)\sin(\delta_n) + \delta_n} \qquad (P4.1c)$$

---

[*] For a derivation of this equation, see Section 10.2.

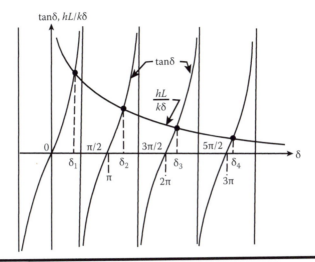

Figure P4.1b   Intersection of $hL/k\delta$ and tan δ curves.

A plot of TRATIO vs. time for several different values of x/L should appear as shown in Figure P4.1c.

Finally, the heat transfer ratio, *Qratio*, from the plate to the bath in time t is given by

$$QRATIO = \frac{Q(t)}{Q_0} = \frac{2hL}{k} \sum_{n=1}^{\infty} \frac{\sin \delta_n \cos \delta_n}{\delta_n^2 [\sin \delta_n \cos \delta_n + \delta_n]} \left[1 - e^{-\alpha t \delta_n^2 / L^2} \right] \qquad \text{(P4.1d)}$$

where
   $Q(t)$ = the amount of heat transferred from the plate to the bath in time t.
   Q_0 = the amount of heat transferred from the plate to the bath in infinite time, which equals the change in internal energy in infinite time.

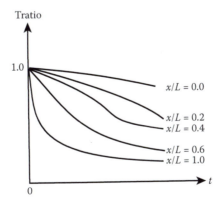

Figure P4.1c   Typical *Tratio* vs. t curves.

Table P4.1 Temperature Ratio, TRATIO

Time(s)	X/L					
	0.0	0.2	0.4	0.6	0.8	1.0
0	1.0	1.0	1.0	1.0	1.0	1.0
10	—	—	—	—	—	—
20	—	—	—	—	—	—
200	—	—	—	—	—	—

1. Write a computer program that will solve for the roots $\delta_1, \delta_2, \ldots, \delta_{50}$ using the fzero function in MATLAB. Print out the δ values in 10 rows and 5 columns. Also print out the functional values at the roots, i.e.; $f(\delta_n)$.
   Note: Only 50 δ values were asked to be computered.
2. Solve Equation (P4.1c) for TRATIO for x/L = 0.0, 0.2, 0.4, 0.6, 0.8, 1.0, and t = 0, 10, 20,...200 seconds. Print out results in table form as shown in Table P4.1.
   Also use MATLAB to produce a plot as shown in Figure P4.1c.
3. Construct a table for *QRATIO* vs. *t* for times 0, 10, 20, 30,..., 200 seconds.
4. Use MATLAB to produce a plot of *QRATIO* vs. *t*. Use the following values for the parameters of the problem:

$T_0 = 300°C$, $T_\infty = 30°C$, $h = 45$ w/m²-°C
$k = 10.0$ w/m-°C, $L = 0.03$ m, $a = 0.279 \times 10^{-5}$ m²/s

Project 4.2

We wish to consider the temperature distribution in a semi-infinite slab (see Figure P4.2), initially at a uniform temperature, whose surface is suddenly subjected to convective heat transfer from the surrounding air. The temperature, T, in the slab will be a function of position and time; that is, $T = T(x,t)$. It will also depend on the parameters: h, T_i, T_∞, k, and α, where h = convective heat transfer coefficient, T_i = the initial temperature of the slab, T_∞ is the air temperature, and k and α are the thermal conductivity and diffusivity of the slab material, respectively. The problem can be solved by Laplace Transforms. The solution is

$$TR(x,t) = 1 - erf\left(\frac{x}{2\sqrt{\alpha t}}\right) - e^{\left(\frac{hx}{k} + \frac{h^2\alpha t}{k^2}\right)} \times \left[1 - erf\left(\frac{x}{2\sqrt{\alpha t}} + \frac{h\sqrt{\alpha t}}{k}\right)\right] - \frac{T - T_i}{T_\infty - T_i} = 0$$

Given: $T_i = 10°C$, $T_\infty = 70°C$, $k = 386.0$ W/m-°C, $\alpha = 1.1234 \times 10^{-4}$ m²/s, and $h = 100$ w/m²-°C. We wish to determine the time, t, when the temperature in the

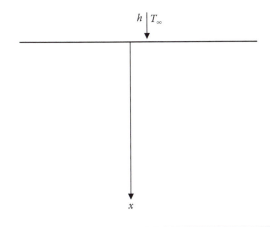

$h \mid T_{\infty}$

x

Figure P4.2   Semi-infinite slab undergoing heat transfer.

slab reaches the following temperatures and at the following positions:

$T = [15, 20, 25, 30]°C$ and $x = [\ 0.1, 0.2, 0.3, 0.4, 0.5, 0.6, 0.7, 0.8, 0.9, 1.0]$ m

Use MATLAB to solve for the *time* for each condition and construct a table as shown by Table P4.2.

Use the search method to find an interval in which the function $TR(x,t)$ changes sign.

Assume that $0 \le time \le 60000\ s$ and subdivide the time domain into 200 intervals.

Also create plots of *time* (*h*) vs. *x* for the four temperature cases listed in Table P4.2. All four plots should be on the same graph.

Table P4.2   Time in Hours to Reach Specified x and T

x(m)	T(C)			
	15	20	25	30
0.1	..	..	..	..
0.2	..	..	..	..
0.3	..	..	..	..
..	..	..	..	..
..	..	..	..	..
1.0	..	..	..	..

Table P4.3a   Equation of State Variables for Air, Oxygen, and Carbon Dioxide

Gas #	Gas	$a\left(\dfrac{N\,m^4\,K^{1/2}}{kmol^2}\right)$	$b\left(\dfrac{m^3}{kmol}\right)$	$\bar{R}\left(\dfrac{N\,m}{K\,kmol}\right)$
1	Air	15.989×10^5	0.02541	8314
2	Oxygen	17.22×10^5	0.02197	8314
3	Carbon dioxide	64.43×10^5	0.02963	8314

Project 4.3

The equation of state for a substance is a relationship between pressure (p), temperature (T), and specific volume ($\bar{v}$). Many gases at low pressures and moderate temperatures behave approximately as an ideal gas. The ideal gas equation of state with p in N/m², $\bar{v}$ in m³/kmol, T in K, and $\bar{R}$ in (N m)/(K kmol) is

$$p = \frac{\bar{R}T}{\bar{v}}$$

where $\bar{R}$ is the universal gas constant. As temperature decreases and pressure increases, gas behavior deviates from ideal gas behavior. The Redlich–Kwong's equation of state is often used to approximate nonideal gas behavior. Redlich–Kwong's equation of state is [1]

$$p = \frac{\bar{R}T}{\bar{v} - b} - \frac{a}{\bar{v}(\bar{v} + b)T^{1/2}}$$

The values for $\bar{R}$, a, and b for three gases are tabulated in Table P4.3a.

We wish to determine the percent error in the specific volume by using the ideal gas relationship while assuming that Redlich–Kwong's equation of state is the correct equation of state for the three gases listed in Table P4.3a. Vary the temperature from 350 K to 700 K in steps of 50 K, while holding the pressure constant at 1.0132×10^7 N/m². Using the specified temperatures and pressure determine the specific volumes, $\bar{v}$, by both the ideal gas equation and the Redlich–Kwong's equation and determine the percent error in the specific volume resulting from the use of the ideal gas equation. Take the percent error in the specific volume to be

$$\% \text{ error} = \frac{\left| \bar{v}_{\text{ideal gas}} - \bar{v}_{\text{Redlich–Kwong}} \right|}{\bar{v}_{\text{Redlich–Kwong}}} \times 100$$

Table P4.3b   Table for Gas 1 (Air)

		Ideal Gas	Redlich-Kwong Eq	% Error in $\bar{v}$
$T(°K)$	$p(N/m^2)$	$\bar{v}$ $(m^3/kmol)$	$\bar{v}$ $(m^3/kmol)$	
350	$1.0132×10^7$	—	—	—
400	$1.0132×10^7$	—	—	—
—	—	—	—	—
—	—	—	—	—
700	$1.0132×10^7$	—	—	—

Write a MATLAB program utilizing the fzero function to calculate the specific volume by Redlich–Kwong's equation. Assume that $\bar{v}$ varies between 0.1 and 1.1 $m^3/kmol$. Use 50 subdivisions on the $\bar{v}$ domain. Construct a table as shown by Table P4.3b.

Repeat table for gases 2 and 3. In your program, use a *for* loop to select all three gases. Use an *if-elseif* ladder to select the proper constants for the gas.

Project 4.4

Repeat Project 4.3, but replace the Redlich–Kwong's equation with van der Waals' equation [1], which is

$$p = \frac{\bar{R}T}{\bar{v} - b} - \frac{a}{\bar{v}^2}$$

The constants a and b are tabulated in Table P4.4.

Table P4.4   Tabulation of Constants a and b

Gas #	Gas	$a\left(\dfrac{N m^4}{kmol^2}\right)$	$b\left(\dfrac{m^3}{kmol}\right)$	$\bar{R}\left(\dfrac{N m}{K\,kmol}\right)$
1	Air	$1.368 × 10^5$	0.0367	8314
2	Oxygen	$1.369 × 10^5$	0.0317	8314
3	Carbon dioxide	$3.647 × 10^5$	0.0428	8314

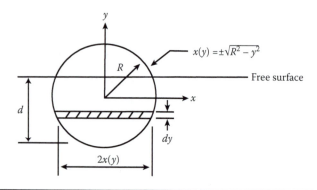

Figure P4.5 Floating log having a circular cross-section.

Project 4.5

A wood circular cylinder, having a specific gravity, S, of 0.54, floats in water as shown in Figure P4.5. For a floating body, the weight of the floating body equals the weight of fluid displaced, thus

$$S\gamma_w \pi R^2 L = \gamma_w V \qquad\qquad (\text{P4.5a})$$

where
  S = the specific gravity of the wood.
  γ_w = the specific weight of water.
  R = the radius of the cylinder.
  L = the cylinder length.
  V = the volume of water displaced.

$$dV = LdA = 2Lx(y)dy = 2L\sqrt{R^2 - y^2}\,dy$$

$$V = 2L \int_{-R}^{d-R} \sqrt{R^2 - y^2}\,dy = 2L\left\{\frac{1}{2}\left(y\sqrt{R^2 - y^2} + R^2 \sin^{-1}\frac{y}{R}\right)\right\}_{-R}^{d-R} \qquad (\text{P4.5b})$$

The integral was obtained from integral tables. Substituting the limits of integration gives

$$V = L\left\{(d-R)\sqrt{R^2 - (d-R)^2} + R^2 \sin^{-1}\frac{d-R}{R} - R^2 \sin^{-1}(-1)\right\} \qquad (\text{P4.5c})$$

$$\sin^{-1}(-1) = -\pi/2.$$

Substituting Equation (P4.5c) into Equation (P4.5a) and rearranging terms and dividing by R^2 gives

$$f\left(\frac{d}{R}\right) = \left(\frac{d}{R} - 1\right)\sqrt{2\frac{d}{R} - \left(\frac{d}{R}\right)^2} + \sin^{-1}\left(\frac{d}{R} - 1\right) - (S - 0.5)\pi = 0$$

If $R = 1$ ft, determine d using the fzero function.

Project 4.6

Do parts (a) and (b) of Project 2.12 and then, using MATLAB's fzero function, determine

  a. The time when the velocity of the piston, described in that project, reaches ½ of its maximum velocity.
  b. The time when the acceleration of the piston, described in that project, reaches ½ of its maximum acceleration.

Project 4.7

In order to solve the temperature distribution of a thick rod having a circular cross-section, initially at a uniform temperature, T_0, and which is suddenly immersed in a huge bath at a temperature T_∞, one needs to determine the roots of $J_1(x)$, where J_1 is the Bessel function of the first kind of order 1. However, in this project, the roots of J_0, where J_0 is the Bessel function of the first kind of order 0, is also to be determined.

MATLAB has functions that evaluate the Bessel functions. The MATLAB functions for $J_1(x)$ and $J_0(x)$ are *besselj*(1,x) and *besselj*(0,x), respectively.

  1. Create vectors for $J_0(x)$ and $J_1(x)$ for $0 \le x \le 40$, subdividing the x domain into 400 subdivisions ($dx = 0.1$).
  2. Create plots of $J_0(x)$ and $J_1(x)$ vs. x on two separate graphs.
  3. Using MATLAB's fzero function, determine the roots of $J_0(x)$ and $J_1(x)$ for $0 \le x \le 40$. Print out the roots in two separate tables.

Project 4.8

The temperature distribution of a thick rod having a circular cross-section, initially at a uniform temperature, T_0, and which is suddenly immersed in a huge bath at a temperature T_∞, is given by

$$TR(r,t) = \frac{T(r,t) - T_\infty}{T_0 - T_\infty} = \sum_{n=1}^{\infty} \frac{2\lambda_n R}{(\lambda_n R)^2 + \left(\frac{hR}{k}\right)^2} \times \frac{J_1(\lambda_n R)J_0(\lambda_n r)}{[J_0(\lambda_n R)]^2} e^{-a\lambda_n^2 t} \qquad \text{(P4.8a)}$$

where

J_1 and J_0 are Bessel functions of the first kind.
h = the convective heat transfer coefficient.
k = the thermal conductivity of the rod material.
R = the radius of the rod.
a = the thermal diffusivity of the rod material.
$\lambda_n R$ = the nth root of Equation (P4.8b):

$$F(\lambda R) = \frac{J_0(\lambda R)}{J_1(\lambda R)} - \lambda R \frac{k}{hR} = 0 \qquad \text{(P4.8b)}$$

MATLAB has functions that evaluate the Bessel functions $J_0(x)$ and $J_1(x)$. These are

$J_0(x) = besselj(0,x)$ and $J_1(x) = besselj(1,x)$. Plots of $J_0(x)$ and $J_1(x)$ are shown in Figure P4.8a and Figure P4.8b, respectively. It can be seen that both functions have an infinite number of zeros. But $F(\lambda R)$ is singular wherever $J_1(\lambda R) = 0$. Before we can evaluate $TR(r, t)$ we need to determine the values of λR that satisfy Equation (P4.8b). The project is to determine the values of λR that satisfy Equation (P4.8b). Designate these values as $\lambda_n R$, for $n =$ 1,2,3, ... To accomplish this, first determine the zeros of J_1 by the fzero function; designate these values as $(\lambda R)_{j1,1}, (\lambda R)_{j1,2}, \ldots$ etc.; then, knowing that

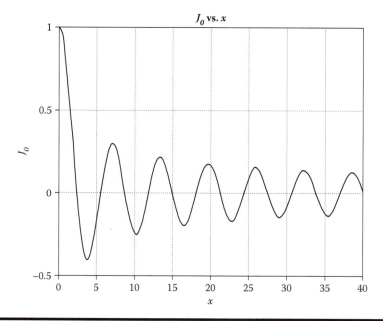

Figure P4.8a **Plot of $J_0(x)$ function.**

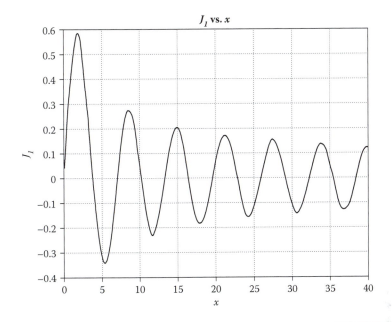

Figure P4.8b **Plot of $J_1(x)$ function.**

the roots of Equation (P4.8b) lie between the zeros of J_1, determine $\lambda_n R$ for $n = 1, 2, 3, \ldots , -30$ by the fzero function. For example: $(\lambda_1 R)$ lies between $(\lambda R)_{j1,1} + \varepsilon$ and $(\lambda R)_{j1,2} - \varepsilon$, $(\lambda_2 R)$ lies between $(\lambda R)_{j1,2} + \varepsilon$ and $(\lambda R)_{j1,3} - \varepsilon$, etc. The plus and minus ε is used because J_1 is singular at the zeros of J_1.

Use the following values:

$$h = 890.0 \frac{\text{w}}{\text{m}^2 - {}^\circ\text{C}}, k = 35.0 \frac{\text{w}}{\text{m} - {}^\circ\text{C}}, R = 0.12 \text{ m}$$

Print out a table of the first 30 values of $\lambda_n R$ in columns of five. Also print out $F(\lambda_n R)$ in columns of five, but in e format.

Reference

1. Moran, M. J. and Shapiro, H. N., *Fundamentals of Thermodynamics,* John Wiley & Sons, Hoboken, NJ, 2004.

Chapter 5

Numerical Integration

5.1 Numerical Integration and Simpson's Rule

We want to evaluate the integral, I, where

$$I = \int_{A}^{B} f(x)\ dx \tag{5.1}$$

- Subdivide the x axis from $x = A$ to $x = B$ into n subdivisions, where n is an even integer.
- Simpson's rule consists of connecting groups of three points on the curve $f(x)$ by second-degree polynomials (parabolas) and summing the areas under the parabolas to obtain the approximate area under the curve (see Figure 5.1).

Expand $f(x)$ in a Taylor Series about x_i using three terms; that is,

$$f(x) = a(x - x_i)^2 + b(x - x_i) + c \tag{5.2}$$

Then,

$$A_{2\ \text{strips}} = \int_{x_{i-1}}^{x_{i+1}} f(x)dx = \int_{x_i - \Delta x}^{x_i + \Delta x} [a\,(x - x_i)^2 + b\,(x - x_i) + c]\ dx \tag{5.3}$$

Let $\xi = x - x_i$

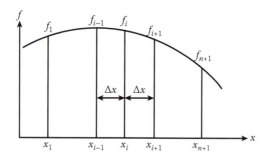

Figure 5.1 Arbitrary three points on the curve f(x).

Then

$$d\xi = dx$$

and when $x = x_i - \Delta x$, $\xi = -\Delta x$, and when $x = x_i + \Delta x$, $\xi = \Delta x$
Making these substitutions into Equation (5.3) gives

$$A_{2\,strips} = \int_{x_{i-1}}^{x_{i+1}} f(x)\,dx = \int_{-\Delta x}^{\Delta x} [a\xi^2 + b\xi + c]\,d\xi$$

$$= \left[\frac{a\xi^3}{3} + \frac{b\xi^2}{2} + c\xi\right]_{-\Delta x}^{\Delta x}$$

$$= \frac{a}{3}(\Delta x)^3 - \frac{a}{3}(-\Delta x)^3 + \frac{b}{2}(\Delta x)^2 - \frac{b}{2}(-\Delta x)^2 + c\Delta x) - c(-\Delta x)$$

Collect like powers of Δx gives

$$A_{2\,strips} = \frac{2a}{3}(\Delta x)^3 + 2c\,\Delta x$$

Now $f(x_i) = f_i = c$

$$f(x_{i+1}) = f_{i+1} = a(\Delta x)^2 + b\,\Delta x + c$$

$$f(x_{i-1}) = f_{i-1} = a(-\Delta x)^2 + b(-\Delta x) + c$$

Adding the two above equations gives: $f_{i+1} + f_{i-1} = 2a\,\Delta x^2 + 2c$

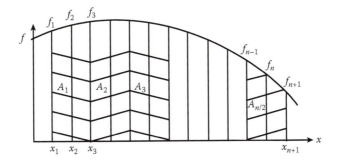

Figure 5.2   Integration areas.

Solving for a gives:

$$a = \frac{1}{2\Delta x^2} [f_{i+1} + f_{i-1} - 2 f_i]$$

Then,

$$A_{2\,strips} = \frac{2}{3} \frac{1}{2\Delta x^2} [f_{i+1} + f_{i-1} - 2f_i] (\Delta x^3) + 2f_i \Delta x$$

$$= \frac{\Delta x}{3} [f_{i+1} + f_{i-1} - 2f_i + 6f_i]$$

or

$$A_{2\,strips} = \frac{\Delta x}{3} [f_{i-1} + 4f_i + f_{i+1}] \tag{5.4}$$

To obtain an approximation for the integral, I, we need to sum all the two-strip areas under the curve from x = A to x = B (see Figure 5.2); that is,

$$A_1 = \frac{\Delta x}{3} [f_1 + 4f_2 + f_3]$$

$$A_2 = \frac{\Delta x}{3} [f_3 + 4f_4 + f_5]$$

$$A_3 = \frac{\Delta x}{3} [f_5 + 4f_6 + f_7]$$

.

.

$$A_{n2} = \frac{\Delta x}{3} [f_{n-1} + 4f_n + f_{n+1}]$$

Thus,

$$I = \int_{x_1=A}^{x_{n+1}=B} f(x)\,dx = \frac{\Delta x}{3}[f_1 + 4f_2 + 2f_3 + 4f_4 + 2f_5 + \cdots + 4f_n + f_{n+1}] \qquad (5.5)$$

This is Simpson's rule for integration.

5.2 Improper Integrals

Example 5.1

$$I = \int_0^1 \frac{\log(1+x)}{x}\,dx$$

The above integral is improper since both the numerator and denominator are zero at the lower limit (x = 0). The exact value of *I* can be obtained by Residue Theory in Complex Variables. It is $I = \dfrac{\pi^2}{12} = 0.822467$

Let

$$I = \int_0^1 \frac{\log(1+x)}{x}\,dx = I_1 + I_2$$

where

$$I_1 = \int_\epsilon^1 \frac{\log(1+x)}{x}\,dx$$

and

$$I_2 = \int_0^\epsilon \frac{\log(1+x)}{x}\,dx$$

To evaluate I_2, expand $\log(1+x)$ in a Taylor Series about x = 0, giving

$$\log(1+x) = x - \frac{1}{2}x^2 + \frac{1}{3}x^3 - \frac{1}{4}x^4 + \frac{1}{5}x^5 - +\cdots$$

then

$$\frac{\log(1+x)}{x} = 1 - \frac{1}{2}x + \frac{1}{3}x^2 - \frac{1}{4}x^3 + \frac{1}{5}x^4$$

$$I_2 = \int\limits_0^\epsilon \left\{ 1 - \frac{1}{2}x + \frac{1}{3}x^2 - \frac{1}{4}x^3 + \frac{1}{5}x^4 - \frac{1}{6}x^5 + \frac{1}{7}x^6 - + \right\} dx$$

$$I_2 = \epsilon - \frac{1}{4}\epsilon^2 + \frac{1}{9}\epsilon^3 - \frac{1}{16}\epsilon^4 + \frac{1}{25}\epsilon^5 - \frac{1}{36}\epsilon^6 + \frac{1}{49}\epsilon^7 - +$$

Evaluate I_1 by Simpson's rule.

To obtain Taylor Series expansion of $\log(1+x)$ use

$$f(x) = f(0) + f'(0)x + f''(0)\frac{x^2}{2!} + f'''(0)\frac{x^3}{3!} + +$$

$$f(0) = \log(1) = 0$$

$$f' = \frac{d}{dx}\log(1+x) = \frac{1}{1+x} \quad ; \quad f'(0) = 1$$

$$f'' = -\frac{1}{(1+x)^2} \qquad ; \quad f''(0) = -1$$

$$f''' = +\frac{2}{(1+x)^3} \qquad ; \quad f'''(0) = 2$$

$$f^{IV} = -\frac{3 \cdot 2}{(1+x)^4} \qquad ; \quad f^{IV}(0) = 3 \cdot 2$$

$$f^V = +\frac{4 \cdot 3 \cdot 2}{(1+x)^5} \qquad ; \quad f^V(0) = 4 \cdot 3 \cdot 2$$

etc.

$$\log(1+x) = 0 + x - \frac{1}{2}x^2 + \frac{2}{3 \cdot 2}x^3 - \frac{3 \cdot 2}{4 \cdot 3 \cdot 2}x^4 + \frac{4 \cdot 3 \cdot 2}{5 \cdot 4 \cdot 3 \cdot 2}x^5$$

$$\log(1+x) = x - \frac{x^2}{2} + \frac{x^3}{3} - \frac{x^4}{4} + \frac{x^5}{5} - \frac{x^6}{6} + -$$

5.3 MATLAB's Quad Function

The MATLAB function for evaluating an integral is quad. A description of the function can be obtained by typing help quad in the command window.

```
Q = quad(FUN,A,B) tries to approximate the integral of function FUN
from A to B to within an error of 1.e-6 using recursive adaptive Simpson
quadrature. The function Y = FUN(X) should accept a vector argument X and
return a vector result Y, the integrand evaluated at each element of X.
   Q = quad(FUN,A,B,TOL) uses an absolute error tolerance of TOL instead
of the default, which is 1.e-6. Larger values of TOL result in fewer
function evaluations and faster computation, but less accurate results.
quad(FUN,A,B,TOL,TRACE,P1,P2,...) provides for additional arguments P1,
P2,... to be passed directly to function FUN, FUN(X,P1,P2,...). Pass empty
matrices for TOL or TRACE to use the default values.
   Use array operators .*, ./, and .^ in the definition of FUN so that it
can be evaluated with a vector argument.
   Function quadl may be more efficient with high accuracies and smooth
integrands.

Example
FUN can be specified as:
An inline object:
   F = inline('1./(x.^3-2*x-5)');
   Q = quad(F,0,2);
A function handle:
   Q = quad(@myfun,0,2);
   where myfun.m is an M-file:
      function y = myfun(x)
      y = 1./(x.^3-2*x-5);
See also quadl, inline, @.
```

It has been found that the quad function is able to evaluate certain improper integrals (see Exercises 5.1d, 5.1e, and 5.1f).

Example 5.2

```
% integralk.m
% This program evaluates the integral of function fk between a & b
% by MATLAB's integration program QUAD
clear; clc;
a=0.0; b=10.0;
integr=quad('fk',a,b);
fprintf('Evaluation of the integration of fk over interval a,b \n');
fprintf('by MATLABs integration program \n\n');
fprintf('fk=x^3+3.2*x^2-5.4*x+20.2 \n\n');
fprintf('integral=%f \n\n',integr);
```

```
% fk.m
% This function is used in an integration program named integralk
% function is f1=x^3+3.2*x^2-5.4*x+20.2
function f1=fk(x)
f1=x.^3+3.2*x.^2-5.4*x+20.2;
```

Example 5.3

```
% integralg.m
% This program evaluates the integral of function f1 between a & b
% by MATLAB's integration program quadl
clear; clc;
integr=quad('funct2',a,b,1.0e-5);
fprintf('\n\n integration of f1 over interval a,b \n');
fprintf('by MATLABs integration program \n\n');
fprintf('funct2=t/(t^3+t+1) \n\n');
fprintf('integral=%f \n\n',integr);
```

```
% funct2.m
% This function is used in an integration program
% named integralg. The function is f1=t/(t^3+t+1)
function f1=f(t)
f1=t./(t.^3+t+1.0);
```

Example 5.4

```
% integralc.m
% This program evaluates the integral of function f1 between a & b
% by MATLAB's integration program quad or quad8
clear; clc;
eps=0.001; a=eps; b=1.0;
integral1=quad('funct3',a,b);
integral2=eps-0.25*eps^2+1.0/9.0*eps^3-1.0/16*eps^4+1.0/25.0*eps^5.0
-1.0/35.0*eps^6+1.0/49.0*eps^7;
integral=integral1+integral2;
fprintf('Evaluation of the integration of f1 over interval a,b \n');
fprintf('by MATLABs integration program \n\n');
fprintf('functc=log(1+x)/x \n\n');
fprintf('integral1=%f integral2=%f \n\n',integral1,integral2);
fprintf('integral=%f \n\n',integral)
```

```
% funct3.m
% This function is used in an integration program
% named integralb
% function is f1=log(1.0+x)/x
function f1=f(x)
f1=log(1.0+x)./x;
```

5.4 MATLAB's DBLQUAD Function

The MATLAB function for numerically evaluating a double integral is DBLQUAD. A description of the function can be obtained by typing *help DBLQUAD* in the command window. The description follows:

```
Q = DBLQUAD(FUN,XMIN,XMAX,YMIN,YMAX) evaluates the double integral of
FUN(X,Y) over the rectangle XMIN <= X <= XMAX, YMIN <= Y <= YMAX. FUN is a
function handle. The function Z=FUN(X,Y) should accept a vector X and a
scalar Y and return a vector Z of values of the integrand.
```

Example

```
Q = dblquad(@ integrnd, pi, 2*pi, 0, pi)
```

where integrnd is the M-file function:

```
function z = integrnd(x, y)
z = y*sin(x)+x*cos(y);
```

Note the integrand can be evaluated with a vector x and a scalar y. Nonsquare regions can be handled by setting the integrand to zero outside of the region.

A program to evaluate the volume of a hemisphere follows.

Example 5.5

```
% two_D_integral.m
% This program calculates the volume of a hemisphere of radius =1
% The dblquad function calculates a double integral over a rectangular
% region XMIN <= X <= XMAX, YMIN <= Y <= YMAX.
Clear; clc;
V=dblquad('fun2D',-1,1,-1,1);
fprintf('V=%10.4f \n',V);

%fun2D
function z=fun2D(x,y)
if (1-(x.^2+y.^2)>=1)
    z=0;
else
    z=sqrt(1-(x.^2+y.^2));
end
```

Exercises

Exercise 5.1

Use MATLAB's quad function to evaluate the following integrals (note that integrals d, e, and f are improper integrals).

a. $I = \int\limits_0^3 \dfrac{dx}{5e^{3x} + 2e^{-3x}}$

b. $I = \int\limits_{-\pi/2}^{\pi/2} \dfrac{\sin x\, dx}{\sqrt{1 - 4\sin^2 x}}$

c. $I = \int\limits_{-\pi}^{\pi} \sinh x \cdot \cos x\, dx$

d. $I = \int\limits_0^1 \dfrac{3e^x\,dx}{\sqrt{1-x^2}}$

e. $I = \int\limits_0^1 \dfrac{\log x\,dx}{(1-x)}$

f. $I = \int\limits_0^1 \dfrac{\log x\,dx}{(1-x^2)}$

Projects

Project 5.1

The solution for the displacement, $Y(x,t)$, from the horizontal of a vibrating string (see Section 10.2) is given by

$$Y(x,t) = \sum_{n=1}^{\infty} a_n \sin\frac{n\pi x}{L}\cos\frac{n\pi ct}{L}$$

where

$$a_n = \frac{2}{L}\int_0^L f(x)\sin\frac{n\pi x}{L}\,dx$$

Use MATLAB's quad function to determine a_n, for $n = 1, 2,\ldots, 10$. Create a table and a plot of a_n vs. n. Take $L = 1.0$ m and

$$f(x) = \begin{cases} 0.4x, & 0 <= x <= 0.75L \\ 0.12 - 0.12x, & 0.75L <= x <= L \end{cases}$$

Project 5.2

In determining the temperature distribution in a cylinder of radius R (see Unsteady Heat Transfer II in Section 10.2.3), the following integral involving the Bessel function, J_0, arises:

$$I_m = \int_0^R r\,J_0(\lambda_m r)\,dr \qquad\qquad (P5.2a)$$

where λ_m is determined by the equation

$$\frac{J_0(\lambda R)}{J_1(\lambda R)} - \lambda R \frac{k}{hR} = 0 \qquad \text{(P5.2b)}$$

where h is the convective heat transfer coefficient and k is the thermal conductivity of the cylinder material. In Project 10.3 the values in Table P5.2 of λ_m were obtained.

Use MATLAB's quad function to determine I_m. Take $R = 0.12$ m. Construct a table of I_m vs. index m.

Table P5.2 λ_m **vs.** *m*

Index *m*	λ_m
1	14.96759
2	37.26053
3	37.26053
4	61.80180
5	87.17614

Chapter 6

Numerical Integration of Ordinary Differential Equations

6.1 The Initial Value Problem

- *Initial Value Problem*—The values of the dependent variable and the necessary derivatives are known at the point at which the integration begins.
- *Modified Euler Method (Self-Starting Method)*—Given the differential equation and initial condition

$$y' = \frac{dy}{dx} = f(x, y)$$

$$y(0) = y_0$$

(6.1)

Subdivide the x domain into N subdivisions. Method involves marching in the x direction.

- Since the initial condition is known, we can assume that for some arbitrary position x_i, the variable y_i is known. We wish to predict y_{i+1}.

Suppose we use a Taylor Series expansion about x_i and only use the first two terms, then

$$y_{i+1}^{P_1} = y_i + y_i' h$$

(6.2)

where $y_i' = y'(x_i)$ and $h = x_{i+1} - x_i$ (see Figure 6.1).

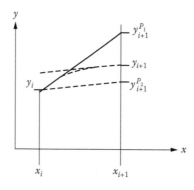

Figure 6.1 Estimate of y_{i+1}.

For the configuration shown in Figure 6.1, $y_{i+1}^{P_1}$ overshoots the true value of y_{i+1}. Suppose in Equation (6.2) we use y_{i+1}' instead of y_i' giving

$$y_{i+1}^{P_2} = y_i + (y_{i+1}')^P h$$

where

$$(y_{i+1}')^P = f\left(x_{i+1}, y_{i+1}^{P_1}\right) \qquad (6.3)$$

For the configuration shown in Figure 6.1, $y_{i+1}^{P_2}$ undershoots the true value of y_{i+1}. A better estimate for y_{i+1} is obtained by using the average of the two derivatives. The corrected value for y_{i+1}, denoted $y_{i+1}^{C_1}$, is given by

$$y_{i+1}^{C_1} = y_i + \frac{h}{2}\left[y_i' + (y_{i+1}')^P \right] \qquad (6.4)$$

Equation (6.4) is known as the corrector equation.

We now substitute Equation (6.4) into Equation (6.1) and obtain

$$(y_{i+1}')^{C_1} = f\left(x_{i+1}, y_{i+1}^{C_1}\right) \qquad (6.5)$$

Now substitute Equation (6.5) into Equation (6.4), giving

$$y_{i+1}^{C_2} = y_i + \frac{h}{2}\left[y_i' + (y_{i+1}')^{C_1} \right] \qquad (6.6)$$

If $|y_{i+1}^{C_2} - y_{i+1}^{C_1}| < \epsilon$ stop iteration for y_{i+1} and move on to the next step to determine y_{i+2}, y_{i+2}', etc. If $|y_{i+1}^{C_2} - y_{i+1}^{C_1}| > \epsilon$ continue iteration; that is, substitute $y_{i+1}^{C_2}$ into

Equation (6.5), obtaining $(y'_{i+1})^{C_2}$ then substitute $(y'_{i+1})^{C_2}$ into Equation (6.4), obtaining $y_{i+1}^{C_3}$, etc.

$$\text{Error estimate, } E = -\frac{1}{12}\, y'''(\xi)h^3, \quad \text{where } x_i < \xi < x_{i+1}.$$

■ The easiest way to determine the accuracy of your answer is to double the number of subdivisions and compare answers. If the desired accuracy is not obtained, continue doubling the number of subdivisions until the desired accuracy is obtained.

6.2 The Fourth-Order Runge–Kutta Method

The fourth-order Runge–Kutta method uses a weighted average of derivative estimates within the interval of interest.

Suppose we are given the equations

$$y' = \frac{dy}{dx} = f(x, y) \quad \text{and} \quad y(0) = y_0$$

In the Euler method we use

$$y_{i+1} = y_i + \frac{h}{2}\left[y'_i + (y'_{i+1})^{C_j} \right]$$

In the Runge–Kutta method, we use

$$y_{i+1} = y_i + \frac{h}{6}[k_1 + 2k_2 + 2k_3 + k_4]$$

where
$$k_1 = f(x_i, y_i) = y'_i \qquad \rightarrow \text{value of } y' \text{ at } x_i$$

$$k_2 = f\left(x_i + \frac{h}{2}, y_i + \frac{h}{2}k_1\right) \rightarrow \text{estimate of } y' \text{ at } x_i + \frac{h}{2}$$

$$k_3 = f\left(x_i + \frac{h}{2}, y_i + \frac{h}{2}k_2\right) \rightarrow \text{a second estimate of } y' \text{ at } x_i + \frac{h}{2}$$

$$k_4 = f(x_i + h, y_i + hk_3) \qquad \rightarrow \text{estimate of } y' \text{ at } x_{i+1}$$

6.3 System of Two First-Order Equations

Consider the following two first-order ordinary differential equations:

$$\frac{d\upsilon}{dx} = f(x, y, \upsilon); \quad y(0) = y_0$$

$$\frac{dy}{dx} = g(x, y, \upsilon); \quad \upsilon(0) = \upsilon_0$$

Take

$$y_{i+1} = y_i + \frac{h}{6}[l_1 + 2l_2 + 2l_3 + l_4]$$

$$\upsilon_{i+1} = \upsilon_i + \frac{h}{6}[k_1 + 2k_2 + 2k_3 + k_4]$$

where

$$k_1 = f(x_i, y_i, \upsilon_i)$$

$$l_1 = g(x_i, y_i, \upsilon_i)$$

$$k_2 = f\left(x_i + \frac{h}{2}, y_i + \frac{h}{2}l_1, \upsilon_i + \frac{h}{2}k_1\right)$$

$$l_2 = g\left(x_i + \frac{h}{2}, y_i + \frac{h}{2}l_1, \upsilon_i + \frac{h}{2}k_1\right)$$

$$k_3 = f\left(x_i + \frac{h}{2}, y_i + \frac{h}{2}l_2, \upsilon_i + \frac{h}{2}k_2\right)$$

$$l_3 = g\left(x_i + \frac{h}{2}, y_i + \frac{h}{2}l_2, \upsilon_i + \frac{h}{2}k_2\right)$$

$$k_4 = f(x_i + h, y_i + hl_3, \upsilon_i + hk_3)$$

$$l_4 = g(x_i + h, y_i + hl_3, \upsilon_i + hk_3$$

where $h = \Delta x$.

6.4 A Single Second-Order Equation

For a single second-order ordinary differential equation, the method of solution is to reduce the equation to a system of two first-order equations.

Given the following single second equation with the initial conditions:

$$y'' = \frac{d^2 y}{dx^2} = f(x, y, y')$$

$$y(0) = y_0 \quad \text{and} \quad y'(0) = y_0'$$

Let $y' = \upsilon$, then $y'' = \dfrac{d\upsilon}{dx} = \upsilon' = f(x, y, \upsilon)$ and $y' = g(x, y, \upsilon) = \upsilon$. Then for the notation used for a system of two first-order equations (see Section 6.3),

$$l_1 = \upsilon_i, \quad l_2 = \upsilon_i + \frac{h}{2} k_1, \quad l_3 = \upsilon_i + \frac{h}{2} k_2, \quad l_4 = \upsilon_i + h k_3$$

Example 6.1

Solve the following two first-order ordinary differential equations by the Runge–Kutta method

$$\frac{dr}{dt} = 2re^{-0.1t} - 2ry = f(t, r, y)$$

$$\frac{dy}{dt} = -y + ry = g(r, y)$$

$$0 \le t \le 10$$

$$r(0) = 1.0$$

$$y(0) = 3.0$$

$$k - f \ or \ \frac{dr}{dt}$$

$$l - g \ or \ \frac{dy}{dt}$$

The problem follows:

```
% runge2.m
% This program solves a system of 2 first order ordinary differential
% equations by the Runge-Kutta method. The equations are:
% dr/dt=2*r*exp(-0.1*t) -2*r*y=f(t,r,y)
% dy/dt=-y+ry=g(r,y)
fo=fopen('output.dat','w');
fprintf(fo,'RUNGE-KUTTA PROBLEM \n');
fprintf(fo,'    t                   r                   y \n');
fprintf(fo,'---------------------------------------------\n');
n=1000; dt=0.01; r(1)=1.0; y(1)=3.0; t(1)=0.0;
for i=1:n
    t(i+1)=i*dt;
    k(1)=rprimef(t(i),r(i),y(i));
    L(1)=yprimef(r(i),y(i));
```

```
      k(2)=rprimef(t(i)+dt/2,r(i)+dt/2*k(1),y(i)+dt/2*L(1));
      L(2)=yprimef(r(i)+dt/2*k(1),y(i)+dt/2*L(1));
      k(3)=rprimef(t(i)+dt/2,r(i)+dt/2*k(2),y(i)+dt/2*L(2));
      L(3)=yprimef(r(i)+dt/2*k(2),y(i)+dt/2*L(2));
      k(4)=rprimef(t(i)+dt,r(i)+dt*k(3),y(i)+dt*L(3));
      L(4)=yprimef(r(i)+dt*k(3),y(i)+dt*L(3));
      r(i+1)=r(i)+dt/6*(k(1)+2*k(2)+2*k(3)+k(4));
      y(i+1)=y(i)+dt/6*(L(1)+2*L(2)+2*L(3)+L(4));
end
for i=1:10:n+1
      fprintf(fo,'%10.2f %16.4f %16.4f \n',t(i),r(i),y(i));
end
fclose(fo);
plot(t,r,t,y,'--'), xlabel('t'), ylabel('r,y'), title('r & y vs t'),
grid;
```

```
% rprimef.m
% This function is used with runge2.m program
function drdt=rprimef(t,r,y)
drdt=2.0*r*exp(-0.1*t) -2.0*r*y;
```

```
% yprimef.m
% This function is used with runge2.m program
function dydt=yprimef(r,y)
dydt=-y+r*y;
```

6.5 MATLAB's ODE Function

MATLAB has an ODE function named ODE45. A description of the function follows. This description can be obtained by typing "help ODE45" in the command window.

ODE45 Solve nonstiff differential equations, medium-order method.

```
[T,Y] = ODE45(ODEFUN,TSPAN,Y0) with TSPAN = [T0 TFINAL] integrates the
system of differential equations y' = f(t,y) from time T0 to TFINAL with
initial conditions Y0. Function ODEFUN(T,Y) must return a column vector
corresponding to f(t,y). Each row in the solution array Y corresponds to a
time returned in the column vector T. To obtain solutions at specific
times T0,T1,...,TFINAL (all increasing or all decreasing), use TSPAN = [T0
T1 ... TFINAL].
      [T,Y] = ODE45(ODEFUN,TSPAN,Y0,OPTIONS) solves as above with default
integration properties replaced by values in OPTIONS, an argument created
with the ODESET function. See ODESET for details. Commonly used options
are scalar relative error tolerance 'RelTol' (1e-3 by default) and vector
of absolute error tolerances 'AbsTol' (all components 1e-6 by default).
```

```
[T,Y] =ODE45(ODEFUN,TSPAN,Y0,OPTIONS,P1,P2...) passes the additional
parameters P1,P2,... to the ODE function as ODEFUN(T,Y,P1,P2...), and to all
functions specified in OPTIONS. Use OPTIONS = [] as a place holder if no
options are set.
```

Example 6.2

```
% ode45_ex.m
% This program solves a system of 3 differential equations
% by using ode45 function
% y1'=y2*y3*t, y2'=-y1*y3, y3'=-0.51*y1*y2
% y1(0)=0, y2(0)=1.0, y3(0)=1.0
clear; clc;
initial=[0.0 1.0 1.0];
tspan=0.0:0.1:10.0;
options=odeset('RelTol',1.0e-6,'AbsTol',[1.0e-6 1.0e-6 1.0e-6]);
[t,Y]=ode45(@dydt3,tspan,initial,options);
P=[t Y];
disp(P);
t1=P(:,1);
y1=P(:,2);
y2=P(:,3);
y3=P(:,4);
fid=fopen('output.txt','w');
fprintf(fid,'      t               y(1)          y(2)                  y(3) \n');
fprintf(fid,'-------------------------------------------------------------\n');
for i=1:2:101
    fprintf(fid,' %6.2f       %10.2f  %10.2f      %10.2 \ n', ...
    t1(i),y1(i),y2(i),y3(i))
end
fclose(fid);
plot(t1,Y(:,1),t1,Y(:,2),'-.',t1,Y(:,3),'--'), xlabel('t'),
ylabel('Y(1),Y(2),Y(3)'),title('Y vs. t'),
grid,
text(6.0, -1.2,'y(1)'), text(7.7, -0.25,'y(2)'), text(4.2,0.85,'y(3)');
```

```
% dydt3.m
% functions for example problem
% y1'=y2*y3*t, y2'=-y1*y3, y3'=-0.51*y1*y2
function Yprime=dydt3(t,Y)
Yprime=zeros(3,1);
Yprime(1)=Y(2)*Y(3)*t;
Yprime(2)= -Y(1)*Y(3);
Yprime(3)= -0.51*Y(1)*Y(2);
```

Example 6.3

```
% ode_vib.m
% This program solves the motion of a spring-dashpot system. The governing
% equation is a Second Order Ordinary Differential Equation (x vs. t)by
% MATLAB's ode45 function.
% m=10 kg, k= 4 N/m, c= 2.0 N-s/m, alpha=0.05 N, omega= 2 rad/s
% y(1)= x, y(2)= xdot=v, yprime1=y(2);
% yprime(2)=alpha*sin(omega *t)-c/m*v-k/m*y(1)
clear; clc;
initial=[0.5 0];
tspan=0:0.1:50;
[t,y]=ode45('dydt_vib',tspan,initial);
P=[t y];
t1=P(:,1);
x=P(:,2);
v=P(:,3);
```

```
fprintf('      t      x      v   \n');
for i=1:5:501
    fprintf(' %10.2f    %10.4f    %10.4f   \n',t1(i),x(i),v(i));
end
plot(t1,x), xlabel('t'), ylabel('x'), title('x vs. t'), grid;
figure;
plot(t1,v), xlabel('t'), ylabel('v'), title('v vs. t'), grid;
```

```
% dydt_vib.m
% m=10 kg, k= 4 N/m, c= 2.0 N-s/m, alpha=0.05 N, omega= 2 rad/s
% y(1)= x, y(2)= xdot=v, yprime1=y(2);
% yprime(2)=alpha*sin(omega *t)-c/m*v-k/m*y(1)
function yprime=dydt_vib(t,y)
yprime=zeros(2,1);
m=10; k=4; c=2.0; alpha=0.05; omega=2;
yprime(1)=y(2);
yprime(2)=alpha*sin(omega*t)-c/m*y(2)-k/m*y(1);
```

6.6 Ordinary Differential Equations That Are Not Initial Value Problems

When an ordinary differential equation involves boundary conditions, instead of initial conditions, then it is convenient to use a numerical approach to solving the problem. An example of this type of problem is the deflection of a beam where boundary conditions at both ends of the beam are specified. With certain types of boundary conditions, the numerical method will reduce to solving a set of linear equations that fall into the category of a tri-diagonal matrix. The solution of a set of linear algebraic equations that are classified as a tri-diagonal system involves fewer calculations than the solution of the set by Gauss Elimination. This is only important if the system of equations is large. The solution of a tri-diagonal system is discussed next.

6.7 Solution of a Tri-Diagonal System of Linear Equations

Suppose we have a system of equations of the form

$$
\begin{bmatrix}
1 & -a_1 & 0 & 0 & 0 \\
-b_2 & 1 & -a_2 & 0 & 0 \\
0 & -b_3 & 1 & -a_3 & 0 \\
0 & 0 & -b_4 & 1 & -a_4 \\
0 & 0 & 0 & -b_5 & 1
\end{bmatrix}
\begin{bmatrix}
x_1 \\ x_2 \\ x_3 \\ x_4 \\ x_5
\end{bmatrix}
=
\begin{bmatrix}
c_1 \\ c_2 \\ c_3 \\ c_4 \\ c_5
\end{bmatrix}
\tag{6.7}
$$

This system is designated as a tri-diagonal system. The set of equations becomes

$$x_1 - a_1 x_2 = c_1 \tag{6.8}$$

$$-b_2 x_1 + x_2 - a_2 x_3 = c_2 \tag{6.9}$$

$$-b_3 x_2 + x_3 - a_3 x_4 = c_3 \tag{6.10}$$

$$-b_4 x_3 + x_4 - a_4 x_5 = c_4 \tag{6.11}$$

$$-b_5 x_4 + x_5 = c_5 \tag{6.12}$$

Concept:

1. One can solve Equation (6.8) for x_1 and substitute it into Equation (6.9), giving an equation involving x_2 and x_3, which is designated as Equation (6.9').
2. One can then solve Equation (6.9') for the x_2 in terms of x_3 and substitute into Equation (6.10). This gives an equation just involving x_3 and x_4, which is designated as Equation (6.10').
3. This process is continued until the last equation. When x_4 is substituted into Equation (6.12), an equation only involving x_5 is obtained. Thus, x_5 can be determined.
4. Then by back substitution, one can obtain all the other x_i values.

Method:

1. Put the set of equations in the general form shown in Equation (6.13):

$$x_i = a_i x_{i+1} + b_i x_{i-1} + c_i \tag{6.13}$$

Note: $b_1 = 0$ and for m equations, $a_m = 0$.

2. By the substitution procedure outlined above, one obtains a set of equations of the form

$$x_i = d_i + e_i x_{i+1} \tag{6.14}$$

Note: For i = m, $e_i = 0$.
Then,

$$x_m = d_m$$

$$x_{m-1} = d_{m-1} + e_{m-1} x_m$$

$$\vdots$$

$$x_1 = d_1 + e_1 x_2$$

Therefore, if a general expression for d_i and e_i can be obtained, one could solve the system for x_i. We start with the assumption that we can put the (i–1) equation in the form

$$x_{i-1} = d_{i-1} + e_{i-1} x_i \qquad \text{(we showed that we can do this for i = 2)}$$

Then, the equation

$$x_i = a_i x_{i+1} + b_i x_{i-1} + c_i$$

becomes

$$x_i = a_i x_{i+1} + b_i (d_{i-1} + e_{i-1} x_i) + c_i$$

$$(1 - b_i e_{i-1}) x_i = (c_i + b_i d_{i-1}) + a_i x_{i+1}$$

or

$$x_i = \frac{c_i + b_i d_{i-1}}{1 - b_i e_{i-1}} + \frac{a_i}{1 - b_i e_{i-1}} x_{i+1} = d_i + e_i x_{i+1}$$

Thus,

$$d_i = \frac{c_i + b_i d_{i-1}}{1 - b_i e_{i-1}}; \qquad e_i = \frac{a_i}{1 - b_i e_{i-1}} \qquad (6.15)$$

valid for $i = 2, 3,\ldots, m$.

Note: The very first equation in the system is already in the form $x_i = d_i + e_i x_{i-1}$ and also that

$$a_m = 0$$

Thus,

$$d_1 = c_1, \quad e_1 = a_1, \quad \text{and} \quad b_1 = 0 \qquad (6.16)$$

Then, $x_m = d_m$ and by back substitution

$$x_{m-1} = d_{m-1} + e_{m-1} x_m$$

$$\vdots$$

$$\qquad (6.17)$$

$$\vdots$$

$$x_1 = d_1 + e_1 x_2$$

Summary:

Set up the equations in the form of Equation (6.13), that is,

$$x_i = a_i x_{i+1} + b_i x_{i-1} + c_i$$

establishing values for a_i, b_i, and c_i.
Determine d_1 and e_1 from Equations (6.16).
Determine d_i and e_i from Equations (6.15), for $i = 2, 3,\ldots,m$.
Determine x_m.
Detemine x_i, for $i = m-1, m-2,\ldots,x_1$, from Equations (6.17).

An example of the use of the tri-diagonal method for solving a system of linear differential equations using the finite difference method is shown in Section 6.9. Finite difference formulas are developed by the use of just a few terms in a Taylor Series expansion. This is shown in the next section.

6.8 Difference Formulas

Difference formulas obtained by Taylor Series expansion are useful in reducing differential and partial differential equations to a set of algebraic equations.

Given $y = f(x)$, a Taylor Series expansion about point x_i is

$$y(x_i + h) = y(x_i) + y'(x_i)h + \frac{y''(x_i)}{2!}h^2 + \frac{y'''(x_i)}{3!}h^3 + \cdots$$

where $h = \Delta x$.

Let

$$y(x_i + h) = y_{i+1} \quad \text{and} \quad y(x_i) = y_i, \quad y'(x_i) = y_i', \quad \text{etc.}$$

Then the Taylor Series expansion equation can be written as

$$y_{i+1} = y_i + y_i'h + \frac{y_i''\,h^2}{2!} + \frac{y_i'''\,h^3}{3!} + \frac{y_i^{IV}\,h^4}{4!} + \cdots \tag{6.18}$$

Also, for equally spaced points on the x axis,

$$y(x_i - h) = y_{i-1} = y_i + y_i'(-h) + \frac{y_i''\,(-h)^2}{2!} + \frac{y_i'''\,(-h)^3}{3!} + \frac{y_i^{IV}\,(-h)^4}{4!} + \cdots$$

or

$$y_{i-1} = y_i - y_i'h + \frac{y_i''\,h^2}{2!} - \frac{y_i'''\,h^3}{3!} + \frac{y_i^{IV}\,h^4}{4!} - + \cdots \tag{6.19}$$

Returning to Equation (6.18) and using two terms in the expansion gives

$$y_{i+1} = y_i + y_i'h$$

Solving for y_i',

$$y_i' = \frac{y_{i+1} - y_i}{h} \tag{6.20}$$

Using $y_i = \dfrac{y_{i+1} - y_i}{h}$ involves an error or order h. This is the forward difference formula for y_i' of order h. Similarly from Equation (6.19), using only two terms in the expansion gives

$$y_i' = \frac{y_i - y_{i-1}}{h} \qquad (6.21)$$

Using $y_i' = \dfrac{y_i - y_{i-1}}{h}$ involves an error of order h. This is the backward difference formula for y_i' of order h.

Now suppose we subtract Equation (6.19) from Equation (6.18), keeping only three terms in each equation. This gives

$$y_{i+1} - y_{i-1} = 2 y_i' h$$

Solving for y_i' gives

$$y_i' = \frac{y_{i+1} - y_{i-1}}{2h} \qquad (6.22)$$

Using $y_i' = \dfrac{y_{i+1} - y_{i-1}}{2h}$ involves an error of order h^2. This is the central difference formula for y_i' of order h^2.

If we add Equation (6.18) and Equation (6.19), keeping only three terms in each equation, we obtain

$$y_{i+1} + y_{i-1} = 2 y_i + y_i'' h^2$$

Solving for y_i'' gives

$$y_i'' = \frac{y_{i+1} + y_{i-1} - 2 y_i}{h^2} \qquad (6.23)$$

Using $y_i'' = \dfrac{y_{i+1} + y_{i-1} - 2 y_i}{h^2}$ involves an error of order h^2. This is the central difference formula for y_i'' of order h^2.

Sometimes in numerical analysis, when a boundary condition involves the first derivative, y', one might wish to use a one-sided estimate for y' of order h^2. This can be accomplished by writing

$$y_{i+2} = y_i + y_i'(2h) + \frac{1}{2!} y_i'' (2h)^2 + \frac{1}{3!} y_i''' (2h)^3 + \cdots \qquad (6.24)$$

By Equation (6.18)

$$y_{i+1} = y_i + y_i' h + \frac{1}{2!} y_i'' h^2 + \frac{1}{3!} y_i''' h^3 + \cdots \qquad (6.25)$$

Using only three terms in Equations (6.24) and (6.25) and multiplying Equation (6.25) by (−4) gives

$$-4\,y_{i+1} = -4\,y_i - 4\,y_i'h - \frac{4\,y_i''}{2!}\,h^2 \tag{6.26}$$

The three-term equation of Equation (6.24) is

$$y_{i+2} = y_i + y_i'(2h) + \frac{y_i''}{2!}(2h)^2 \tag{6.27}$$

Adding Equations (6.26) and (6.27) gives

$$y_{i+2} - 4\,y_{i+1} = -3\,y_i - 2\,y_i'h$$

Solving for y_i' gives

$$y_i' = \frac{-y_{i+2} + 4\,y_{i+1} - 3\,y_i}{2h} \tag{6.28}$$

Equation (6.28) involves an error of order h^2. This is the one-sided forward difference formula for y_i' of order h^2.

This formula can be useful in applying a boundary condition involving y' at x_1, the starting point of the x domain. Similarly,

$$y_{i-2} = y_i + y_i'(-2h) + \frac{1}{2!}y_i''(-2h)^2 = y_i - 2\,hy_i' + \frac{4}{2!}h^2 y_i'' \tag{6.29}$$

Again use only three terms in Equation (6.19) and Equation (6.29), and multiply Equation (6.19) by (−4) and add the result to Equation (6.29); that is,

$$-4\,y_{i-1} = -4\,y_i + 4\,hy_i' - \frac{4}{2!}h^2 y_i''$$

$$y_{i-2} = y_i - 2\,hy_i' + \frac{4}{2!}h^2 y_i''$$

Adding gives

$$y_{i-2} - 4\,y_{i-1} = -3\,y_i + 2\,hy_i'$$

Solving for y_i',

$$y_i' = \frac{y_{i-2} - 4\,y_{i-1} + 3\,y_i}{2h} \tag{6.30}$$

Equation (6.30) involves an error at order h^2. This is the one-sided backward difference formula for y_i' of order h^2.

This formula can be useful in applying a boundary condition involving y' at x_{N+1}, the end point in the x domain that is subdivided into N subdivisions.

6.9 Deflection of a Beam

The governing equation for the deflection of a beam is (for a derivation of this equation see Appendix A)

$$\frac{d^2 y}{dx^2} = \frac{M(x)}{EI(x)}$$ (6.31)

where
  y = deflection of the beam at x.
  M = internal bending moment.
  E = modulus of elasticity of the beam material.
  I = moment of inertia of the cross-sectional area.

Consider the beam shown in Figure 6.2. To obtain the finite difference form of the governing equation, subdivide the x axis into N subdivisions, giving x_1, x_2, x_3,..., x_{N+1}. Let the deflections at these points be

$$y_1, y_2, y_3, ..., y_{N+1}$$

The finite difference for $\dfrac{d^2 y}{dx^2}$, as discussed in Section 6.8, is given by

$$\frac{d^2 y}{dt^2}(x_n) = \frac{y_{n+1} + y_{n-1} - 2y_n}{\Delta x^2}$$ (6.32)

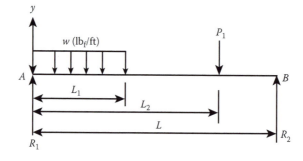

Figure 6.2 Beam loading.

Thus, the governing differential equation becomes

$$\frac{y_{n+1} + y_{n-1} - 2y_n}{\Delta x^2} = \frac{M_n}{EI_n}, \qquad \text{for } n = 2,3,4,\ldots,N$$

or

$$y_n = \frac{1}{2} y_{n+1} + \frac{1}{2} y_{n-1} - \frac{M_n \Delta x^2}{2EI_n} \qquad (6.33)$$

Boundary conditions:

$$y_1 = 0 \qquad (6.34)$$

$$y_{N+1} = 0 \qquad (6.35)$$

The system of equations is a tri-diagonal system.

To obtain an expression for the bending moment M_n, first solve for the reactions R_1 and R_2.

$$\sum M_A = 0 = R_2 L - wL_1 \times \frac{L_1}{2} - P_1 L_2$$

Solving for R_2 gives

$$R_2 = \frac{wL_1^2}{2L} + \frac{P_1 L_2}{L} \qquad (6.36)$$

$$\sum M_B = 0 = P_1 (L - L_2) + wL_1 \left(L - \frac{L_1}{2} \right) - R_1 L$$

Solving for R_1 gives

$$R_1 = P_1 \left(1 - \frac{L_2}{L} \right) + wL_1 \left(1 - \frac{1}{2} \frac{L_1}{L} \right) \qquad (6.37)$$

Internal bending moments are taken about the neutral axis of the section at x. For $0 \le x \le L_1$ (see Figure 6.3),

$$M(x) + wx \cdot \frac{x}{2} - R_1 x = 0$$

Solving for $M(x)$ and expressing the equation in finite difference form gives

$$M_n = R_1 x_n - \frac{w}{2} x_n^2 \qquad (6.38)$$

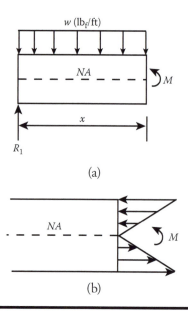

(a)

(b)

Figure 6.3 Sketch indicating (a) internal moment at section x, where $0 \leq x \leq L_1$; (b) stress distribution.

For $L_1 < x \leq L_2$ (see Figure 6.4),

$$M(x) + wL_1\left(x - \frac{L_1}{2}\right) - R_1 x = 0$$

Solving for $M(x)$ and expressing the equation in finite difference form gives

$$M_n = R_1 x_n - wL_1\left(x_n - \frac{L_1}{2}\right) \tag{6.39}$$

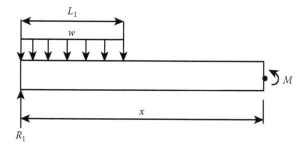

Figure 6.4 Sketch indicating internal moment at section x, where $L_1 < x \leq L_2$.

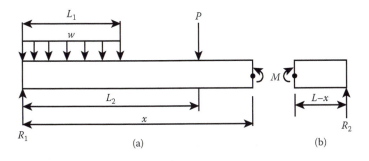

Figure 6.5   Sketch indicating internal moment at section *x*, where, $L_2 < x \le L$, (a) as seen from the left side and (b) as seen from the right side.

For $L_2 < x \le L$ (see Figure 6.5a)

$$M(x) + P_1(x - L_2) + wL_1\left(x - \frac{L_1}{2}\right) - R_1 x = 0$$

Solving for $M(x)$ and expressing the equation in finite difference form gives

$$M_n = R_1 x_n - P_1(x_n - L_2) - wL_1\left(x_n - \frac{L_1}{2}\right) \tag{6.40}$$

For this section, it is more convenient to select the section from the right side of the beam (see Figure 6.5b).

$$-M(x) + R_2(L - x) = 0$$

Solving for $M(x)$ and expressing the equation in finite difference form gives

$$M_n = R_2(L - x_n) \tag{6.41}$$

Equation (6.41) is equivalent to Equation (6.40) for M_n for the region $L_2 < x \le L$.

The system of equations is tri-diagonal and thus can be solved by the method described in Section 6.7. (See Projects 6.8 and 6.9.)

Projects

Project 6.1

An airplane flying horizontally at 50 m/s and at an altitude of 300 m is to drop a food package weighing 2000 N to a group of people stranded in an inaccessible area

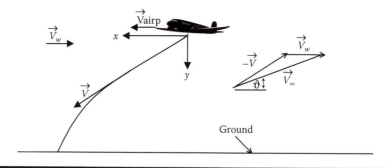

Figure P6.1   Airplane dropping food package.

resulting from an earthquake. A wind velocity, V_w, of 20 m/s flows horizontally in the opposite airplane direction (see Figure P6.1). A drag force, $\vec{D}$, acts on the package in the direction of the free stream, $\vec{V}_\infty$, as seen from the package (see Figure P6.1). We wish to determine the (t,x,u,v) values when the package hits the ground, where

(x,y) = the position of the package at time t.
(u,v) = the horizontal and vertical components of the package velocity, respectively.

Governing equations:

$$M\frac{d\vec{V}}{dt}=Mg\,\vec{j}+\vec{D}$$

$$\vec{V}=u\vec{i}+v\,\vec{j}=\frac{dx}{dt}\vec{i}+\frac{dy}{dt}\vec{j}$$

$$\vec{D}=C_d\rho\frac{V_\infty^2}{2}A\vec{e}$$

$$V_\infty=\sqrt{(u+V_w)^2+v^2}$$

where $\vec{i}$ and $\vec{j}$ are unit vectors in the x and y directions, respectively, $\vec{e}$ is a unit vector in the direction of the free stream velocity as seen from the package, C_d is the drag coefficient, ρ is the air density, and A is the frontal area of the package.
Equations reduce to

$$\frac{du}{dt}=-\frac{C_d\rho V_\infty^2 A}{2M}\cos\vartheta \tag{P6.1a}$$

$$\frac{dv}{dt}=g-\frac{C_d\rho V_\infty^2 A}{2M}\sin\vartheta \tag{P6.1b}$$

$$\frac{dx}{dt} = u \tag{P6.1c}$$

$$\frac{dy}{dt} = v \tag{P6.1d}$$

$$\cos \vartheta = \frac{u + V_w}{V_\infty} \quad \text{and} \quad \sin \vartheta = \frac{v}{V_\infty} \tag{P6.1e}$$

Initial conditions:

$$x(0) = 0, \quad y(0) = 0, \quad u(0) = 50 \, \text{m/s}, \quad v(0) = 0$$

Use the following parameters:

$$C_d = 0.8, \quad \rho = 1.225 \, \text{kg/m}^3, \quad A = 1.0 \, \text{m}^2$$

Using the MATLAB function ODE45 to obtain values for (t, x, y, u, v) at intervals of 0.10 seconds for $0 \le t \le 10.0$ seconds.

(a) Create plots of x and y vs. t both on the same graph.
(b) Create plots of u and v vs. t both on the same graph.
(c) Create a table containing (t, x, y, u, v) at intervals of 0.10 seconds. Stop printing the table the first time $y > 300$ m.
(d) Use MATLAB's function interp1 to interpolate for the (t, x, u, v) values when the package hits the ground. Print out these values.

Project 6.2

A small rocket with an initial mass of 350 kg, including a mass of 100 kg of fuel, is fired from a rocket launcher (see Figure P6.2). The rocket leaves the launcher at

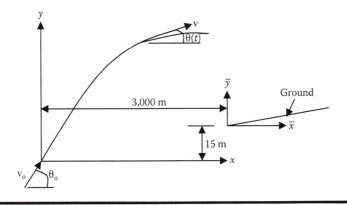

Figure P6.2   Rocket trajectory.

velocity v_o and at an angle of θ_o with the horizontal. Neglect the fuel consumed inside the rocket launcher. The rocket burns fuel at the rate of 10 kg/s, and develops a thrust $T = 6000$ N. The thrust acts axially along the rocket and lasts for 10 seconds. Assume that the drag force also acts axially and is proportional to the square of the rocket velocity. The governing differential equations describing the position and velocity components of the rocket are as follows:

$$\frac{dv_x}{dt} = \frac{T}{m}\cos\theta - \frac{Kv^2}{m}\cos\theta \tag{P6.2a}$$

$$\frac{dv_y}{dt} = \frac{T}{m}\sin\theta - \frac{Kv^2}{m}\sin\theta - g \tag{P6.2b}$$

$$\frac{dx}{dt} = v_x \tag{P6.2c}$$

$$\frac{dy}{dt} = v_y \tag{P6.2d}$$

where

$$v^2 = v_x^2 + v_y^2$$

θ is the angle the velocity vector makes with the horizontal.
m is the mass of the rocket (varies with time).
v_x, v_y are the x and y components of the rocket's velocity relative to the ground.
K is the drag coefficient.
g is the gravitational constant.
(x, y) are the position of the rocket relative to the ground.
t is the time of rocket flight.
$\cos\theta = v_x/v$ and $\sin\theta = v_y/v$.

Substituting for $\cos\theta$ and $\sin\theta$ in Equations (P6.2a) and (P6.2b) they become

$$\frac{dv_x}{dt} = \frac{v_x T}{m\sqrt{v_x^2 + v_y^2}} - \frac{v_x K\sqrt{v_x^2 + v_y^2}}{m} \tag{P6.2e}$$

$$\frac{dv_y}{dt} = \frac{v_y T}{m\sqrt{v_x^2 + v_y^2}} - \frac{v_y K\sqrt{v_x^2 + v_y^2}}{m} - g \tag{P6.2f}$$

The target lies on ground that has a slope of 5%. The ground elevation, y_g, relative to the origin of the coordinate system of the rocket is given by

$$y_g = 15 + 0.05(x - 3000) \tag{P6.2g}$$

Using Equations (P6.2e), (P6.2f), (P6.2c), and (P6.2d), write a computer program in MATLAB using the fourth-order Runge–Kutta method described in Sections 6.3 and 6.4 that will solve for x, y, v_x, and v_y for $0 \leq t \leq 60$ seconds. Use Equation (P6.2g) to solve for y_g.

Use a fixed time step of 0.01 second. Take $x(0) = 0$, $y(0) = 0$, $v_x(0) = v_o \cos \theta_o$, $v_y(0) = v_o \sin \theta_o$, $v_o = 150$ m/s, $K = 0.045$ N–s²/m², $g = 9.81$ m/s², $\theta_o = 60°$, and

(a) Print out a table for x, y, y_g, v_x, v_y at every 1.0 seconds. Run the program for $0 \leq t \leq 60$ seconds.
(b) Use MATLAB to plot x, y, and y_g vs. t and v_x, v_y vs. t.
(c) Assume a linear trajectory between the closest two data points where the rocket hits the ground. The intersection of the two straight lines gives the (x,y) position of where the rocket hits the ground.

Project 6.3

Repeat Project 6.2, but this time use MATLAB's ODE45 function to solve the problem. Use a tspan = [0: 1:60] seconds.

Project 6.4

We wish to examine the time temperature variation of a fluid, T_f, enclosed in a container with a heating element and a thermostat. The walls of the container are pure copper. The fluid is engine oil, which has a temperature T_f that varies with time. The thermostat is set to cut off power from the heating element when the T_f reaches 65°C and to resume supplying power when T_f reaches 55°C.

Wall properties:

$$k = 386.0 \text{ w/m-°C}, c = 0.3831 \text{ kJ/kg-°C}, \rho = 8954 \text{ kg/m}^3$$

Engine oil properties:

$$k = 0.137 \text{ w/m-°C}, c = 2.219 \text{ kJ/kg-°C}, \rho = 840 \text{ kg/m}^3$$

The inside size of the container is (0.5 m × 0.5 m × 0.5 m). The wall thickness is 0.01 m. Thus,

Inside surface area, $A_{s,i} = 1.5$ m².
Outside surface area, $A_{s,o} = 1.5606$ m².
Engine oil volume, $V_{oil} = 0.125$ m³.
Wall volume, $V_{wall} = 0.0153$ m³.

The power, Q, of the heating element = 10,000 w.
The inside convective heat transfer coefficient, $h_i = 560$ w/m²-C.
The outside convective heat transfer coefficient, $h_o = 110$ w/m²-C.

Using a lump parameter analysis (assume that the engine oil is well mixed) and the First Law of Thermodynamics, the governing equations describing the time temperature variation of both materials are as follows:

$$\frac{d\theta_f}{dt} = -a_1(\theta_f - \theta_w) + a_5 \tag{P6.4a}$$

$$\frac{d\theta_w}{dt} = a_2(\theta_f - \theta_w) - a_3\theta_w = a_2\theta_f - (a_2 + a_3)\theta \tag{P6.4b}$$

where

$$\theta_f = T_f - T_\infty$$

$$\theta_w = T_w - T_\infty$$

$$a_1 = \frac{h_i A_{s,i}}{m_f c_f}, \quad a_2 = \frac{h_i A_{s,i}}{m_w c_w}, \quad a_3 = \frac{h_o A_{s,o}}{m_w c_w}, \quad a_4 = a_2 + a_3, \quad a_5 = \frac{Q}{m_f c_f}$$

Initial conditions:

$$T_f(0) = T_w(0) = 15°C$$

$$T_\infty = 15°C$$

Using ODE45 function, construct a simulation of this system. Run the time for 3600 seconds. Print out values of T_f and T_w vs. t at every 100 seconds. Construct plots of T_f and T_w vs. t.

Project 6.5

We wish to determine the altitude and velocity of a helium-filled spherically shaped balloon as it lifts off from its mooring. We will assume that atmospheric conditions can be described by the U.S. Standard Atmosphere. We will assume that there is no change in the balloon's volume. The governing equation describing the motion of the balloon is

$$M\frac{d^2z}{dt^2} = (B - W - \text{sgn}* D) \tag{P6.5a}$$

where

  z = altitude of the centroid of the balloon.
  B = buoyancy force acting on the balloon.
  M = the total mass of the balloon material, ballast, and the gas.
  W = the total weight of the balloon material, ballast, and the gas = Mg.
  D = the drag on the balloon.
  sgn = +1, if $\frac{dz}{dt} \geq 0$ and sgn = −1, if $\frac{dz}{dt} < 0$.

The U.S. Standard Atmosphere as applied to this balloon problem consists of the following governing equations:

$$\frac{dp}{dz} = -\gamma = -\rho g, \qquad \frac{dp}{dz} = \frac{dp}{dt}\frac{dt}{dz} = \frac{dp}{dt}\frac{1}{v} = -\rho g$$

or

$$\frac{dp}{dt} = -\rho g v \qquad\qquad\qquad (P6.5b)$$

$$T = T_i - \lambda z, \qquad \rho = \frac{p}{RT} \qquad\qquad (P6.5c)$$

where

  p = the outside air pressure at the centroid of the balloon.
  g = the gravitational constant that varies with altitude.
  R = the gas constant for air.
  T = the outside air temperature at the centroid of the balloon.
  v = the vertical velocity of the balloon.
  T_i = the air temperature at the earth's surface = 288.15 K.
  λ = the lapse rate.

The second-order ordinary differential equation, Equation (P6.5a), can be reduced to two first-order differential equations, by letting

$$\frac{dz}{dt} = v \qquad\qquad\qquad (P6.5d)$$

Then

$$\frac{dv}{dt} = \frac{1}{M}(B - W - \text{sgn} * D) \qquad\qquad (P6.5e)$$

The three equations—Equations (P6.5e), (P6.5d), and (P6.5b)—represent three coupled ordinary differential equations that can be solved using MATLAB's ODE45 function. The buoyancy force, B, is given by

$$B = \rho g \, \forall \qquad\qquad\qquad (P6.5f)$$

and

$$g = g_0 \left(\frac{r_e}{z + r_e} \right) \qquad\qquad (P6.5g)$$

where

  ρ = the air density at the centroid of the balloon.
  $\forall$ = the volume of the balloon.

r_b = the radius of the balloon.
r_e = the radius of the earth.
g_0 = the gravitational acceleration near the earth's surface.
g = the gravitational acceleration at an elevation of the centroid of the balloon.

For low Reynolds number, Re, less than 0.1, the drag force D is given by the Stokes formula, which is

$$D = 6\pi\mu v r_b \qquad \text{(P6.5h)}$$

For flow speeds with Re > 0.1, use

$$D = C_d \frac{\rho}{2} v^2 A \qquad \text{(P6.5i)}$$

where
C_d = the drag coefficient.
A = the frontal area of the balloon $= \pi r_b^2$.
μ = the fluid viscosity.

$$\text{Re} = \frac{2\rho v r_b}{\mu} \qquad \text{(P6.5j)}$$

The fluid viscosity, μ, can be determined by the Sutherland formula, which is

$$\mu = \mu_0 \left(\frac{T}{T_0}\right)^{1.5} \left(\frac{T_0 + S}{T + S}\right) \qquad \text{(P6.5k)}$$

For air, $S = 110.4\ K$, $\mu_0 = 1.71\text{e-}5$ N-s/m², $T_0 = 273$ K.
The drag coefficient, C_d, is given by

$$C_d = \frac{24}{\text{Re}} + \frac{6}{1.0 + \sqrt{\text{Re}}} + 0.4 \qquad \text{(P6.5l)}$$

Write a computer program, using MATLAB's ODE45 function, that will determine the balloon's altitude as a function of time. Create a table of 100 lines giving the balloon's altitude, velocity, and the pressure of the atmosphere at the balloon's centroid. Also, create plots of z vs. t, v vs. t, and p vs. t. Use the following values:

$M = 2200$ kg, $r_b = 7.816$ m, $\forall = 4/3\,\pi r_b^3$ m³, $R = 287$ J/(kg-K), $T_i = 288.15$ K, $\lambda = 0.0065$ K/m, $g_0 = 9.81$ m/s², $r_e = 6371\text{e+}3$ m.

Use a tspan = 0.0:0.1:1000 and the following initial conditions:

$$z(0) = r_b, \quad v(0) = 0, \quad p(0) = 1.0132\text{e+}5$$

Project 6.6

A small tank with its longitudinal axis in a vertical position is connected to a pressurized air supply system as shown in Figure P6.6. The tank contains two gate valves, one that controls the pressurization of the tank and the other that controls the discharge of the tank through a converging nozzle. The tank is instrumented with a copper-constantan thermocouple and a pressure transducer. We wish to predict the temperature and pressure time histories of the air inside the tank as it is being discharged. We shall assume that the air properties inside the tank are uniform. Due to its large heat capacity, as compared to the air inside the tank, we shall assume that the wall temperature remains nearly constant during the discharge phase of the problem. We will also assume that the change in kinetic and potential energy inside the tank is negligible. Applying the energy equation to a control volume enclosing the interior of the tank gives

$$\frac{d}{dt}(m_a u_a) = -\dot{m}_e\left(u + \frac{p}{\rho} + \frac{V^2}{2}\right)_e + \left(\frac{\delta Q}{\delta t}\right)_{w \to a} \tag{P6.6a}$$

where
  m = mass
  u = internal energy
  p = pressure
  ρ = density
  V = velocity
  $\dot{m}_e$ = the mass flow rate of air exiting the tank
  $\left(\dfrac{\delta Q}{\delta t}\right)_{w \to a}$ = the rate of heat transfer from the wall to the air inside the tank
  subscript a = air
  subscript w = wall
  subscript e = conditions at the exit

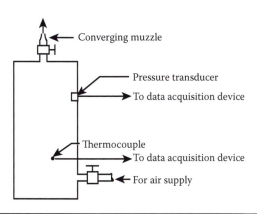

Figure P6.6   Air pressurized tank.

$$u_a = c_{v,a} T_a \tag{P6.6b}$$

$$\left(u + \frac{p}{\rho}\right)_e = h_e = c_{p,e} T_e \tag{P6.6c}$$

where

h = enthalpy.
c_p = specific heat at constant pressure.
c_v = specific heat at constant volume.
T = temperature.

$$\left(\frac{\delta Q}{\delta t}\right)_{w \to a} = \bar{h} A_{s,i} (T_w - T_a) \tag{P6.6d}$$

where

$\bar{h}$ = the convective heat transfer coefficient.
$A_{s,i}$ = interior surface area of the tank.

The equation describing the rate of change of mass in the tank is

$$\frac{dm_a}{dt} = -\dot{m}_e \tag{P6.6e}$$

Substituting Equations (P6.6b) through (P6.6e) into Equation (P6.6a) gives

$$\frac{dT_a}{dt} = \frac{T_a}{m_a}\dot{m}_e + \frac{A_{s,i}\bar{h}_i}{c_{v,a} m_a}(T_w - T_a) - \frac{\dot{m}_e}{m_a}k_a T_e - \frac{\dot{m}_e V^2}{2 c_{v,a} m_a} \tag{P6.6f}$$

where

$c_{v,a}$ = air specific heat at constant volume.

k_a = ratio of specific heats for air $= \dfrac{c_{p,a}}{c_{v,a}}$.

We have assumed that $c_{v,a}$ and $c_{p,a}$ do not vary significantly in the temperature range of the problem. The functional relation for $\dot{m}_e$ in terms of the other variables is obtained from one-dimensional compressible flow-through nozzles. Two possible cases exist, depending on the ratio of $\dfrac{p_b}{p_a}$. The pressure p_b is the back pressure, which is the surrounding air pressure.

Case 1:

$$\frac{p_b}{p_a} > 0.528$$

Then

$$p_e = p_b$$

$$M_e = \frac{2}{k_a - 1}\left[1 - \left(\frac{p_e}{p_a}\right)^{(k_a - 1)/k_a}\right]$$

$$T_e = \frac{T_a}{1 + \dfrac{k_a - 1}{2}M_e^2}$$

$$\bar{c}_e = \sqrt{k_a R_a T_e}$$

$$V_e = M_e \bar{c}_e$$

$$\dot{m}_e = \frac{p_e}{R_a T_e} A_e V_e$$

Case 2:

$$\frac{p_b}{p_a} <= 0.528$$

Then

$$p_e = 0.528 \, p_a$$

$$M_e = 1.0$$

$$T_e = \frac{T_a}{1 + \dfrac{k_a - 1}{2}}$$

$$V_e = \bar{c}_e = \sqrt{k_a R_a T_e}$$

$$\dot{m}_e = \frac{p_e}{R_a T_e} A_e V_e$$

where $\bar{c}$ is the speed of sound and M is the Mach number.

From these relations it can be seen that $\dot{m}_e$ is an implicit function of the variables T_a, m_a, and t. These equations, along with Equations (P6.6e) and (P6.6f), form two coupled differential equations of the form

$$\frac{dT_a}{dt} = f_1(t, T_a, m_a)$$

$$\frac{dm_a}{dt} = f_2(t, T_a, m_a)$$

This system of equations can be solved using the ODE45 solver in MATLAB. However, before this can be done, one needs to determine $\bar{h}$. We shall assume that the heat transfer from the wall to the air inside the tank occurs by natural convection. We will also assume that the wall temperature remains nearly constant, since it has a much larger heat capacity than the air. The empirical relation for natural convection for vertical plates and cylinders is

$$\bar{h} = \frac{k_{t,a}}{L}\left[0.825 + \frac{0.387(Gr\,Pr)^{1/6}}{\left[1 + \left(\frac{0.492}{Pr}\right)^{9/16}\right]^{8/27}}\right]^2 \qquad \text{(P6.6g)}$$

where

$$Gr = \text{Grashof number} = \frac{g\beta(T_w - T_a)L^3}{(\mu_a/\rho_a)^2} \qquad \text{(P6.6h)}$$

Pr is the Prandtl number, $k_{t,a}$ is the thermal conductivity of air, L is the cylinder length, μ_a is the viscosity of air, β is the coefficient of expansion, and g is the gravitational constant. The properties of Pr, $k_{t,a}$, μ_a are evaluated at the film temperature, T_f, which is

$$T_f = 0.5(T_w + T_a) \quad \text{and} \quad \rho_f = \left(\frac{p}{RT}\right)_f$$

For an ideal gas, $\beta = \frac{1}{T}$. Since β is a fluid property, take $\beta = \frac{1}{T_f}$. Air property values for these variables are given in Table P6.6.

Using MATLAB's ODE45 function, determine the air temperature, T_a, air pressure, p_a, and mass, m_a, in the tank as a function of time. Use a tspan = 0:0.1:70 seconds.

Print out a table of these values at every 1 second. Also, create plots of T_a vs. t, p_a vs. t, and m_a vs. t. Use the following values for the program:

D = 0.07772 m, L = 0.6340 m, d_e = 0.001483 m, $c_{p,a}$ = 1.006e+3 J/(kg-K),

$c_{v,a}$ = 0.721e+3 J/(kg-K), k_a = 1.401, g = 9.807 m/s², $p_b = p_{atm}$ = 1.013e+5,

$p_{a,I}$ = 704648 N/m², $T_{a,I}$ = 294.5 K, T_w = 294.5 K, and R_a = 287.2 J/(kg-K)

Table P6.6   Air Properties vs. Temperature

T (K)	μ (N-s/m²)	k (W/(m-K))	Pr
100	6.9224e-6	0.009246	0.77
150	1.0283e-5	0.013735	0.753
200	1.3289e-5	0.01809	0.739
250	1.4880e-5	0.02227	0.722
300	1.9830e-5	0.02624	0.708
350	2.075e-5	0.03003	0.697

where

D = the inside diameter of the tank.

L = the inside length of the tank.

D_e = the diameter of the nozzle.

$T_{a,I}$ = the initial air temperature inside the tank.

$P_{a,I}$ = the initial air pressure inside the tank.

Project 6.7

Low-speed rear-end collisions between two vehicles are very common in urban congested areas. In this kind of collision very little or no damage is observed after the accident. This indicates that most of the energy involved in the impact is absorbed by damping mechanisms attached to the dampers (piston, honeycombs, etc.). Therefore, one may use a linear model in which each vehicle can be represented as a lumped mass with a spring-dashpot energy-absorbing bumper. A linear three-degree-of-freedom model for such a collision is shown in Figure P6.7. The mass and position of the struck vehicle are m_1 and x_1, respectively; the mass and position of

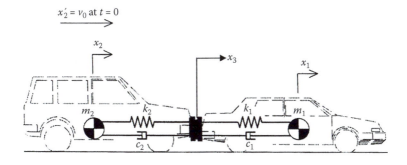

Figure P6.7   Linear three-degree-of-freedom model for low-speed car collision. (SAE Paper 980360. Copyright 1998 SAE International. With permission.)

the striking vehicle are m_2 and x_2, respectively; and the mass and position of the two bumpers are m_3 and x_3, respectively. It is assumed that the two bumpers remain in contact during the investigation period.

The governing dynamic equations for the system are given in Equations (P6.7a) through (P6.7c).

$$\frac{d^2 x_1}{dt^2} = -\frac{C_1}{m_1}\left(\frac{dx_1}{dt} - \frac{dx_3}{dt}\right) - \frac{k_1}{m_1}(x_1 - x_3) \pm \mu g \qquad (P6.7a)$$

$$\frac{d^2 x_2}{dt^2} = \frac{C_2}{m_2}\left(\frac{dx_3}{dt} - \frac{dx_2}{dt}\right) + \frac{k_2}{m_2}(x_3 - x_2) \pm \mu g \qquad (P6.7b)$$

$$\frac{d^2 x_3}{dt^2} = \frac{k_1}{m_3}(x_1 - x_3) - \frac{k_2}{m_3}(x_3 - x_2) + \frac{C_1}{m_3}\left(\frac{dx_1}{dt} - \frac{dx_3}{dt}\right) - \frac{C_2}{m_3}\left(\frac{dx_3}{dt} - \frac{dx_2}{dt}\right) \qquad (P6.7c)$$

For the μ term in Equation (P6.7a), use the $-$ sign if $\frac{dx_1}{dt} > 0$, and the $+$ sign if $\frac{dx_1}{dt} < 0$.

Similarly, for the μ term in Equation (P6.7b), use the $-$ sign if $\frac{dx_2}{dt} > 0$, and the $+$ sign if $\frac{dx_2}{dt} < 0$.

The initial conditions are

$$x_1(0) = 0, \quad x_2(0) = 0, \quad x_3(0) = 0 \qquad (P6.7d)$$

$$\frac{dx_1}{dt}(0) = 0, \quad \frac{dx_2}{dt}(0) = V_0, \quad \frac{dx_3}{dt}(0) = V_0 \qquad (P6.7e)$$

where k_1, k_2, C_1, and C_2 are the springs' and the dampers' constants of the vehicles. V_0 is the speed of the striking vehicle and μ is the coefficient of friction.

The linear spring and dashpot constants are deduced from an analysis of the time history data coming from destructive tests. Use the following constants:

$$C_1 = C_2 = 20000 \text{ N s/m}$$

$$k_1 = k_2 = 6600 \text{ kN/m}$$

$$m_1 = 930 \text{ kg}, \ m_2 = 960 \text{ kg}, \ m_3 = 150 \text{ kg}$$

$$V_0 = 2.77 \text{ m/s}$$

For Equation (P6.7a),

$$\text{take } \mu = 0.04 \quad \text{if } \left|\frac{dx_1}{dt}\right| > 0 \quad \text{and} \quad \mu = 0 \quad \text{if } \frac{dx_1}{dt} = 0.$$

For Equation (P6.7b),

$$\text{take } \mu = 0.04 \quad \text{if } \left|\frac{dx_2}{dt}\right| > 0 \quad \text{and} \quad \mu = 0 \quad \text{if } \frac{dx_2}{dt} = 0.$$

Develop a MATLAB program that will solve the collision model problem using the ODE45 function for $0 \leq t \leq 0.4\,s$. Use a tspan = $[0:001:0.4]$.

(a) Print out a table for x_1, x_2, x_3, $\dfrac{dx_1}{dt}$, $\dfrac{dx_2}{dt}$, $\dfrac{dx_3}{dt}$, $\dfrac{d^2x_1}{dt^2}$, $\dfrac{d^2x_2}{dt^2}$, and $\dfrac{d^2x_3}{dt^2}$ every 0.01 second.

(b) Plot x_1, x_2, and x_3 on one graph, $\dfrac{dx_1}{dt}$, $\dfrac{dx_2}{dt}$, and $\dfrac{dx_3}{dt}$ on a second graph, and $\dfrac{d^2x_1}{dt^2}$, $\dfrac{d^2x_2}{dt^2}$, and $\dfrac{d^2x_3}{dt^2}$ on a third graph.

Project 6.8

A small fin is used to increase the heat loss from an electronic element. A sketch of the fin is shown in Figure P6.8.

The temperature distribution in the fin is governed by the following equation:

$$\frac{d}{dx}\left(A\frac{dT}{dx} \right) = \frac{hP}{k}(T - T_\infty)$$

or

$$\frac{dA}{dx}\frac{dT}{dx} + A\frac{d^2T}{dx^2} = \frac{hP}{k}(T - T_\infty)$$

where T is the temperature in the fin at position x, T_∞ is the surrounding air temperature, A is the fin cross-sectional area, h is the corrective heat transfer coefficient, k is the thermal conductivity of the fin, and P is its perimeter.

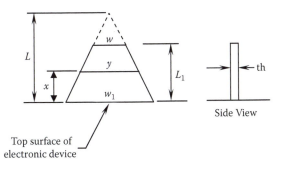

Figure P6.8   Fin attached to an electronic device.

$$A(x) = y(th), \quad P(x) = 2y + 2th$$

$$\frac{y}{w_1} = \frac{L-x}{L}, \quad \frac{w_2}{w_1} = \frac{L-L_1}{L}$$

$$A = \frac{w_1(th)}{L}(L-x), \quad \frac{dA}{dx} = -\frac{w_1(th)}{L}$$

The Governing Differential Equation is

$$\frac{w_1(th)}{L}(L-x)\frac{d^2T}{dx^2} - \frac{w_1(th)}{L}\frac{dT}{dx} = \frac{h}{k}\left[\frac{2w_1}{L}(L-x) + 2(th)\right](T - T_\infty) \quad \text{(P6.8a)}$$

The boundary conditions are

$$T(0) = T_w \quad \text{(P6.8b)}$$

To obtain the second boundary condition write:

The rate that heat leaves the fin at $(x = L_1)$ per unit surface area
   = the rate that heat is carried away by convection per unit surface area.

The above statement can be written mathematically as

$$-k\frac{dT}{dx}(L_1) = h\,[T(L_1) - T_\infty] \quad \text{(P6.8c)}$$

We wish to solve this problem numerically using the finite difference method. First subdivide the x axis into I subdivision giving $x_1, x_2, x_3,\ldots,x_{I+1}$. Take the temperature at x_i to be T_i. The finite difference formulas for $\frac{d^2T}{dx^2}$ and $\frac{dT}{dx}$ (see Section 6.8) are

$$\frac{d^2T}{dx^2}(x_i) = \frac{T_{i+1} + T_{i-1} - 2T_i}{\Delta x^2}$$

$$\frac{dT}{dx}(x_i) = \frac{T_{i+1} - T_i}{\Delta x}$$

The finite differential form of Equation (P6.8a) is

$$\frac{w_1(th)}{L}(L-x_i)\left(\frac{T_{i+1} + T_{i-1} - 2T_i}{\Delta x^2}\right) - \frac{w_1(th)}{L}\left(\frac{T_{i+1} - T_i}{\Delta x}\right)$$

$$= \frac{h}{k}\left[\frac{2w_1}{L}(L-x_i) + 2(th)\right](T_i - T_\infty)$$

Solving for T_i gives

$$T_i = \left\{ \frac{2w_1(th)}{L}(L - x_i) - \frac{w_1(th)}{L}\Delta x + \frac{h\Delta x^2}{k}\left(\frac{2w_1}{L}(L - x_i) + 2(th) \right) \right\}^{-1}$$

$$\times \left\{ \left[\frac{w_1(th)}{L}(L - x_i) - \frac{w_1(th)}{L}\Delta x \right]T_{i+1} + \frac{w_1(th)}{L}(L - x_i)T_{i-1} \right.$$

$$\left. + \frac{h\Delta x^2}{k}\left[\frac{2w_1}{L}(L - x_i) + 2(th) \right]T_\infty \right\} \tag{P6.8d}$$

Equation (P6.8d) is valid for $i = 2, 3, \ldots,$ I.

The finite difference form for Equation (P6.8c) is

$$-k\frac{T_{I+1} - T_I}{\Delta x} = h\left[T_{I+1} - T_\infty \right]$$

Solving for T_{I+1} gives

$$T_{I+1} = \frac{1}{1 + \frac{h\Delta x}{k}}\left\{ T_I + \frac{h\Delta x}{k}T_\infty \right\}$$

Also, $T_1 = T_W$.

The rate of heat loss, Q, through the fin is given by

$$Q = -\frac{kA_1}{\Delta x}(T_2 - T_1)$$

where A_1 is the cross-sectional area of the fin at $x = 0+$.

Using the method described in Section 6.7 for a tri-diagonal system of equations, write a computer program that will solve for the temperature distribution in the fin and the rate of heat loss through the fin. Use the following values:

$T_w = 200°C, \quad T_\infty = 40°C, \quad k = 204$ W/m-°C, $\quad h = 60$ W/m²-°C,

$I = 80, \quad w_1 = 2$ cm, $\quad L = 6$ cm, $\quad L_1 = 4$ cm, $\quad th = 0.2$ cm.

The output of your program should include the values of h, k, T_w, T_∞, Q, and a table of T vs. x at every 0.1 cm.

Project 6.9

We wish to obtain the reactions, the bending moment, and the deflection of the statically indeterminate beam as shown in Figure P6.9a. The problem can be solved by the method of superposition. First solve for the deflection $y(x)$ by the finite

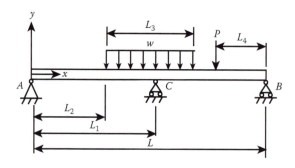

Figure P6.9a Indeterminate beam structure.

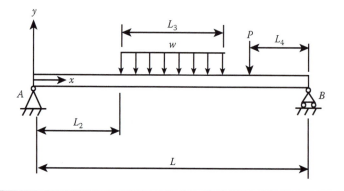

Figure P6.9b Determinate beam structure.

difference method utilizing the tri-diagonal method to obtain a solution for the statically determinate structure shown in Figure P6.9b. Then determine $y(L_1)$ for the structure shown in Figure P6.9b. Next determine the value of F in the structure shown in Figure P6.9c that would cause the deflection at L_1 in that structure to be $-y(L_1)$. You may use the following formula to determine the F value that would give the required deflection at $x = L_1$ (see Figure P6.9d).

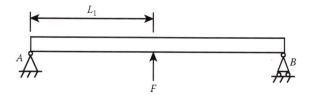

Figure P6.9c Beam with single concentrated load.

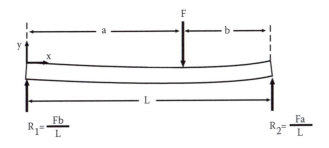

Figure P6.9d   Beam loaded with a single concentrated load.

$$y(x) = \frac{Fbx}{6LEI}[x^2 - (L^2 - b^2)] \qquad (x \le a)$$

$$= \frac{Fb}{6LEI}\left[x^3 - (L^2 - b^2)x - \frac{L}{b}(x - a)^3 \right] \quad (x \ge a)$$

Finally, superimpose both solutions to give the true values for the reactions, R_A, R_B, and R_C, the bending moment $M(x)$ and the deflection $y(x)$. Print out the final reactions, R_A, R_B, and R_C. Print out a table of $M(x)$ and $y(x)$ vs. x at every other node. Use MATLAB to plot $M(x)$ and $y(x)$.

Let $w = 40$ kN/m, $EI = 1.5 \times 10^3$ kN-m², $P = 35$ kN, $L = 3$ m, $L_1 = 1.3$ m, $L_2 = 0.5$ m, $L_3 = 1.5$ m, $L_4 = 0.5$ m. Let the number of subdivisions on the x axis be 150.

Chapter 7

Simulink

7.1 Introduction

Simulink* is used with MATLAB® to model, simulate, and analyze dynamic systems. With Simulink, models can be built from scratch, or additions can be made to existing models. Simulations can be made interactive, so a change in parameters can be made while running the simulation. Simulink supports linear and nonlinear systems, modeled in continuous time, sample time, or a combination of the two.

Simulink provides a graphical user interface (GUI) for building models as block diagrams, using click-and-drag mouse operations.

The program includes a comprehensive library of components (blocks). Using scopes and other display blocks, simulation results can be seen while the simulation is running.

7.2 Creating a Model in Simulink

1. Click on the Simulink icon on the menu bar in the MATLAB command window or type Simulink in the MATLAB command window. This brings up the Simulink library browser window (see Figure 7.1).
2. Click on *file* in the Simulink library browser window.

---

* MATLAB®, Simulink®, and Stateflow® are trademarks of The MathWorks, Inc. and are used with permission. The MathWorks does not warrant the accuracy of the text or exercises in this book. This book's use or discussion of MATLAB, Simulink, and Stateflow software or related products does not constitute endorsement or sponsorship by The MathWorks of a particular pedagogical approach or particular use of the MATLAB, Simulink, and Stateflow software.

Figure 7.1 Simulink windows. (From MATLAB. With permission.)

3. Select a "new model" (for a new model) or "open" for an existing model. This will bring up an untitled model window (for the case of a new model) or an existing model window.

4. To create a new model you need to copy blocks from the library browser window into the new model window. This can be done by highlighting a particular block and dragging it into the model window. To simplify the connections of blocks, you may need to rotate a block 90° or 180°. To do this highlight the block and click on format in the menu bar; select rotate block (for 90°) or flip block (for 180°).

Simulink has many categories for distributing the library blocks; those of interest for this book are commonly used blocks, continuous, discontinuities, math operations, ports and subsystems, signal routing, sinks, sources, and user-defined functions.

The blocks that will be used in this chapter are constant, product, gain, sum, integrator, scope, display, relay, switch, to workspace, mux, and fcn. The product, gain, and sum are in the math category, the integrator is in the continuous category, the scope, display, and to workspace are in the sink category, the constant, step, and sine wave are in the source category, the relay is in the nonlinear category, the mux and switch are in the signal routing category, and the fcn is in the user-defined functions category.

7.3 Typical Building Blocks in Constructing a Model

1. Addition of two constants (see Figure 7.2). To set the value for the constants, highlight the block, click on the right button to open a dialog box, select update parameters, and type in the value of the constant. To run the program click on simulation in the menu bar and select start.
2. Subtraction (see Figure 7.3).
3. Product of two blocks (see Figure 7.4).
4. Division of two blocks (see Figure 7.5).
5. Integrate a sine wave (see Figure 7.6). In sine wave block parameters set frequency to 2 rads/second.

$$\int_0^t \sin 2x \, dx = -\frac{1}{2}[\cos 2x]_0^t = -\frac{1}{2}[\cos 2t - 1] = \frac{1}{2}[1 - \cos 2t] \quad (\text{range } 0 \rightarrow 1)$$

(7.1)

6. Solution of a simple ordinary first-order differential equation. The method of solution is illustrated in Example 7.1, which considers the temperature change of a small, good heat-conducting object that is suddenly immersed in a fluid of temperature T_∞. The temperature, T, of the object varies with time. The governing equation is given by Equation (7.2).

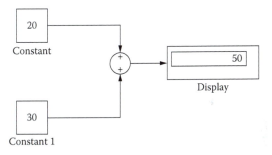

Figure 7.2 Constants, summation, and display blocks for addition.

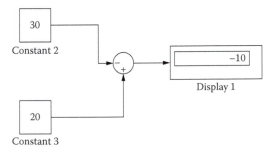

Figure 7.3 Constants, summation, and display blocks for subtraction.

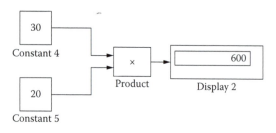

Figure 7.4   Constants, product, and display blocks for multiplication.

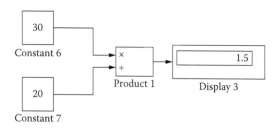

Figure 7.5   Constants, product, and display blocks for division.

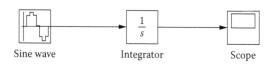

Figure 7.6   Sine wave, integration, and scope blocks.

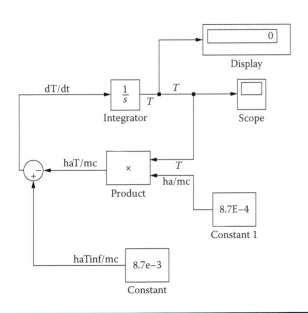

Figure 7.7   Block diagram for solving Equation (7.2).

Example 7.1

First put the equation in the form

$$\frac{dT}{dt} = f(T,t)$$

$$\frac{dT}{dt} = \frac{hA_s}{mc}T_\infty - \frac{hA_s}{mc}T \qquad (7.2)$$

where
    m = the mass of the object.
    A_s = the surface area.
    c = the specific heat of the object.
    h = the convective heat transfer coefficient.

The following parameters for Equation (7.2) were used:

$$\frac{hA_s}{mc} = 8.7 \times 10^{-4}\frac{1}{s}$$

$$T_\infty = 10°C$$

Block diagram for Equation (7.2) is shown in Figure 7.7.

7.4 Constructing and Running the Model

1. To connect lines from the output of a block to the input of a second block, place the pointer on the output (or input) of the block, right click on the mouse, and drag the line to the input (or output) of the second block.
2. To connect a point on a line to the input of a block, place the pointer on the line and right click on the mouse and drag the line to the input of the block.
3. To add alphanumerical information above a line double click (left button) on the line. A small box will appear on the line. Type in desired info and click elsewhere.
4. To view the results on the scope, use the right mouse button and click on the scope. Select open block. A graph appears. To select the graph axis, right click on graph and select axis properties or autoscale. In most cases selecting autoscale is sufficient.
5. To select the initial condition, right click on integrator box, select integrator parameters, and type in the initial condition; select OK.
6. To select start and stop run times, click on Simulation in menu bar, select simulation parameter, and edit start and stop time boxes.
7. To run the simulation, click on simulation in the menu bar and click on start.

7.5 Constructing a Subsystem

Suppose we are building a large system consisting of many blocks and we wish to reduce the number of blocks appearing in the overall block diagram. This can be done by creating a subsystem. The subsystem will appear as a single block. To create a subsystem, place the pointer in the vicinity of the region that is to become a subsystem and right click the mouse. This produces a small dashed box that can be enlarged by dragging the mouse over the region enclosing the number of blocks to be included in the subsystem. When the mouse button is released, a drop-down menu appears. Select create a subsystem. In the above example, the constant, constant 1, and product blocks have been included in the subsystem. Note that the subsystem will have one input and two outputs (see Figure 7.8). These three blocks will appear as a single subsystem block (see Figure 7.9). A block diagram of the subsystem is shown in Figure 7.10.

7.6 Using the mux and fcn Blocks

An alternative to the block diagram for Example 7.1 is shown in Figure 7.11. In this solution the mux and fcn blocks are used to solve the problem. The mux block allows you to select the number of inputs (right click on the block, select mux

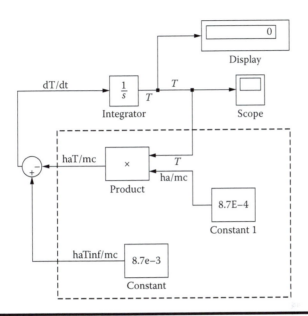

Figure 7.8   Constructing a subsystem.

parameters, and type in the number of inputs). The uppermost input is designated as u [1], the one below is designated as u [2], etc. The output from the mux block should go to the input of the fcn block. The math expression in the fcn block needs to be in terms of the u [ ]'s. See block sketch, Figure 7.11.

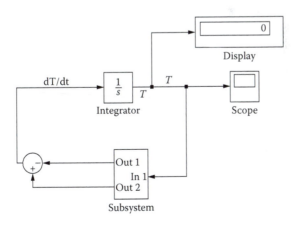

Figure 7.9   Block diagram for solving Equation (7.2).

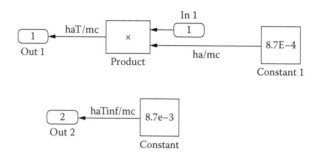

Figure 7.10   Sketch of subsystem.

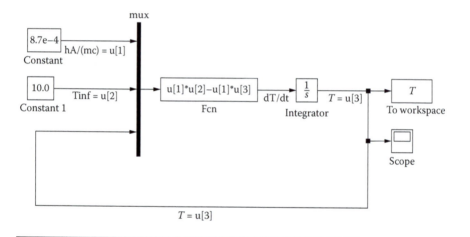

Figure 7.11   Block diagram for solving Equation (7.2) using the mux and fcn blocks.

7.7 The Relay Block

Suppose a model involves the addition of a heating/cooling element to a system controlled by a thermostat. A relay may be used to simulate the thermostat behavior. This concept can be represented by this simple differential equation in which y is to go from 0 to 10 and then fluctuate between 5 and 10.

$$\frac{dy}{dt} = c \quad \text{where } c = \begin{cases} 20 & \text{if} \quad y \leq 5 \\ -20 & \text{if} \quad y \geq 10 \end{cases}$$

The block diagram for this system consists of an integrator, a constant, a relay, a product, and a scope. The relay parameters are

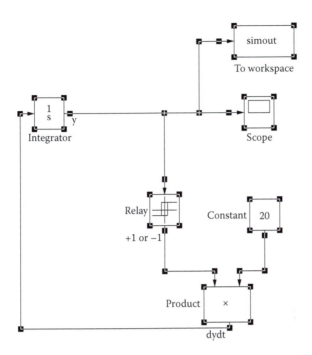

Figure 7.12 Block diagram using the relay block.

Switch point on = 10
Switch point off = 5
Output when on = –1
Output when off = +1

Note: Switch point on has to be greater than switch point off.

The block diagram for this problem is shown in Figure 7.12.

Start by setting the initial condition for y to 0. At this point, $y \leq 5$, thus the relay switch is off and the output of the relay is +1, causing y to increase. The relay output will remain +1 until y reaches 10, then the relay switch will turn on and the relay output will be –1, causing y to decrease. The relay output will be –1, until y reaches 5; then the relay switch is off and the output is +1. The process will continue until the simulation end time is reached.

7.8 The Switch Block

Some problems may involve a function that varies in time for $0 <= t <= t_1$ and the remains are constant for $t_1 < t <= t_2$. This type of function can best be modeled with the switch.

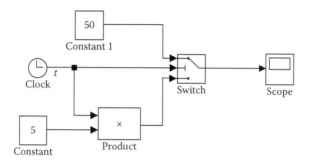

Figure 7.13 Block diagram using the switch block.

Suppose

$$y = \begin{cases} 5t & for \quad 0 <= t <= 10 \\ 50 & for \quad 10 < t <= 20 \end{cases}$$

The Simulink model for this problem is shown in Figure 7.13.

Switch parameters (u1 is the top input, u2 is the middle input, and u3 is the bottom input). Switch parameters for this example follow.

Criteria for passing first input: u2 > = Threshold
Threshold: 10
Sample time (–1 for inherited): –1

7.9 Trigonometric Function Blocks

Functions such as sine, cosine, tangent, etc. can be found in math operations under trigonometric function. The input to the trig function block is the argument to the trig function. If the argument involves the independent variable, *t*, the output of the clock gives *t*. This is shown in the following example (see Figure 7.14).

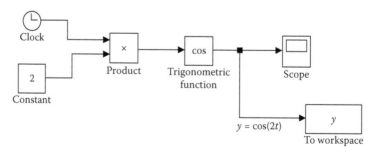

Figure 7.14 Block diagram using the clock, the trig function, and the workspace blocks.

Example 7.2 Governing Equation for a Spring-Dashpot System

Given a simple spring-dashpot system subjected to an oscillatory force. The governing equation is

$$\ddot{x} + \frac{c}{m}\dot{x} + \frac{k}{m}x = \frac{F}{m}\sin(\omega t), \quad x(0) = 5 \text{ m}, \quad \dot{x}(0) = 0 \text{ m/s} \qquad (7.3)$$

The following Simulink program (see Figure 7.15) gives the solution. The values used are

$$k = 1 \text{ N/m}, \quad c = 0.5 \text{ kg/s}, \quad \omega = 20 \text{ l/s}, \quad F = 1 \text{ N}, \quad m = 10 \text{ kg}$$

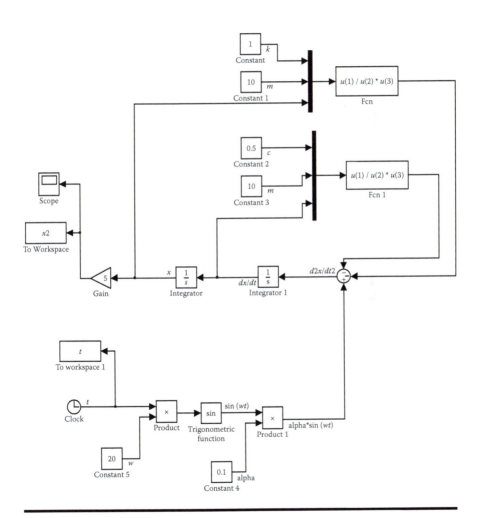

Figure 7.15   Block diagram for solving Equation (7.3).

To obtain output values in table form, one needs to send the variables to workspace by the workspace block as shown in Figure 7.15. After the simulation is run, those variables become available for use in any MATLAB program. Typical simulation and workspace parameters follow:

```
Simulation time
Start time: 0.0           Stop time: 100.0
Solver Options
Type: fixed-step          Solver: ode4 (Runge-Kutta)
Periodic sample time constraint: unconstrained
Fixed-step size (fundamental sample time): 0.1
Tasking mode for periodic sample time: auto

Workspace Parameters
Variable name: x
Limit data points to last: inf
Decimation: 1
Sample time (-1 for inherited): -1
Save format: array
```

Projects

Project 7.1

A small rocket with an initial mass of 350 kg, including a mass of 100 kg of fuel, is fired from a rocket launcher. The rocket leaves the launcher at velocity v_o and at an angle of θ_o with the horizontal. Neglect the fuel consumed inside the rocket launcher. The rocket burns fuel at the rate of 10 kg/s, and develops a thrust $T = 6000$ N. The thrust acts axially along the rocket and lasts for 10 seconds. Assume that the drag force also acts axially and is proportional to the square of the rocket velocity. The governing differential equations describing the position and velocity components of the rocket (see Figure P6.2) are as follows:

$$\frac{d^2x}{dt^2} = \frac{T}{m}\cos\theta - \frac{Kv^2}{m}\cos\theta$$

$$\frac{d^2y}{dt^2} = \frac{T}{m}\sin\theta - \frac{Kv^2}{m}\sin\theta - g$$

$$\frac{dx}{dt} = v_x$$

$$\frac{dy}{dt} = v_y$$

where

$$v^2 = v_x^2 + v_y^2$$

θ is the angle the velocity vector makes with the horizontal.
m is the mass of the rocket (varies with time).
v_x, v_y are the x and y components of the rocket's velocity relative to the ground.
K is the drag coefficient.
g is the gravitational constant.
(x, y) are the position of the rocket relative to the ground.
t is the time of rocket flight.
$\cos \theta = v_x/v$ and $\sin \theta = v_y/v$.

Substituting for cos θ, sin θ, and v in the above set, the equations become

$$\frac{d^2x}{dt^2} = \frac{v_x T}{m\sqrt{v_x^2 + v_y^2}} - \frac{v_x K \sqrt{v_x^2 + v_y^2}}{m} \qquad \text{(P7.1a)}$$

$$\frac{d^2y}{dt^2} = \frac{v_y T}{m\sqrt{v_x^2 + v_y^2}} - \frac{v_y K \sqrt{v_x^2 + v_y^2}}{m} - g \qquad \text{(P7.1b)}$$

(a) Develop a SIMULINK model that will solve for x, y, v_x, v_y for $0 \le t \le 60$ seconds. Use a fixed time step of 0.1 second. Run the program for $t \le 60$ seconds.

Take $x(0) = 0$, $y(0) = 0$, $v_x(0) = v_o \cos \theta_o$, $v_y(0) = v_o \sin \theta_o$, $v_o = 150$ m/s, $\theta_o = 60°$, $K = 0.045$ N $-s^2/m^2$, and $g = 9.81$ m/s². Send the results to workspace.

(b) From the workspace print out a table for t, x, y, v_x, v_y every 1 second and create plots of (x and y vs. t) and (v_x and v_y vs. t).

(c) The target lies on the ground, which has a slope of 5%. The ground elevation, y_g, relative to the origin of the coordinate system of the rocket is given by

$$y_g = 15 + 0.05(x - 3000) \qquad \text{(P7.1c)}$$

In the workspace program determine where the rocket hits the ground. Assume a linear trajectory between the closest two data points. The intersection of the two straight lines gives the (x,y) position where the rocket hits the ground. Print out these (x,y) values.

Project 7.2

Repeat Project 6.4, but this time use Simulink to construct a simulation of this system. Scope output should be for T_f and T_w vs. t. Set the end time to 1 hour and print out the block diagram.

Project 7.3

Repeat Project 6.5, but this time use Simulink to construct a simulation of this system. Scope output should be for z, v, and p vs. t. Send z, v, p, and t to the workspace and create a table of 50 lines giving z, v, and p vs. t. Also create plots of z vs. t, v vs. t, and p vs. t.

Chapter 8

Curve Fitting

8.1 Curve-Fitting Objective

There are many occasions in engineering that require an experiment to determine the behavior of a particular phenomenon. The experiment may produce a set of data points that represents a relationship between the variables involved in the phenomenon. The engineer may then wish to express this relationship analytically. This analytical expression is designated as the approximating function to the data. There are two approaches.

1. The approximating function passes through all data points (see Section 8.5). If there is some scatter in the data points, the approximating function may not be satisfactory.
2. The approximating function graphs as a smooth curve. In this case the approximating curve will not, in general, pass through all the data points. A plot of the data on regular graph paper, semilog, or log-log paper may suggest an appropriate form for the approximating function.

8.2 Method of Least Squares

Best-fit straight line:

Given a set of data points: $(x_1, y_1), (x_2, y_2), \ldots, (x_n, y_n)$
We wish to represent the approximating curve, y_c, as:

$$y_c = c_1 x + c_2 \tag{8.1}$$

Let

$$D = \sum_{i=1}^{n} [y_i - y_c(x_i)]^2 = \sum_{i=1}^{n} [y_i - (c_1 x_i + c_2)]^2$$

$$= [y_1 - (c_1 x_1 + c_2)]^2 + [y_2 - (c_1 x_2 + c_2)]^2 + \cdots + [y_n - (c_1 x_n + c_2)]^2 \qquad (8.2)$$

To obtain the best-fit straight line approximating function, take $\dfrac{\partial D}{\partial c_1} = 0$ and $\dfrac{\partial D}{\partial c_2} = 0$

$$D = \sum_{i=1}^{n} [y_i - (c_1 x_i + c_2)]^2 \qquad (8.3)$$

$$\frac{\partial D}{\partial c_1} = \sum_{i=1}^{n} 2[y_i - (c_1 x_i + c_2)][-x_i] = 0$$

$$0 = \sum_{i=1}^{n} x_1 y_i - c_1 \sum_{i=1}^{n} x_i^2 - c_2 \sum_{i=1}^{n} x_i$$

or

$$\left(\sum_{i=1}^{n} x_i^2 \right) c_1 + \left(\sum_{i=1}^{n} x_i \right) c_2 = \sum_{i=1}^{n} x_i y_i \qquad (8.4)$$

$$\frac{\partial D}{\partial c_2} = 0 = \sum_{i=1}^{n} 2[y_i - (c_1 x_i + c_2)][-1]$$

$$0 = \sum_{i=1}^{n} y_i - c_1 \sum_{i=1}^{n} x_i - nc_2$$

or

$$\left(\sum_{i=1}^{n} x_i \right) c_1 + nc_2 = \sum_{i=1}^{n} y_i \qquad (8.5)$$

Solving by method of determinates:

$$c_1 = \frac{\begin{vmatrix} \Sigma x_i y_i & \Sigma x_i \\ \Sigma y_i & n \\ \Sigma x_i^2 & \Sigma x_i \\ \Sigma x_i & n \end{vmatrix}}{\begin{vmatrix} \Sigma x_i^2 & \Sigma x_i \\ \Sigma x_i & n \end{vmatrix}} = \frac{n\Sigma x_i y_i - (\Sigma x_i)(\Sigma y_i)}{n\Sigma x_i^2 - (\Sigma x_i)(\Sigma x_i)} \tag{8.6}$$

$$c_2 = \frac{\begin{vmatrix} \Sigma x_i^2 & \Sigma x_x y_i \\ \Sigma x_i & \Sigma y_i \end{vmatrix}}{n\Sigma x_i^2 - (\Sigma x_i)(\Sigma x_i)} = \frac{(\Sigma y_i)(\Sigma x_i^2) - (\Sigma x_i)(\Sigma x_i y_i)}{n\Sigma x_i^2 - (\Sigma x_i)(\Sigma x_i)} \tag{8.7}$$

For an m degree polynomial fit, take the approximating curve, y_c, to be

$$y_c = c_1 + c_2 x + c_3 x^2 + c_4 x^3 + \cdots + c_{m+1} x^m$$

where $m \le n - 1$ and n = number of data points.

Measured values are (x_i, y_i), $i = 1, 2,\ldots, n$
Let $y_{c,i} = y_c(x_i)$
Then

$$D = \sum_{i=1}^{n} [y_i - y_{c,i}]^2 = \sum_{i=1}^{n} \left[y_i - \left(c_1 + c_2 x_i + c_3 x_i^2 \ldots c_{m+1} x_i^m \right) \right]^2$$

To minimize D, take

$$\frac{\partial D}{\partial c_1} = 0, \quad \frac{\partial D}{\partial c_2} = 0, \text{ etc.}$$

$$\frac{\partial D}{\partial c_1} = 0 = \sum_{i=1}^{n} 2\left[y_i - \left(c_1 + c_2 x_i + \cdots + c_{m+1} x_i^m \right) \right][-1]$$

$$\frac{\partial D}{\partial c_2} = 0 = \sum_{i=1}^{n} 2\left[y_i - \left(c_1 + c_2 x_i + \cdots + c_{m+1} x_i^m \right) \right][-x_i]$$

$$\frac{\partial D}{\partial c_3} = 0 = \sum_{i=1}^{n} 2\left[y_i - \left(c_1 + c_2 x_i + \cdots + c_{m+1} x_i^m \right) \right]\left[-x_i^2\right]$$

$$\vdots$$

$$\frac{\partial D}{\partial c_{m+1}} = 0 = \sum_{i=1}^{n} 2\left[y_i - \left(c_1 + c_2 x_i + \cdots + c_{m+1} x_i^m \right) \right]\left[-x_i^m\right]$$

The set of equations reduces to

$$nc_1 + \left(\sum x_i\right)c_2 + \left(\sum x_i^2\right)c_3 + \cdots + \left(\sum x_i^m\right)c_{m+1} = \sum y_i$$

$$\left(\sum x_i\right)c_1 + \left(\sum x_i^2\right)c_2 + \left(\sum x_i^3\right)c_3 + \cdots + \left(\sum x_i^{m+1}\right)c_{m+1} = \sum x_i y_i$$

$$\vdots$$ (8.8)

$$\left(\sum x_i^m\right)c_1 + \left(\sum x_i^{m+1}\right)c_2 + \cdots + \left(\sum x_i^{2m}\right)c_{m+1} = \sum x_i^m y_i$$

Solve by Gauss Elimination

The standard error of the fit, s_{yx}, is defined by

$$s_{yx} = \left\{ \frac{1}{v} \sum_{i=1}^n (y_i - y_{c,i})^2 \right\}^{1/2}$$

where $v = n - (m + 1)$ = number of degrees of freedom.

Note: MATLAB® uses mean square error *(mse)*, which is defined by

$$(mse) = \frac{1}{n} \sum_{i=1}^n (y_i - y_{c,i})^2$$ (8.9)

s_{yx} is a measure of the precision with which the polynomial describes the data. Run the program for several different m values. Use the one with the lowest s_{yx} value.

8.3 Curve Fitting with the Exponential Function

A mathematical analysis of physical systems frequently leads to exponential functions. If experimental data appear to fall into this category, one should use

$$y_c = \alpha_1 e^{-\alpha_2 x}$$ (8.10)

as the approximating function, where α_1 and α_2 are real constants.

For data points (x_1, y_1), (x_2, y_2), ... (x_n, y_n), let $z_i = ln\, y_i$ and $z_c = ln\, y_c = ln\, \alpha_1 - \alpha_2 x$. Letting $c_1 = -\alpha_2$ and $c_2 = ln\, \alpha_1$, the above equation becomes linear in z_c; that is,

$$z_c = c_1 x + c_2$$ (8.11)

For data points (x_1, y_1), (x_2, y_2), ... (x_n, y_n), the new set of data points becomes

$$(x_1, z_1), (x_2 z_2), ... (x_n, z_n)$$

From our previous work, the best-fit approximating straight-line curve by the method of least squares gives

$$c_1 = \frac{n \sum x_i z_i - \left(\sum x_i\right)\left(\sum z_i\right)}{n \sum x_i^2 - \left(\sum x_i\right)^2} \tag{8.12}$$

and

$$c_2 = \frac{\left(\sum z_i\right)\left(\sum x_i^2\right) - \left(\sum x_i\right)\left(\sum x_i z_i\right)}{n \sum x_i^2 - \left(\sum x_i\right)^2} \tag{8.13}$$

Then

$$\alpha_1 = e^{c_2}, \quad \alpha_2 = -c_1$$

and

$$mse = \frac{1}{n} \sum (y_i - y_{c,i})^2 \tag{8.14}$$

where $y_{c,i} = y_c(x_i)$.

The above analysis can be used to determine the damping constant in a mass-spring-dashpot system. This is accomplished by examining the Oscilloscope graph of free damped vibration (see Figure 8.1). The governing equation of the envelope is

$$y = y_0 \exp\left(-\frac{c}{2m}t\right) \tag{8.15}$$

where

c = damping constant.

m = the mass.

y = the mass displacement from the equilibrium position.

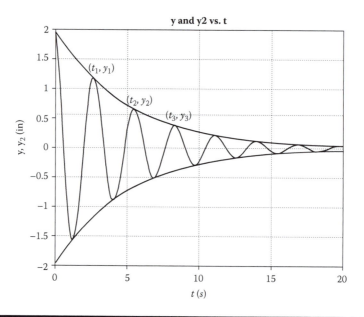

Figure 8.1 Oscilloscope graph of free damped oscillations.

Comparing Equation (8.15) with Equation (8.10) we see that

$$\alpha_1 = y_0 \text{ and } \alpha_2 = \frac{c}{2m} \text{ with } t \text{ replacing } x$$

Therefore, $c = 2m\alpha_2$ (see Project 2.10).

8.4 MATLAB's Curve-Fitting Function

MATLAB calls curve fitting with a polynomial by the name "Polynomial Regression." The function polyfit (x, y, m) returns a vector of $(m + 1)$ coefficients that represent the best-fit polynomial of degree m for the (x_i, y_i) set of data points. The coefficient order corresponds to decreasing powers of x; that is,

$$y_c = a_1 x^m + a_2 x^{m-1} + a_3 x^{m-2} + \ldots a_m x + a_{m+1} \tag{8.16}$$

To obtain y_c at $(x_1, x_2, \ldots, x_n)$ use the MATLAB function polyval (a, x). The function polyval (a,x) returns a vector of length n giving $y_{c,i}$, where

$$y_{c,i} = a_1 x_i^m + a_2 x_i^{m-1} + a_3 x_i^{m-2} + \ldots a_m x_i + a_{m+1} \tag{8.17}$$

The mean square error (*mse*) is a measure of the precision of the fit. The *mse* is defined as follows:

$$mse = \frac{1}{n}\sum_{i=1}^{m}(y_i - y_{c,i})^2 \tag{8.18}$$

where n is the number of data points.

Example 8.1

```
% polyfit_text.m
% This program determines the best-fit polynomial curve passing
% through a given set of data points
clear; clc;
x=-10:2:10;
y=[-980 -620 -70 80 100 90 0 -80 -90 10 220];
x2=-10:0.5:10;
mse=zeros(4);
for n=2:5
    fprintf('n= %i \n',n);
    coef=zeros(n+1);
    coef=polyfit(x,y,n);
    yc2=polyval(coef,x2);
    yc=polyval(coef,x);
    mse(n)=sum((y-yc).^2)/11;
    if n==2
    fprintf('    x        y         yc \n');
    fprintf('-----------------------------------\n');
    for i=1:11
        fprintf('%5.0f %5.0f %8.2f \n',x(i),y(i),yc(i));
    end
    fprintf('\n\n');
end
if n==3
    fprintf('    x        y         yc \n');
    fprintf('-----------------------------------\n');
    for i=1:11
        fprintf('%5.0f %5.0f %8.2f \n',x(i),y(i),yc(i));
    end
    fprintf('\n\n');
end
if n==4
    fprintf('    x        y         yc \n');
    fprintf('-----------------------------------\n');
    for i=1:11
        fprintf('%5.0f %5.0f %8.2f \n',x(i),y(i),yc(i));
    end
    fprintf('\n\n');
end
if n==5
    fprintf('    x        y         yc \n');
    fprintf('-----------------------------------\n');
    for i=1:11
```

```
        fprintf('%5.0f    %5.0f    %8.2f \n',x(i),y(i),yc(i));
    end
    fprintf('\n\n');
end
subplot(2,2,n-1),plot(x2,yc2,x,y,'o'),
xlabel('x'), ylabel('y'), grid,
if n==2
    title('Second Degree Polynomial Fit')
end
if n==3
    title('Third Degree Polynomial Fit')
end
if n==4
    title('Fourth Degree Polynomial Fit')
end
if n==5
    title('Fifth Degree Polynomial Fit')
end
end
fprintf(' n mse            \n')
fprintf('------------------------\n');
for n=2:5
    fprintf(' %g            %6.2f \n',n,mse(n))
end
```

8.5 Cubic Splines

Suppose for a given set of data points, all attempted degree polynomial approximating curves produced points that were not allowed. For example, suppose it is known that a particular property represented by the data (such as absolute pressure or absolute temperature) must be positive and all the attempted polynomial approximating curves produced some negative values. For this case, the polynomial approximating function would not be satisfactory. The method of cubic splines eliminates this problem.

Given a set of $(n + 1)$ data points (x_i, y_i), $i = 1, 2,\ldots, (n + 1)$, the method of cubic splines develops a set of n cubic functions, such that $y(x)$ is represented by a different cubic in each of the n intervals and the set of cubics passes through the $(n + 1)$ data points.

This is accomplished by forcing the slopes and curvatures to be the same for each pair of cubics that join at a data point.

$$\text{Note:} \quad \text{Curvature,} \quad K = \frac{\pm \dfrac{d^2 y}{dx^2}}{\left[1+\left(\dfrac{dy}{dx}\right)^2\right]^{3/2}} \tag{8.19}$$

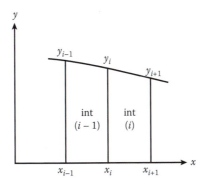

Figure 8.2 Two adjacent arbitrary intervals.

This is accomplished by the following equations:

$$[y(x_i)]_{\text{int } i-1} = [y(x_i)]_{\text{int } i}$$

$$[y'(x_i)]_{\text{int } i-1} = [y'(x_i)]_{\text{int } i} \qquad (8.20)$$

$$[y''(x_i)]_{\text{int } i-1} = [y''(x_i)]_{\text{int } i}$$

In interval $(i-1), (x_{i-1} \leq x \leq x_i)$ (see Figure 8.2).

$$y(x) = A_{i-1} + B_{i-1}(x - x_{i-1}) + C_{i-1}(x - x_{i-1})^2 + D_{i-1}(x - x_{i-1})^3 \qquad (8.21)$$

In interval $i, (x_i \leq x \leq x_{i+1})$

$$y(x) = A_i + B_i(x - x_i) + C_i(x - x_i)^2 + D_i(x - x_i)^3 \qquad (8.22)$$

This gives $(n-1)$ equations in $(n+1)$ unknowns.

Thus, values for $\dfrac{d^2 y}{dx^2}$ at x_1 and x_{n+1} must be assumed.

Several alternatives exist:

1. Assume $y''(x_1) = y''(x_{n+1}) = 0$
   Widely used—forces splines to approach straight lines at end points.
2. Assume $y''(x_{n+1}) = y''(x_n)$ and $y''(x_1) = y''(x_2)$. This forces the splines to approach parabolas at the end points.

MATLAB's built-in function *interp1* interpolates between data points by the cubic spline method.

8.6 The Function Interp1 for Cubic Spline Curve Fitting

$$y_i = \text{interp1}\,(x, y, x_i, \text{'}spline\text{'})$$

where x,y are the set of data points and x_i is the set of x values at which the set of values, y_i, is to be returned.

Example 8.2

```
% cubic_spline
% This program uses interpolation by cubic splines to determine
% overpressure resulting from a blast.
clear;
clc;
dist=0.2:0.2:2.6;
press=[24.0 14.0 10.0 7.6 5.4 4.0 3.1 2.5 2.0 1.7 1.5 1.3 1.1];
d=0.2:0.1:2.6;
p=interp1(dist,press,d,'spline');
fid=fopen('output1.dat','w');
fprintf(fid,'PEAK OVERPRESSURE VS. DISTANCE FROM BLAST \n');
fprintf(fid,'CUBIC SPLINE FIT \n');
fprintf(fid,' dist     over-press \n');
fprintf(fid,' (miles)   (psi) \n');
for n=1:25
    fprintf(fid,' %5.1f %10.3f \n',d(n),p(n));
end
plot(d,p,dist,press,'o'),xlabel('miles from ground zero'),
    ylabel('overpressure(psi)'),axis([0.0,3.0,0.0,25.0]),
    grid,title('peak overpressure vs. distance from blast')
fclose(fid);
```

8.7 Curve Fitting with Fourier Series

Suppose an experimental data set produced a plot as shown in Figure 8.3 and it was desired to obtain an analytical expression that comes close to fitting the data. Let us assume that the $-L \le x \le L$. If not, make it so by shifting the origin. Actually the original abscissa data were for $0 \le t \le 10.5$ seconds. The data were shifted by letting $x = t - 5.25$.

The t domain was subdivided into 70 equal spaces, with $\Delta t = 10.5/70 = 0.15$ second. Thus, $x_{i+1} - x_i$ is uniform over the entire domain. An attempt to fit a polynomial approximating curve to these data would not be successful. However, the use of a Fourier series could give a reasonable analytical expression approximating the data. If uc is the approximating curve, then by a Fourier series,

$$uc(x) = a_0 + \sum_{m=1}^{\infty} \left(a_m \cos\left(\frac{m\pi x}{L}\right) + b_m \sin\left(\frac{m\pi x}{L}\right) \right) \tag{8.23}$$

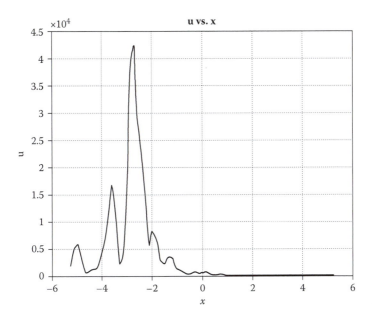

Figure 8.3   Experimental data of *u* vs. x.

where

$$a_0 = \frac{1}{2L} \int_{-L}^{L} u(x)\, dx$$

$$a_m = \frac{1}{L} \int_{-L}^{L} u(x) \cos\left(\frac{m\pi x}{L}\right) dx$$

$$b_m = \frac{1}{L} \int_{-L}^{L} u(x) \sin\left(\frac{m\pi x}{L}\right) dx$$

Using 30 terms in the series and Simpson's rule on integration an approximating curve as shown in Figure 8.4 was obtained.

The $a_m \cos\left(\frac{m\pi x}{L}\right) + b_m \sin\left(\frac{m\pi x}{L}\right)$ terms can be put into the following form by the trigonomic identity $a\cos\beta + b\sin\beta = c\sin(\beta - \phi)$, where c represents the amplitude. The amplitude, c, is given by

$$c = \sqrt{(a^2 + b^2)}$$

A plot of amplitude vs. $\frac{m\pi}{L}$ is shown in Figure 8.5.

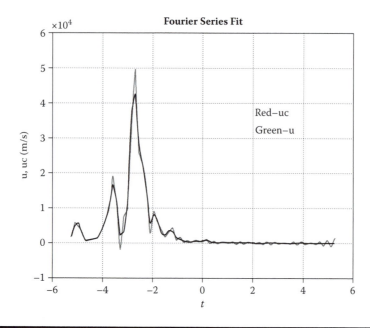

Figure 8.4 Fourier series fit of the data. (See color insert following page 334.)

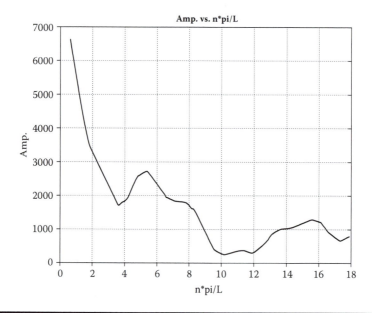

Figure 8.5 Fourier series coefficient amplitudes vs. $\dfrac{n\pi}{L}$.

Projects

Project 8.1

A formula describing the fluid level, h_{eq}, in a tank, as a function of time, as the fluid discharges through a small orifice, is

$$\sqrt{h_{eq}} = \sqrt{h_{eq,o}} - \frac{C_d A_0}{2 A_T} \sqrt{2g}\, t \qquad \text{(P8.1a)}$$

where

C_d = the discharge coefficient.
$h_{eq,o}$ = the fluid level in the tank at time, $t = 0$.
A_0 = the area of the orifice.
A_T = the cross-sectional area of the tank.

An experiment consisting of a cylindrical tank with a small orifice was used to determine C_d for that particular orifice and cylinder. The tank walls were transparent and a ruler was pasted to the wall allowing for the determination of the fluid level in the tank. The procedure was to fill the tank with water while the orifice was plugged. The plug was then removed and the water was allowed to flow through the orifice. The water level in the tank, h_{exp} in meters, was recorded as a function of time, t. The experimental data are shown in Table P8.1.

Table P8.1 h_{exp} **vs. time**

h_{exp} (m)	t(s)		h_{exp} (m)	t(s)
0.288	0		0.080	110
0.258	10		0.065	120
0.234	20		0.053	130
0.215	30		0.041	140
0.196	40		0.031	150
0.178	50		0.022	160
0.160	60		0.013	170
0.142	70		0.006	180
0.125	80		0.002	190
0.110	90		0.000	200
0.095	100			

The diameters of the orifice and the tank are $D_o = 0.0055$ *m* and $D_t = 0.146$ *m*, respectively. The free surface elevation, $h_{eq,o}$, at $t = 0$ is 0.288 *m*. The gravitational constant, $g = 9.81$ *m/s²*.

Use the *mse* as defined by Equation (8.12) to determine the value for C_d that best fits the data. Vary C_d from 0.3 to 0.9 in steps of 0.01 and evaluate the *mse* for each C_d selected, where

$$mse = \frac{1}{N} \sum_{i=1}^{N} [h_{eq}(t_i) - h_{exp}(t_i)]^2 \tag{P8.1b}$$

where

N = the number of data points.

$h_{eq}(t_i)$ = the water level in the tank at t_i as determined by Equation (P8.1a).

$h_{exp}(t_i)$ = the water level in the tank at t_i as determined by experiment.

For the C_d with the lowest *mse*, create a plot of h_{eq} vs. t (solid line) and superimpose h_{exp} vs. t as little x's onto the plot of h_{eq} vs. t. Also print out the value of C_d that gives the lowest *mse*.

Also create a 30-line table of *mse* vs. C_d.

Project 8.2

This project involves determining the best fit polynomial approximating curve to the (H vs. Q) data obtained from a pump manufacturer's catalog (units changed to SI units). The data points of the (H vs. Q) curve are shown in Table P8.2.

Table P8.2 H vs. Q Data from the Pump Manufacturer

Q	H		Q	H
(m³/h)	(m)		(m³/h)	(m)
3.3	43.3		61.6	40.8
6.9	43.4		68.5	39.6
13.7	43.6		75.3	38.7
20.5	43.6		82.2	37.2
27.4	43.3		89.0	36.3
34.2	43.0		95.8	34.4
41.1	42.7		102.7	32.6

Try degree polynomials of 2 through 4 to determine which degree polynomial will give the smallest *mse*. Use MATLAB's function *polyfit*, which returns the coefficients for each of the three polynomials. Then use MATLAB's function *polyval* to create for each polynomial:

(a) A table containing Q, H_c, and H, where H_c is the approximating curve for H vs. Q.

(b) A plot of H_c vs. Q (solid line) and H vs. Q (small circles), all plots on the same page.

Chapter 9

Optimization

9.1 Introduction

The objective of optimization is to maximize or minimize some function f. The function f is called the *object function*. For example, suppose there is an electronics company that manufactures several different types of circuit boards. Each circuit board must pass through several different departments (such as drilling, component assembly, testing, etc.) before shipping. The time required for each circuit board to pass through the various departments is also known. There is a minimum production quantity per month that the company must produce. However, the company is capable of producing more than the minimum production requirement for each type of circuit board each month. The profit the company will make on each circuit board it produces is known. The problem is to determine the production amount of each type of circuit board per month that will result in the maximum profit. A similar type of problem may be one in which the object is to minimize the cost of producing a particular product. These types of optimization problems are discussed in greater detail later in this chapter. In most optimization problems, the object function, f, will depend on several variables—x_1, x_2, x_3, ..., x_n. These are called the *control variables* because their values can be selected. Optimization theory develops methods for selecting optimal values for the control variables, x_1, x_2, x_3, ..., x_n, that either maximizes (or minimizes) the objective function f. In many cases, the choice of values for x_1, x_2, x_3, ..., x_n is not entirely free, but is subject to some *constraints*.

9.2 Unconstrained Optimization Problems

In calculus it is shown that a necessary (but not sufficient) condition for f to have a maximum or minimum at point P, is that at point P, each of the first partial derivatives of f, be zero; that is,

$$\frac{\partial f}{\partial x_1}(P) = \frac{\partial f}{\partial x_2}(P) = \cdots = \frac{\partial f}{\partial x_n}(P) = 0$$

If $n = 1$, say, $y = f(x)$, then a necessary condition for an extremum (maximum or minimum) at x_0 is for $y'(x_0) = 0$.

For y to have a local minimum at x_0, $y'(x_0) = 0$ and $y''(x_0) > 0$.
For y to have a local maximum at x_0, $y'(x_0) = 0$ and $y''(x_0) < 0$.

For f involving several variables, the condition for f to have a relative minimum is more complicated. First, Equation (9.1)

$$\frac{\partial f}{\partial x_1}(P) = \frac{\partial f}{\partial x_2}(P) = \cdots = \frac{\partial f}{\partial x_n}(P) = 0 \tag{9.1}$$

must be satisfied. Second, the quadratic form (Equation 9.2)

$$Q = \sum_{i=1}^{n}\sum_{j=1}^{n} \frac{\partial^2 f}{\partial x_i \partial x_j}(P)(x_i - x_i(P))(x_j - x_j(P)) \tag{9.2}$$

must be positive for all choices of x_i and x_j in the vicinity of point P, and $Q = 0$ only when $x_i = x_i(P)$ for $i = 1, 2, \ldots, n$. This condition comes from a Taylor Series expansion of $f(x_1, x_2, \ldots, x_n)$ about point P using only terms up to $\frac{\partial^2 f}{\partial x_i \partial x_j}(P)$. This gives

$$f(x_1, x_2, \ldots, x_n) = f(P) + \sum_{i}^{n} \frac{\partial f}{\partial x_i}(P)(x_i - x_i(P))$$

$$+ \sum_{i=1}^{n}\sum_{j=1}^{n} \frac{\partial^2 f}{\partial x_i \partial x_j}(P)(x_i - x_i(P))(x_j - x_j(P))$$

If $f(x_1, x_2, \ldots, n)$ has a relative minimum at point P, then $\frac{\partial f}{\partial x_i}(P) = 0$ for $i = 1$, $2, \ldots, n$ and $f(x_1, x_2, \ldots, x_n) - f(P) > 0$ for all $(x_1, x_2, \ldots, x_n)$ in the vicinity of point P. But $f(x_1, x_2, \ldots, x_n) - f(P) = Q$. Thus, for $f(x_1, x_2, \ldots, x_n)$ to have a relative minimum at point P, Q must be positive for all choices of x_i and x_j in the vicinity of point P.

Since the above analysis is quite complicated when f is a function of several variables, an iterative scheme is frequently used as a method of solution. One such method is the method of steepest descent. In this method one needs to guess for a point where an extremum exists. Using a grid to evaluate the function at different values of the control variables can be helpful in establishing a good starting point for the iteration process.

9.3 Method of Steepest Descent

Consider a function, f, of three variables (x, y, z). From calculus, we know that the gradient of f, written as ∇f, is given by

$$\nabla f = \frac{\partial f}{\partial x}\hat{i} + \frac{\partial f}{\partial y}\hat{j} + \frac{\partial f}{\partial z}\hat{k}$$

where $\hat{i}$, $\hat{j}$, and $\hat{k}$ are unit vectors in the x, y, and z directions, respectively.

At (x_0, y_0, z_0), we also know that $\nabla f(x_0, y_0, z_0)$ points in the direction of the maximum rate of change of f with respect to distance.

A unit vector, $\hat{e}_g$, which points in this direction, is

$$\hat{e}_g = \frac{\nabla f}{|\nabla f|}$$

where

$$|\nabla f| = \sqrt{\left(\frac{\partial f}{\partial x}\right)^2 + \left(\frac{\partial f}{\partial y}\right)^2 + \left(\frac{\partial f}{\partial z}\right)^2}$$

To find a relative minimum, the method of steepest descent is frequently used. This method starts at some initial point and moves in small steps in the direction of steepest descent, which is $(-\hat{e}_g)$. Let $(x_{n+1}, y_{n+1}, z_{n+1})$ be the new position on the nth iteration and (x_n, y_n, z_n) the old position; then

$$x_{n+1} = x_n - \frac{\frac{\partial f}{\partial x}(x_n, y_n, z_n)}{|\nabla f(x_n, y_n, z_n)|}\Delta s$$

$$y_{n+1} = y_n - \frac{\frac{\partial f}{\partial y}(x_n, y_n, z_n)}{|\nabla f(x_n, y_n, z_n)|}\Delta s$$

$$z_{n+1} = z_n - \frac{\frac{\partial f}{\partial z}(x_n, y_n, z_n)}{|\nabla f(x_n, y_n, z_n)|}\Delta s$$

where Δs is some small length.

Example 9.1

Given: $f = 4 + 4.5x_1 - 4x_2 + x_1^2 + 2x_2^2 - 2x_1x_2 + x_1^4 - 2x_1^2x_2$.

Determine: The minimum of f by the method of steepest descent starting at point $(x_1, x_2) = (6, 10)$. Use a $\Delta s = 0.1$ and 100 iterations.

Instead of starting at some arbitrary point as specified above, one might wish to first use a grid program to establish a good starting point. A sample grid program follows:

```
% grid2.m
% This program determines the functional values of a specified
% function of 2 variables
% for determining a good starting point for the method of steepest
% decent
clear; clc;
x1min=-10.0; x1max=10.0;
x2min=-10.0; x2max=10.0;
dx1=2.0; dx2=2.0;
for i=1:11
    x1(i)=x1min+(i-1)*dx1;
    for j=1:11
        x2(j)=x2min+(j-1)*dx2;
        f(i,j)=fxf(x1(i),x2(j));
    end
end
fprintf('==========================================================\n');
fprintf('                functional values of f(x1,x2) \n');
fprintf('==========================================================\n');
fprintf(' x2 | x1 %6.1f %6.1f %6.1f %6.1f %6.1f %6.1f \n',...
        x1(1),x1(2),x1(3),x1(4),x1(5),x1(6));
fprintf('==========================================================\n');
for j=1:11
    fprintf('%6.1f ',x2(j));
    for i=1:6
        fprintf('%10.1f',f(i,j));
    end
    fprintf('\n');
end
fprintf('\n\n\n');
fprintf('==========================================================\n');
fprintf('                functional values of f(x1,x2) \n');
fprintf('==========================================================\n');
fprintf(' x2 | x1 %6.1f  %6.1f  %6.1f  %6.1f  %6.1f  \n',...
        x1(7),x1(8),x1(9),x1(10),x1(11));
fprintf('==========================================================\n');
for j=1:11
    fprintf('%6.1f ',x2(j));
    for i=7:11
        fprintf('%10.1f',f(i,j));
    end
    fprintf('\n');
end
end
```

Next use a steepest descent program. A sample program follows:

```
% steep_descent.m
% This program determines a relative minimum by the
% method of steepest decent.
% The function is:
% f(x1,x2)=4+4.5*x1-4*x2+x1^2+2*x2^2-2*x1*x2+x1^4-2*x1^2*x2
% Relative minimum points are known. They occur at
% (x1,x2)=(1.941,3.854) and (-1.053,1.028). The minimum functional
% values are 0.9855 & -0.5134 respectively
clear; clc;
% First guess
x1=6;
    x2=10;
    ds=0.1;
    fx=fxf(x1,x2);
    fprintf('    x1        x2        fx \n');
        for n=1:100
        dfx1=dfx1f(x1,x2);
        dfx2=dfx2f(x1,x2);
        gradf_mag=sqrt(dfx1^2+dfx2^2);
        x1n=x1-dfx1/gradf_mag*ds;
        x2n=x2-dfx2/gradf_mag*ds;
        fxn=fxf(x1n,x2n);
        fprintf(' %7.4f %7.4f %10.4f \n',x1n,x2n,fxn);
        if(fxn > fx)
            fprintf(' a minimum has been reached \n\n');
            break
        else
            x1=x1n;
            x2=x2n;
            fx=fxn
        end
    end
end
    fmin=fxf(x1,x2);
    fprintf(' The relative minimum occurs at x1=%7.4f  x2=%7.4f\n',x1,x2);
    fprintf(' The minimum value for f=%10.4f \n',fmin);
```

```
% fxf.m
% This function is used in the steep_decent.m program
    function fx=fxf(x1,x2)
    fx=4+4.5*x1-4*x2+x1^2+2*x2^2-2*x1*x2+x1^4-2*x1^2*x2;
```

```
% dfx1f.m
% This function is used in the steep_descent.m program
    function dfx1=dfx1f(x1,x2)
    dfx1=4.5+2*x1-2*x2+4*x1^3-4*x1*x2;
```

```
% dfx2f.m
% This function is used in the steep_descent.m program
    function dfx2=dfx2f(x1,x2)
    dfx2=-4+4*x2-2*x1-2*x1^2;
```

9.4 Optimization with Constraints

In many optimization problems the variables in the function to be maximized or minimized are not all independent, but are related by one or more conditions or constraints.

Suppose we are given the object function $f(x_1, x_2, x_3, ..., x_n)$, in which the variables $x_1, x_2, ..., x_n$ are subject to N constraints, say,

$$\Phi_1(x_1, x_2, x_3, ..., x_n) = 0$$
$$\Phi_2(x_1, x_2, x_3, ..., x_n) = 0$$

.

.

.

$$\Phi_N(x_1, x_2, x_3, ..., x_n) = 0$$

Theoretically, N x's can be solved in terms of the remaining x's. Then these N variables can be eliminated from the objective function f by substitution and the extreme problem can be solved as if there were no constraints. This method is referred to as the implicit method.

■ Lagrange's Multipliers

Suppose $f(x_1, x_2, x_3,..., x_n)$ is to be maximized subject to constraints

$$\Phi_1(x_1, x_2, x_3, ..., x_n) = 0$$
$$\Phi_2(x_1, x_2, x_3, ..., x_n) = 0$$

.

.

.

$$\Phi_N(x_1, x_2, x_3, ..., x_n) = 0$$

Define the Lagrange function F as

$$F(x_1, x_2, x_3, ..., x_n) = f(x_1, x_2, x_3, ..., x_n) + \lambda_1 \Phi_1(x_1, x_2, x_3, ..., x_n)$$
$$+ \lambda_2 \Phi_2(x_1, x_2, x_3, ..., x_n) + \lambda_N \Phi_N(x_1, x_2, x_3, ..., x_n)$$

where λ_j are unknown multipliers to be determined. Set

$$\frac{\partial F}{\partial x_1} = 0, \quad \frac{\partial F}{\partial x_2} = 0, ..., \frac{\partial F}{\partial x_n} = 0$$

$$\Phi_1 = 0, \quad \Phi_2 = 0, ..., \Phi_N = 0,$$

$$(Note: \frac{\partial F}{\partial \lambda_j} = 0 \quad gives \ \Phi_j = 0)$$

This set of $(n + N)$ equations gives all possible extrema of f. Proof is beyond the scope of this textbook [1].

Example 9.2

A silo is to consist of a right circular cylinder with a hemispherical roof (see Figure 9.1). If the silo is to have a specified volume V, find the dimensions that make its surface area a minimum. Assume that the silo has a floor of the same material.

$$\text{Note: } V_{sphere} = \frac{4}{3}\pi R^3, S_{sphere} = 4\pi R^2$$

Take $V = 8400 \text{ m}^3$

Solution:

$$V = \frac{2}{3}\pi R^3 + \pi R^2 L,$$

$$S = 2\pi RL + \pi R^2 + 2\pi R^2 = 2\pi RL + 3\pi R^2$$

$$F = 2\pi RL + 3\pi R^2 + \lambda\left(\frac{2}{3}\pi R^3 + \pi R^2 L - V\right)$$

Variables are R, L, and λ.

$$\frac{\partial F}{\partial R} = 2\pi L + 6\pi R + \lambda(2\pi R^2 + 2\pi RL) = 0$$

$$\frac{\partial F}{\partial L} = 2\pi R + \lambda\pi R^2 = 0 \quad \Rightarrow \quad \lambda R = -2 \quad \text{or} \quad \lambda = -\frac{2}{R}$$

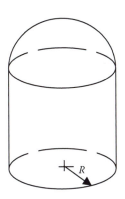

Figure 9.1 Sketch of a silo consisting of a right circular cylinder topped by a hemisphere.

Substituting the value of λ in the above equation gives

$$2\pi L + 6\pi R - \frac{2}{R}(2\pi R^2 + 2\pi RL) = 0$$

The above equation reduces to $R - L = 0$ or $R = L$. Substituting this result into the V equation gives

$$V = \pi R^3 + \frac{2}{3}\pi R^3 = \frac{5}{3}\pi R^3$$

For $V = 8400$ m³,

$$R = \left(\frac{8400 \times 3}{5\pi}\right)^{1/3} = 11.7065 \text{ m}$$

Substituting the values for R and L into the equation for S gives $S = 2152.6$ m².

9.5 MATLAB's Optimization Function

MATLAB's optimization function is fmincon. A description of the function can be obtained by typing *help fmincon* in the command window. A description of the function follows:

```
X=fmincon(FUN,X0,A,B) starts at X0 and finds a minimum X to the function
    FUN, subject to the linear inequalities A*X <= B. FUN accepts input X
    and returns a scalar function value F evaluated at X. X0 may be a
    scalar, vector, or matrix.
X=fmincon(FUN,X0,A,B,Aeq,Beq) minimizes FUN subject to the linear
    equalities Aeq*X = Beq as well as A*X <= B. (Set A=[] and B=[] if no
    inequalities exist.)
X=fmincon(FUN,X0,A,B,Aeq,Beq,LB,UB) defines a set of lower and upper
    bounds on the design variables, X, so that the solution is in the range
    LB <= X <= UB. Use empty matrices for LB and UB if no bounds exist. Set
    LB(i)=-Inf if X(i) is unbounded below; set UB(i)=Inf if X(i) is
    unbounded above.
X=fmincon(FUN,X0,A,B,Aeq,Beq,LB,UB,NONLCON) subjects the minimization to
    the constraints defined in NONLCON. The function NONLCON accepts X and
    returns the vectors C and Ceq, representing the nonlinear inequalities
    and equalities, respectively. Fmincon minimizes FUN such that C(X)< = 0
    and Ceq(X) = 0. (Set LB=[ ] and/or UB=[] if no bounds exist.)
```

X=fmincon(FUN,X0,A,B,Aeq,Beq,LB,UB,NONLCON,OPTIONS) minimizes with the
    default optimization parameters replaced by values in the structure
    OPTIONS, an argument created with the OPTIMSET function. See OPTIMSET
    for details.
X=fmincon(FUN,X0,A,B,Aeq,Beq,LB,UB,NONLCON,OPTIONS,P1,P2,...) passes the
    problem-dependent parameters P1, P2, ... directly to the functions FUN
    and NONLCON.
[X,FVAL]=fmincon(FUN,X0,...) returns the value of the objective function FUN
    at the solution X.

Example 9.3

```
% optimsilo.m
% This program minimizes the material surface area of a silo
% The silo consists of a right cylinder topped by a hemisphere.
% The volume is set at 7000 and 8400 m^3
% Variables are length, L = x(2) and radius, R = x(1)
clear; clc;
% Take a guess at the solution
LB = [0,0];
UB = [];
xo = [10.0 20.0];
% Set optimization options:
% Turn off the large-scale algorithms (the default)
options = optimset('LargeScale','off');
% We have no inequality constraints, so pass [] for those arguments
VT=[7000 8400];
fo=fopen('output.dat','w');
fprintf(fo,' Optimization Problem \n\n');
fprintf(fo,'This program minimizes the material surface area of ...
        a silo \n');
fprintf(fo,' The silo consists of a right cylinder topped by ...
        a hemisphere. \n');
fprintf(fo,' The volume is set at 7000 and 8400 ft^3 \n\n');
for i=1:2
    V=VT(i);
[x,fval]=fmincon(@objfunsilo,xo,[],[],[],[],LB,UB,@confunsilo, ...
        options,V);
fprintf(fo,'V=%6.1f R=%7.3f L=%7.3f minimum surf area=%9.3f \n', ...
        V,x(1),x(2),fval);
end
fclose(fo);
```

```
% objfunsilo.m
function s=objfunsilo(x,V)
s=(2.0*pi*x(1)*x(2)+3.0*pi*x(1)^2);
```

```
% confunsilo.m
function [c, ceq] = confunsilo(x,V)
% Nonlinear equality constraints:
ceq = pi*x(1)^2*x(2)+2.0/3.0*pi*x(1)^3-V;
% No nonlinear inequality constraints:
c = [];
```

Example 9.4

```
% optim_shafts.m
% Two machine shops, Machine Shop A and Machine Shop B, are to
% manufacture two types of shafts, shaft S1 and shaft S2. Each machine
% shop has two Turning Machines, Turning Machine T1 and Turning Machine
% T2. The following table lists the production time for each shaft type
% on each machine and at each location.
%-------------------------------------------------|
%                      TIME IN MINUTES            |
%----------|-----------------||-----------------|
%          | MACHINE SHOP A  || MACHINE SHOP B |
% ---------|-----------------||-----------------|
% TURNING  |     SHAFTS      ||     SHAFTS      |
% MACHINE  |                 ||                 |
%----------|--------|--------||--------|-------|
%          |   S1   |   S2   ||   S1   |   S2  |
% ---------|--------|--------||--------|-------|
%    T1    |   4    |   9    ||   5    |   8   |
% ---------|--------|--------||--------|-------|
%    T2    |   2    |   6    ||   3    |   5   |
% ---------|--------|--------||--------|-------|
%
% Shaft S1 sells for $35 and shaft S2 sells for $85.
% Determine the number of S1 and S2 shafts that should be produced at
% each machine shop and on each machine that will maximize the
% revenue per hour.
% Let:
% x(1)=the number of S1 shafts produced/hr by machine T1 by shop A
% x(2)=the number of S2 shafts produced/hr by machine T1 by shop A
% x(3)=the number of S1 shafts produced/hr by machine T2 by shop A
% x(4)=the number of S2 shafts produced/hr by machine T2 by shop A
% x(5)=the number of S1 shafts produced/hr by machine T1 by shop B
% x(6)=the number of S2 shafts produced/hr by machine T1 by shop B
% x(7)=the number of S1 shafts produced/hr by machine T2 by shop B
% x(8)=the number of S2 shafts produced/hr by machine T2 by shop B
% Z=the total revenue/hr for producing the shafts.
% Z=35*(x(1)+x(3)+x(5)+x(7))+85*(x(2)+x(4)+x(6)+x(8)).
% The problem is to maximize Z subject to the following constraints
% 4*x(1)+9*x(2) <= 60
% 2*x(3)+6*x(4) <= 60
% 5*x(5)+8*x(6) <= 60
% 3*x(7)+5*x(8) <= 60
clear; clc;
fo=fopen('shaftoutput.dat','w');
fprintf(fo,'optim_shafts.m \n');
fprintf(fo,'Shaft Production Problem \n');
fprintf(fo,'This program maximizes the revenue/hr for the ...
        production \n');
fprintf(fo,'of two types of shafts, type S1 and type S2. There ...
        are two \n');
fprintf(fo,'machine shops producing these shafts, ...
        shop A & shop B. \n');
fprintf(fo,'Each shop has two types of turning machines, ...
        T1 and T2, \n');
```

```
fprintf(fo,'capable of producing these shafts. \n');
fprintf(fo,'Shop A: \n');
fprintf(fo,' Machine T1 takes 4 minutes to produce type ...
     S1 shafts \n');
fprintf(fo,' and 9 minutes to produce type S2 shafts. \n')
fprintf(fo,' Machine T2 takes 2 minutes to produce type ...
     S1 shafts \n');
fprintf(fo,' and 6 minutes to produce type S2 shafts. \n');
fprintf(fo,'\n');
fprintf(fo,'Shop B: \n');
fprintf(fo,' Machine T1 takes 5 minutes to produce type ...
     S1 shafts \n');
fprintf(fo,' and 8 minutes to produce type S2 shafts. \n')
fprintf(fo,' Machine T2 takes 3 minutes to produce type S1 shaft \n');
fprintf(fo,' and 5 minutes to produce type S2 shafts. \n');
fprintf(fo,'\n');
fprintf(fo,'Shaft S1 sells for $35 & shaft S2 sells for $85. \n');
fprintf(fo,'\n');
fprintf(fo,'We wish to determine the number of S1 & S2 tanks that \n');
fprintf(fo,'should be produced at each shop and by each machine \n');
fprintf(fo,'that will maximize the revenue/hr for producing ...
     the shafts \n');
fprintf(fo,' Let: \n');
fprintf(fo,'x(1)=the number of S1 shafts produced/hr by ...
     machine T1 at shop A \n');
fprintf(fo,'x(2)=the number of S2 shafts produced/hr by ...
     machine T1 at shop A \n');
fprintf(fo,'x(3)=the number of S1 shafts produced/hr by ...
     machine T2 at shop A \n');
fprintf(fo,'x(4)=the number of S2 shafts produced/hr by ...
     machine T2 at shop A \n');
fprintf(fo,'x(5)=the number of S1 shafts produced/hr by ...
     machine T1 at shop B \n');
fprintf(fo,'x(6)=the number of S2 shafts produced/hr by ...
     machine T1 at shop B \n');
fprintf(fo,'x(7)=the number of S1 shafts produced/hr by ...
     machine T2 at shop B \n');
fprintf(fo,'x(8)=the number of S2 shafts produced/hr by ...
     machine T2 at shop B \n');
fprintf(fo,'\n');
fprintf(fo,'Let Z=the total revenue/hr for producing these tanks \n');
fprintf(fo,' Z=35*(x(1)+x(3)+x(5)+x(7))+75*(x(2)+x(4)+x(6)+x(8)) \n');
% Take a guess at the solution
xo = [0 0 0 0 0 0 0 0];
LB = [0 0 0 0 0 0 0 0];
UB = [];
% We have linear inequality constraints
A=[4 9 0 0 0 0 0 0;
 0 0 2 6 0 0 0 0;
 0 0 0 0 5 8 0 0;
 0 0 0 0 0 0 3 5];
B=[60 60 60 60]';
% We have no linear equality constraints, so pass [] for those
% arguments. We have no equality or inequality nonlinear constraints.
[x, fval] = fmincon(@objshafts,xo,A,B,[],[],LB,UB);
```

```
fprintf(fo,'\n');
fprintf(fo,'The number of S1 shafts produced at shop A ...
      on machine T1=%5.0f\n',x(1));
fprintf(fo,'The number of S2 shafts produced at shop A ...
      on machine T1=%5.0f\n',x(2));
fprintf(fo,'The number of S1 shafts produced at shop A ...
      on machine T2=%5.0f\n',x(3));
fprintf(fo,'The number of S2 shafts produced at shop A ...
      on machine T2=%5.0f\n',x(4));
fprintf(fo,'The number of S1 shafts produced at shop B ...
      on machine T1=%5.0f\n',x(5));
fprintf(fo,'The number of S2 shafts produced at shop B ...
      on machine T1=%5.0f\n',x(6));
fprintf(fo,'The number of S1 shafts produced at shop B ...
      on machine T2=%5.0f\n',x(7));
fprintf(fo,'The number of S2 shafts produced at shop B ...
      on machine T2=%5.0f\n',x(8));
fprintf(fo,'\n');
fprintf(fo,'The revenue for producing these shafts = $%6.0f \n',-fval);
fclose(fo);
```

```
% objshafts.m
% This object function is required for program optimtanks3.m
function Z=objshafts(x)
Z=-(35*(x(1)+x(3)+x(5)+x(7))+85*(x(2)+x(4)+x(6)+x(8)));
```

Exercises

Exercise 9.1

Use Lagrange Multipliers to find the volume of the largest box that can be placed inside the ellipsoid

$$\frac{x^2}{a^2}+\frac{y^2}{b^2}+\frac{z^2}{c^2}=1$$

so that the edges will be parallel to the coordinate axis.

Projects

Project 9.1

A silo consists of a right circular cylinder topped by a right circular cone as shown in Figure P9.1. The radius of the cylinder and the base of the cone are R. The length of the cylinder is L and the height of the cone is H. The cylinder, the cone, and the silo floor are all made of the same material. Write a program in MATLAB using MATLAB's *fmincon* function to determine the values of R, H, and L that will result in the minimum surface silo area for an internal silo volume of 7000 m³.

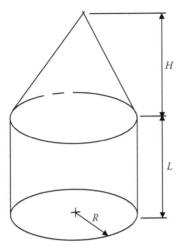

Figure P9.1   Sketch of a silo consisting of a right circular cylinder topped by a right circular cone.

For a right circular cone:

$$V = \frac{\pi R^2 H}{3}$$

$$S = \pi R \sqrt{R^2 + H^2}$$

Project 9.2

A retail store sells computers to the public. There are eight different computer types that the store may carry. Table P9.2 lists the type of computer, the selling price,

Table P9.2   Selling Price and Store Cost of Computer Types

Computer Type	Selling Price ($)	Cost ($)
C1	675	637
C2	805	780
C3	900	874
C4	1025	990
C5	1300	1250
C6	1500	1435
C7	350	340
C8	1000	1030

and the cost to the store. The store plans to spend $20,000 per month purchasing the computers.

The store plans to spend no more than 30% of its costs on computer types C1 and C2, no more than 30% on computer types C3 and C4, no more than 10% on computer types C5 and C6, and no more than 30% on computer types C7 and C8. The store estimates that it can sell 30% more of type C1 than C2, 20% more of type C3 than C4, 20% more of type C5 than C6, and 60% more of type C7 than C8. Use the *fmincon* function in MATLAB to determine the number of each type of computer that will provide the store with the most profit. Print out the number of each type of computer the store should purchase per month, the total profit per month, and the total cost per month to the store.

Project 9.3

The Jones Electronics Corp. has a contract to manufacture four different computer circuit boards. The manufacturing process requires each of the boards to pass through the following four departments before shipping: Etching & Lamination (etches circuits into board), Drilling (drills holes to secure components), Assembly (installs transistors, micro processes, etc.), and Testing. The time requirement in minutes for each unit produced and its corresponding profit value are summarized in Table P9.3a.

Each department is limited to 3 days per week to work on this contract. The minimum weekly production requirement to fulfill the contract is shown in Table P9.3b. Write a MATLAB program that will

(a) Determine the number of each type of circuit board for the coming week that will provide the maximum profit. Assume that there are 8 hours per day, 5 days per week available for factory operations.
   Note: Not all departments work on the same day.
(b) Determine the total profit for the week.
(c) Determine the total number of minutes it takes to produce all the boards.
(d) Determine the total number of minutes spent in each of the four departments.
(e) Print out to a file the requested information.

Table P9.3a Manufacturing Time in Minutes in Each Department

Circuit Board	Etching & Lamination	Drilling	Assembly	Testing	Unit Profit ($)
Board A	15	10	8	15	12
Board B	12	8	10	12	10
Board C	18	12	12	17	15
Board D	13	9	4	13	10

Table P9.3b

Circuit Board	Minimum Production
Board A	10
Board B	10
Board C	10
Board D	10

Project 9.4

The XYZ oil company operates three oil wells (OW1, OW2, OW3) and supplies crude oil to four refineries (refinery A, refinery B, refinery C, refinery D). The cost of shipping the crude oil from each oil well to each of the refineries, the capacity of each of the three oil wells, and the demand (equality constraint) for gasoline at each refinery are tabulated in Table P9.4a. The crude oil at each refinery is distilled into six basic products: gasoline, lubricating oil, kerosine, jet fuel, heating oil, and plastics. The cost of distillation per 100 liters at each refinery from each of the oil wells is given in Table P9.4b. The percentage of each distilled product per 100 liters is tabulated in Table P9.4c. The revenue from each product is tabulated in Table P9.4d.

Using the *fmincon* function in MATLAB, determine the liters of oil to be produced at each oil well and shipped to each of the four refineries that will satisfy the gasoline demand and that will produce the maximum profit. Print out the following items:

(a) The liters produced at each oil well.
(b) The liters of gasoline received at each refinery.
(c) The total cost of shipping and distillation of all products.
(d) The total revenue from the sale of all of the products.
(e) The total profit from all of the products.

Table P9.4a Cost of Shipping per 100 Liters

Oil Well	Refinery A	Refinery B	Refinery C	Refinery D	Oil Well Capacity
OW 1	9	7	10	11	7000 liters
OW 2	7	10	8	10	6100 liters
OW 3	10	11	6	7	6500 liters
Demand (liters of gasoline)	2,000	1,800	2,100	1,900	

Table P9.4b   Cost of Distillation per 100 Liters ($)

Oil Well	Refinery A	Refinery B	Refinery C	Refinery D
OW 1	15	16	12	14
OW 2	17	12	14	10
OW 3	12	15	16	17

Table P9.4c   Distillation Products per Liter of Crude Oil

	Product Percentage per Liter from Distillation (%)					
Oil Well	Gasoline	Lubricating Oil	Kerosene	Jet Fuel	Heating Oil	Plastics
OW 1	43	10	9	15	13	10
OW 2	38	12	5	14	16	15
OW 3	46	8	8	12	12	14

Table P9.4d   Product Revenue per Liter ($)

Gasoline	Lubricating Oil	Kerosene	Jet Fuel	Heating Oil	Plastics
0.40	0.20	0.20	0.50	0.25	0.15

Reference

1. Wylie, C. R., *Advanced Engineering Mathematics*, 4th Ed., McGraw-Hill, New York, 1975.

Chapter 10

Partial Differential Equations

10.1 The Classification of Partial Differential Equations

The mathematical modeling of many types of engineering-type problems involves partial differential equations (PDEs). PDEs of the general form as given by Equation (10.1) fall into one of three categories. These categories are listed below:

$$A\frac{\partial^2 u}{\partial x^2} + B\frac{\partial^2 u}{\partial x \partial y} + C\frac{\partial^2 u}{\partial y^2} = f\left(x, y, u, \frac{\partial u}{\partial x}, \frac{\partial u}{\partial y}\right) \tag{10.1}$$

where A, B, and C are constants.

If $B^2 - 4AC < 0,$ the equation is said to be *elliptic*.

If $B^2 - 4AC = 0,$ the equation is said to be *parabolic*.

If $B^2 - 4AC > 0,$ the equation is said to be *hyperbolic*.

The steady-state heat conduction problem in two dimensions is an example of an elliptic PDE. Laplace's PDE falls into this category. The parabolic PDE is also called the diffusion equation. The unsteady heat conduction problem is an example of a parabolic PDE. The hyperbolic PDE is also called the wave equation. Sound waves and vibration problems, such as the vibrating string, fall into this category. How a PDE is treated numerically depends into which category it falls. However, there are cases in all three categories where a closed-form solution can be obtained by a method called *separation of variables*. This solution method is discussed in the next section.

10.2 Solution by Separation of Variables

10.2.1 The Vibrating String

The first problem to be considered is the vibrating string, such as a violin or a viola string (see Figure 10.1). We will assume that

1. The string is elastic.
   The string motion is vertical.
   The gravitational forces are negligible compared to the tension in the string.
2. The displacement, $Y(x, t)$, from the horizontal is small and the angle that the string makes with the horizontal is small. Then $\dfrac{\partial Y}{\partial t}$ is the vertical velocity of the string and $\dfrac{\partial^2 Y}{\partial t^2}$ is the acceleration of the string at position x.

To obtain the governing equation, select an arbitrary element of the string as shown in Figure 10.2. Taking the sum of the forces in the y direction and applying Newton's second law to this element give

$$M \frac{\partial^2 Y}{\partial t^2} = (T \sin \vartheta)_{x+\Delta x} - (T \sin \vartheta)_x \qquad (10.2)$$

String

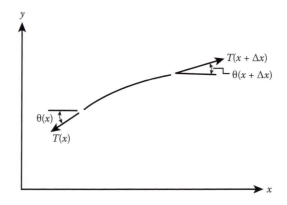

Figure 10.1   A vibrating string.

Figure 10.2   An arbitrary string section.

Since there is no horizontal movement in the string,

$$(T\cos\vartheta)_{x+\Delta x} = (T\cos\vartheta)_x = T_0 \qquad (10.3)$$

Dividing both sides of Equation (10.2) by T_0 (but using the appropriate expression from Equation 10.3) gives

$$\frac{M}{T_0}\frac{\partial^2 Y}{\partial t^2} = \left(\frac{T\sin\vartheta}{T\cos\vartheta}\right)_{x+\Delta x} - \left(\frac{T\sin\vartheta}{T\cos\vartheta}\right)_x = (\tan\vartheta)_{x+\Delta x} - (\tan\vartheta)_x \qquad (10.4)$$

$$M = \rho\Delta x \qquad (10.5)$$

where ρ is the mass per unit length. Dividing both sides by Δx and taking the limit as $\Delta x \to 0$ on both sides of Equation (10.4) gives

$$\frac{\rho}{T_0}\frac{\partial^2 Y}{\partial t^2} = \frac{\partial(\tan\vartheta)}{\partial x} \qquad (10.6)$$

But $\tan\vartheta$ is the slope of the string, which is $\dfrac{\partial Y}{\partial x}$. Thus, Equation (10.6) becomes

$$\frac{1}{c^2}\frac{\partial^2 Y}{\partial t^2} = \frac{\partial^2 Y}{\partial x^2} \qquad (10.7)$$

where

$$c^2 = \frac{T_0}{\rho} \qquad (10.8)$$

Comparing Equation (10.7) with Equation (10.1), we see that Equation (10.7) is a wave equation. Since ϑ is very small, $\cos\vartheta \approx 1$ and, thus, T_0 is essentially the tension in the string. To complete the formulation, two initial conditions are needed (PDE is second order in t) and two boundary conditions (PDE is second order in x). We will assume that the string is deflected at x_0 and then released from rest. Then, the initial conditions become

$$\frac{\partial Y}{\partial t}(x,0) = 0 \qquad (10.9)$$

$$Y(x,0) = f(x) \qquad (10.10)$$

The boundary conditions are

$$Y(0,t) = 0 \tag{10.11}$$

$$Y(L,t) = 0 \tag{10.12}$$

We seek a function Y that satisfies the PDE and initial and boundary conditions. Let us examine the possibility that Y is a product of a pure function of x and a pure function of t, that is,

$$Y = F(x)G(t)$$

Substituting this expression into the PDE gives

$$\frac{F}{c^2}\frac{d^2 G}{dt^2} = G(t)\frac{d^2 F}{dx^2}$$

Dividing both sides by GF gives

$$\frac{1}{c^2 G}\frac{d^2 G}{dt^2} = \frac{1}{F}\frac{d^2 F}{dx^2} \tag{10.13}$$

The left-hand side is a pure function of t and the right-hand side is a pure function of x. Since x and t are independent variables, Equation (10.13) can only be true if both sides equal the same constant, say, $(-\lambda^2)$. The minus sign is selected so that Y does not blow up as $t \to \infty$. Then Equation (10.13) reduces to two ordinary differential equations, which are

$$\frac{d^2 F}{dx^2} + \lambda^2 F = 0 \tag{10.14}$$

and

$$\frac{d^2 G}{dt^2} + c^2 \lambda^2 G = 0 \tag{10.15}$$

The general solution to Equation (10.14) is

$$F = a_1 \cos \lambda x + a_2 \sin \lambda x \tag{10.16}$$

The boundary condition given by Equation (10.11) reduces to

$$F(0)G(t) = 0 \to F(0) = 0 \to a_1 = 0$$

or

$$F = a_2 \sin \lambda x \tag{10.17}$$

The boundary condition given by Equation (10.12) is

$$F(L)G(t) = 0 \to F(L) = 0$$

or

$$a_2 \sin \lambda L = 0 \tag{10.18}$$

There are an infinite number of solutions to Equation (10.18), that is, $\lambda L = n\pi$, where $n = 0, 1, 2,..., \infty$. Then,

$$\lambda = \frac{n\pi}{L} \tag{10.19}$$

The solution to Equation (10.15) is

$$G = b_1 \cos(c\lambda t) + b_2 \sin(c\lambda t) \tag{10.20}$$

and

$$\frac{dG}{dt} = -c\lambda b_1 \sin(c\lambda t) + c\lambda b_2 \cos(c\lambda t) \tag{10.21}$$

Applying the initial condition given by Equation (10.9) gives

$$F(x)\frac{dG}{dt}(0) = 0 \;\to\; \frac{dG}{dt}(0) = 0 \to b_2 = 0$$

Then,

$$G(t) = b_1 \cos(c\lambda t)$$

The b's can be absorbed into the a_n constants giving

$$Y(x,t) = \sum_{n=1}^{\infty} a_n \sin\frac{n\pi x}{L} \cos\frac{n\pi ct}{L} \tag{10.22}$$

Applying the initial condition given by Equation (10.10) gives

$$Y(x,0) = f(x) = \sum_{n=1}^{\infty} a_n \sin\frac{n\pi x}{L} \tag{10.23}$$

The coefficients a_n can be determined by knowing that the $\sin\dfrac{n\pi x}{L}$ functions are orthogonal, that is,

$$\int_0^L \sin\frac{n\pi x}{L}\sin\frac{m\pi x}{L}\,dx = 0, \text{if } m \neq n \text{ and equals} \frac{L}{2}\text{if } m = n \tag{10.24}$$

Multiply Equation (10.23) by $\sin\dfrac{m\pi x}{L}\,dx$ and integrate from 0 to L, giving

$$\int_0^L f(x)\sin\frac{m\pi x}{L}\,dx = \sum_{n=0}^{\infty} a_n \int_0^L \sin\frac{n\pi x}{L}\sin\frac{m\pi x}{L}\,dx = a_m \frac{L}{2} \tag{10.25}$$

Since m is an index from 1 to ∞, we can replace m with n and write

$$a_n = \frac{2}{L}\int_0^L f(x)\sin\frac{n\pi x}{L}\,dx \tag{10.26}$$

Having a value for a_n, Equation (10.22) gives the solution for $Y(x, t)$. See Project 10.2.

10.2.2 Unsteady Heat Transfer I (Bar)

Consider a thick bar, as shown in Figure 10.3, that is initially at a uniform temperature, T_0, that is suddenly immersed in a large bath at temperature T_∞. We wish to determine the temperature time history of the bar, $T(x,t)$, and the amount of heat transferred to the bath (see Appendix B for derivation of the heat conduction equation). Due to symmetry, one only needs to consider $T(x,t)$ for $0 \leq x\ L$. The PDE is

$$\frac{1}{a}\frac{\partial T}{\partial t} = \frac{\partial^2 T}{\partial x^2} \tag{10.27}$$

The initial condition is

$$T(x,0) = T_0 \tag{10.28}$$

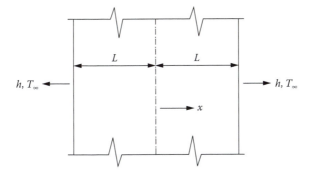

Figure 10.3   A thick bar suddenly immersed in a liquid.

The boundary conditions are

$$\frac{\partial T}{\partial x}(0, t) = 0 \tag{10.29}$$

$$\frac{\partial T}{\partial x}(L,t)+\frac{h}{k}(T(L,t) - T_{\infty}) = 0 \tag{10.30}$$

where
   t = time.
   a = thermal diffusivity of the bar material.
   h = the convective heat transfer coefficient.
   k = the thermal conductivity of the bar material.

Equation (10.29) is a statement that there is no heat transfer in any direction at $x = 0$ (this is due to problem symmetry). Equation (10.30) is a statement that the rate that heat leaves the bar at $x = L$ by conduction is equal to the rate that heat is carried away by convection in the bath. To make the boundary condition at $x = L$ homogeneous, let

$$\vartheta(x,t) = T(x,t)-T_{\infty}$$

Then the PDE and the initial and boundary conditions become

$$\frac{1}{a}\frac{\partial \vartheta}{\partial t} = \frac{\partial^{2}\vartheta}{\partial x^{2}} \tag{10.31}$$

$$\vartheta(x,0) = T_{0}-T_{\infty} \tag{10.32}$$

$$\frac{\partial \vartheta}{\partial x}(0,t) = 0 \tag{10.33}$$

$$\frac{\partial \vartheta}{\partial x}(L,t) + \frac{h}{k}\vartheta(L,t) = 0 \tag{10.34}$$

Comparing Equation (10.31) with Equation (10.1), we see that Equation (10.31) is a diffusion equation. We will assume that

$$\vartheta = F(x)G(t) \tag{10.35}$$

where F is a pure function of x and G is a pure function of t. Substituting Equation (10.35) into Equation (10.31) gives

$$\frac{1}{a}F(x)G'(t) = G(t)F''(x) \tag{10.36}$$

where

$$G'(t) = \frac{dG}{dt} \text{ and } F'' = \frac{d^2 F}{dx^2}$$

Dividing both sides of Equation (10.36) by $F\,G$ gives

$$\frac{1}{a}\frac{G'}{G} = \frac{F''}{F} \tag{10.37}$$

The left-hand side of Equation (10.37) is a pure function of t and the right-hand side is a pure function of x. The only way a pure function of t can equal a pure function of x is for both sides to equal the same constant, say, $(-\lambda)$. Then Equation (10.37) can be expressed as two ordinary differential equations, which are

$$G' + a\lambda^2 G = 0 \tag{10.38}$$

$$F'' + \lambda^2 F = 0 \tag{10.39}$$

The boundary conditions become

$$F'(0)G(t) = 0$$

or

$$F'(0) = 0 \tag{10.40}$$

and

$$F'(L)G(t) + \frac{h}{k}F(L)G(t) = 0$$

or

$$F'(L) + \frac{h}{k}F(L) = 0 \tag{10.41}$$

The function that will satisfy Equation (10.39) is of the form

$$F = A e^{\beta x}$$

Substituting this form into Equation (10.41) gives

$$\beta^2 A e^{\beta x} + \lambda^2 A e^{\beta x} = 0$$

Then,

$$\beta = \pm \lambda i, \quad \text{where} \quad i = \sqrt{-1}$$

Knowing that $e^{ix} = \cos(x) + i\sin(x)$, F becomes

$$F(x) = A\cos(\lambda x) + B\sin(\lambda x) \tag{10.42}$$

and

$$F'(x) = -\lambda A \sin(\lambda x) + \lambda B \cos(\lambda x) \tag{10.43}$$

Applying Equation (10.40) to Equation (10.42) gives $B = 0$. Thus,

$$F(x) = A\cos(\lambda x) \tag{10.44}$$

and

$$F'(x) = -\lambda A \sin(\lambda x) \tag{10.45}$$

Applying Equations (10.44) and (10.45) to Equation (10.41) gives

$$-\lambda A \sin(\lambda L) + \frac{h}{k} A \cos(\lambda L) = 0$$

or

$$\tan(\lambda L) - \frac{hL}{k(\lambda L)} = 0 \tag{10.46}$$

A plot of $\tan(\delta)$ and $\dfrac{hL}{k\delta}$ vs. δ, where $\delta = \lambda L$, is shown in Figure 10.4.

It can be seen that there is an infinite number of roots that satisfy Equation (10.46), say, $\delta_1, \delta_2, \delta_3, ..., \delta_n, = \lambda_1 L, \lambda_2 L, \lambda_3 L, ..., \lambda_n L$, giving an infinite number of solutions, each satisfying the PDE and the boundary conditions. The solution to Equation (10.38) can readily be obtained by separating the variables, giving

$$\frac{dG}{G} = -a\lambda_n^2 \, dt$$

Integrating and taking the antilog gives

$$G = \exp\left(-a\lambda_n^2 t\right) \tag{10.47}$$

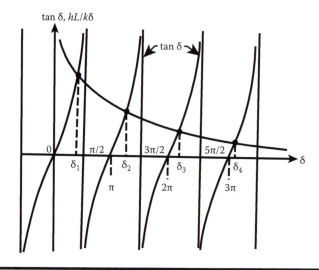

Figure 10.4   The plot of $\tan\delta$ and $\dfrac{hL}{k\delta}$ vs. δ.

Thus the general solution is

$$\vartheta(x,t) = \sum_{n=1}^{\infty} A_n \cos(\lambda_n x) \exp(-a\lambda_n^2 t) \tag{10.48}$$

The initial condition now needs to be applied. Applying Equation (10.32) to Equation (10.48) gives

$$T_0 - T_\infty = \sum_{n=1}^{\infty} A_n \cos(\lambda_n x) \tag{10.49}$$

It can be shown that the functions $\cos(\lambda_n x)$ are orthogonal, that is,

$$\int_0^L \cos(\lambda_n x)\cos(\lambda_m x)\, dx = \begin{cases} 0, & \text{if } m \neq n \\ \dfrac{L}{2} + \dfrac{\sin(\lambda_m L)\cos(\lambda_m L)}{2\lambda_m}, & \text{if } m = n \end{cases}$$

This will be demonstrated later. Now multiply both sides of Equation (10.49) by $\cos(\lambda_m x)\, dx$ and integrate from 0 to L, giving

$$(T_0 - T_\infty)\int_0^L \cos(\lambda_m x)\, dx = \sum_{n=1}^{\infty} A_n \int_0^L \cos(\lambda_n x)\cos(\lambda_m x)\, dx \tag{10.50}$$

The only term in the summation on the right-hand side of Equation (10.50) that is not multiplied by zero is the *m*th term. Thus,

$$(T_0 - T_\infty)\int_0^L \cos(\lambda_m x)\, dx = A_m \left(\frac{L}{2} + \frac{\sin(\lambda_m L)\cos(\lambda_m L)}{2\lambda_m} \right) \tag{10.51}$$

or

$$A_m = \frac{2(T_0 - T_\infty)\sin(\lambda_m L)}{\lambda_m L + \sin(\lambda_m L)\cos(\lambda_m L)} \tag{10.52}$$

Since *m* is an index from 0, 1, 2,…, we can replace *m* with *n*. Substituting Equation (10.52) into Equation (10.48) gives

$$\vartheta(x,t) = 2(T_0 - T_\infty)\sum_{n=1}^{\infty} \left(\frac{\sin(\lambda_n L)\cos(\lambda_n x)}{\lambda_n L + \sin(\lambda_n L)\cos(\lambda_n L)} e^{-a\lambda_n^2 t} \right) \tag{10.53}$$

Now to demonstrate the orthogonality of the $\cos(\lambda_n x)$ functions. Let $f_n = \cos(\lambda_n x)$ and $f_m = \cos(\lambda_m x)$. Each function satisfies the ODE, that is,

$$f_n'' + \lambda_n^2 f_n = 0 \tag{10.54}$$

and

$$f_m'' + \lambda_m^2 f_m = 0 \tag{10.55}$$

Each function satisfies the boundary conditions, that is,

$$f_n' (0) = 0 \quad \text{and} \quad f_m' (0) = 0$$

and

$$f_n' (L) + \frac{h}{k} f_n (L) = 0 \quad \text{and} \quad f_m' (L) + \frac{h}{k} f_m (L) = 0$$

Now multiply Equation (10.54) by f_m and Equation (10.55) by f_n and subtract the second equation from the first, giving

$$\left(f_m f_n'' + \lambda_n^2 f_n f_m \right) - \left(f_n f_m'' + \lambda_m^2 f_n f_m \right) = 0 \tag{10.56}$$

or

$$f_m f_n'' - f_n f_m'' = \left(\lambda_m^2 - \lambda_n^2 \right) f_n f_m \tag{10.57}$$

But

$$\frac{d}{dx}(f_m f_n') = f_m f_n'' + f_m' f_n' \tag{10.58}$$

and

$$\frac{d}{dx}(f_n f_m') = f_n f_m'' + f_m' f_n' \tag{10.59}$$

Then

$$\frac{d}{dx}(f_m f_n' - f_n f_m') = f_m f_n'' - f_n f_m'' \tag{10.60}$$

Substituting Equation (10.60) into Equation (10.57) gives

$$\frac{d}{dx}(f_m f_n' - f_n f_m') = \left(\lambda_m^2 - \lambda_n^2\right) f_n f_m \tag{10.61}$$

Multiplying both sides of Equation (10.61) by dx and integrating from 0 to L gives

$$\int_0^L \frac{d}{dx}(f_m f_n' - f_n f_m')\, dx = \left(\lambda_m^2 - \lambda_n^2\right) \int_0^L f_n f_m\, dx \tag{10.62}$$

or

$$\left(\lambda_m^2 - \lambda_n^2\right) \int_0^L \cos(\lambda_n x)\cos(\lambda_m x)\, dx = f_m(L) f_n'\ (L) - f_n(L) f_m'\ (L)$$

$$- f_m(0) f_n'\ (0) + f_n(0) f_m'\ (0) \tag{10.63}$$

But $f_n'\ (0) = 0$ and $f_m'\ (0) = 0$ and

$$f_m(L) f_n'\ (L) - f_n(L) f_m'\ (L) = -f_m(L)\frac{h}{k} f_n(L) + f_n(L)\frac{h}{k} f_m(L) = 0 \tag{10.64}$$

Thus,

$$\left(\lambda_m^2 - \lambda_n^2\right) \int_0^L \cos(\lambda_n x)\cos(\lambda_m x)\, dx = 0 \tag{10.65}$$

If $m \neq n$, then $\left(\lambda_m^2 - \lambda_n^2\right) \neq 0$ and $\displaystyle\int_0^L \cos(\lambda_n x)\cos(\lambda_m x)\, dx = 0$ (10.66)

If $m = n$, then $(\lambda_m^2 - \lambda_n^2) = 0$ and $\displaystyle\int_0^L \cos^2(\lambda_n x)\, dx$ needs to be evaluated.
From integral tables, we can determine that

$$\int_0^L \cos^2(\lambda_m x)\, dx = \left(\frac{L}{2} + \frac{\sin(\lambda_m L)\cos(\lambda_m L)}{2\lambda_m}\right) \tag{10.67}$$

10.2.3 Unsteady Heat Transfer II (Cylinder)

A similar problem to the one described in the previous section is one in which a cylinder, initially at temperature T_0, is suddenly immersed in a fluid at temperature

T_∞, with $T_0 > T_\infty$. We will assume that end effects are negligible and that h and k are constant, where h is the convective heat transfer coefficient and k is the thermal conductivity of the cylinder material. We wish to determine the temperature distribution and the amount of heat transferred to the bath in time t. The governing PDE is

$$\frac{1}{a}\frac{\partial T}{\partial t} = \nabla^2 T \tag{10.68}$$

In cylindrical coordinates (see Appendix C, Section C.4) with $T = T(r,t)$,

$$\nabla^2 T = \frac{\partial^2 T}{\partial r^2} + \frac{1}{r}\frac{\partial T}{\partial r} \tag{10.69}$$

Thus, Equation (10.69) becomes

$$\frac{1}{a}\frac{\partial T}{\partial t} = \frac{\partial^2 T}{\partial r^2} + \frac{1}{r}\frac{\partial T}{\partial r} \tag{10.70}$$

The initial and boundary conditions are

$$T(r,0) = T_0 \tag{10.71}$$

$$T(0,t) \text{ is finite} \tag{10.72}$$

$$\frac{\partial T}{\partial r}(R,t) + \frac{h}{k}(T(R,t) - T_\infty) = 0 \tag{10.73}$$

To obtain a homogeneous boundary condition at $r = R$, let $\vartheta(r,t) = T(r,t) - T_\infty$; then Equations (10.70), (10.71), (10.72), and (10.73) become

$$\frac{1}{a}\frac{\partial \vartheta}{\partial t} = \frac{\partial^2 r}{\partial r^2} + \frac{1}{r}\frac{\partial \vartheta}{\partial r} \tag{10.74}$$

$$\vartheta(r,0) = T_0 - T_\infty \tag{10.75}$$

$$\vartheta(0,t) \text{ is finite} \tag{10.76}$$

$$\frac{\partial \vartheta}{\partial r}(R,t) + \frac{h}{k}\vartheta(R,t) = 0 \tag{10.77}$$

We wish to see if $\vartheta(r,t) = f(r)g(t)$ can satisfy both the PDE and the initial and boundary conditions. Substituting this form of ϑ into Equation (10.74), it becomes

$$\frac{1}{a} f\, g' = \left(f'' + \frac{1}{r} f' \right) g \tag{10.78}$$

where

$$g' = \frac{dg}{dt}, \quad f'' = \frac{d^2 f}{dr^2}, \quad f' = \frac{df}{dr}$$

Dividing both sides of Equation (10.78) by fg it becomes

$$\frac{1}{a}\frac{g'}{g} = \frac{1}{f}\left(f'' + \frac{1}{r} f' \right) \tag{10.79}$$

The left-hand side of Equation (10.79) is a pure function of t and the right-hand side is a pure function of r, yet r and t are independent variables. The only way that the left-hand side of Equation (10.79) can equal the right-hand side is for both to be equal to the same constant, say, $(-\lambda^2)$. Then Equation (10.79) reduces to two ordinary differential equations, which are

$$g' + a\lambda^2 g = 0 \tag{10.80}$$

$$f'' + \frac{1}{r} f' + \lambda^2 f = 0 \tag{10.81}$$

Equation (10.80) can be immediately solved giving

$$g = B e^{-a\lambda^2 t} \tag{10.82}$$

Multiplying Equation (10.81) by r^2, it becomes

$$r^2 f'' + r f' + \lambda^2 r^2 f = 0 \tag{10.83}$$

By letting $\lambda r = x$, Equation (10.83) can be reduced to the standard form of Bessel's equation of order v, which is

$$x^2 y'' + x y' + (x^2 - v^2) y = 0 \tag{10.84}$$

It is left as a student exercise to show that Equation (10.83) reduces to Equation (10.84) with f replacing y and $v = 0$. The Bessel functions $J_0(\lambda r)$ and $Y_0(\lambda r)$ are two solutions to Equation (10.83). Thus,

$$f(r) = c_1 J_0(\lambda r) + c_2 Y_0(\lambda) \tag{10.85}$$

where J_0 is a Bessel function of the first kind and Y_0 is a Bessel function of the second kind. These functions are oscillatory with a variable frequency and amplitude. They have an infinite number of zeros (see Figure 10.5). However, the Y_0 function is singular at $r = 0$ (see Figure 10.6).

The boundary condition $\vartheta(0,t) = f(0) g(t)$ is finite implies that $f(0)$ is finite. Since Y_0 is singular at $r = 0$, thus, $c_2 = 0$ and

$$f(r) = c_1 J_0(\lambda r) \tag{10.86}$$

The boundary condition described by Equation (10.77) reduces to

$$f'(R) g(t) + \frac{h}{k} f(R) g(t) = 0$$

or

$$f'(R) + \frac{h}{k} f(R) = 0 \tag{10.87}$$

Substituting Equation (10.86) into Equation (10.87) gives

$$\left[\frac{d}{dr}(c_1 J_0(\lambda r)) + \frac{h}{k} c_1 J_0(\lambda r) \right]_{r=R} = 0 \tag{10.88}$$

To utilize the boundary condition expressed by Equation (10.88), we need to turn to a recursion formula involving the Bessel functions (see Equation C.10 in Appendix C). The recursion formula is

$$x \frac{d}{dx}[J_n(x)] = n J_n(x) - x J_{n+1}(x) \tag{10.89}$$

Letting $x = \lambda r$ and using the chain rule to obtain the derivative with respect to r and setting $n = 0$, Equation (10.89) reduces to

$$\frac{d}{dr} J_0(\lambda r) = -\lambda J_1(\lambda r) \tag{10.90}$$

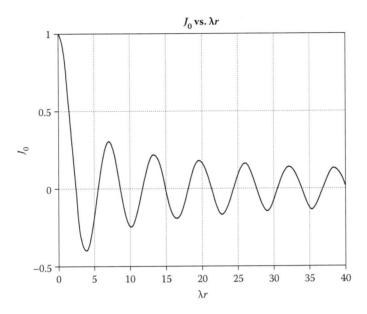

Figure 10.5   The plot of J_0 vs. λr.

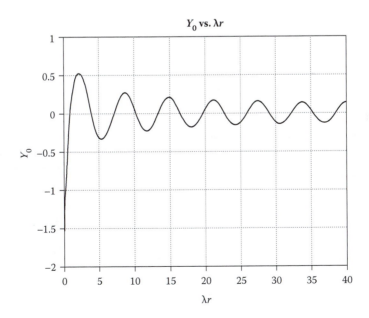

Figure 10.6   The plot of Y_0 vs. λr.

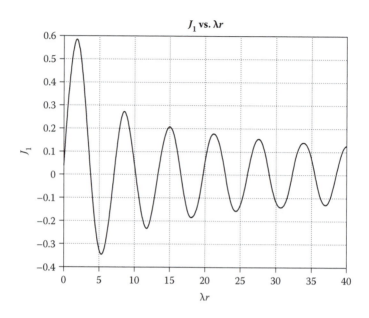

Figure 10.7   The plot of J_1 vs. λr.

Substituting Equation (10.90) into Equation (10.88), we obtain

$$-\lambda J_1(\lambda R) + \frac{h}{k} J_0(\lambda R) = 0$$

or

$$\frac{J_0(\lambda R)}{J_1(\lambda R)} - \lambda R \frac{k}{hR} = 0 \qquad (10.91)$$

The function J_1 has an infinite number of zeros as can be seen from Figure 10.7. As a result there are an infinite number or roots to Equation (10.91) as can be seen in Figure 10.8.

Although only seven $\dfrac{J_0}{J_1}$ curves were plotted in Figure 10.8, it should be understood that there are an infinite number of $\dfrac{J_0}{J_1}$ curves. The intersection of curves $\dfrac{J_0}{J_1}$ and $\dfrac{\lambda k}{hR}$ give the eigen values of λ, say λ_j. Since each of these functions satisfies the differential equation, the general solution is

$$\vartheta(r,t) = \sum_{n=1}^{\infty} A_n e^{-a\lambda_n^2 t} J_0(\lambda_n r) \qquad (10.92)$$

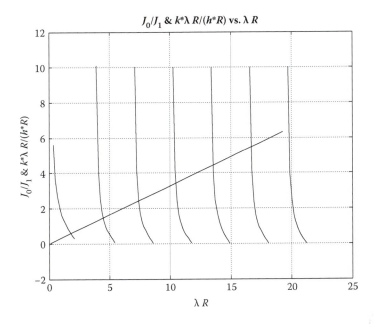

Figure 10.8 The plot of J_0/J_1 and $\lambda\, R\, k/hR$ vs. λR.

The initial condition still needs to be satisfied, that is,

$$\vartheta(r,0)=T_0-T_\infty=\sum_{n=1}^{\infty}A_n J_0(\lambda_n r) \tag{10.93}$$

The constants A_n can be determined because the J_0 functions are orthogonal, that is,

$$\int_0^{R}r\,J_0(\lambda_m r)\,J_0(\lambda_n r)\,dr=\delta_{mn}\frac{\lambda_m^2 R^2+\left(\dfrac{hR}{k}\right)^2}{2\lambda_m^2}[J_0(\lambda_m R)]^2 \tag{10.94}$$

where

$$\delta_{mn}=\begin{cases}1, & \text{if } m=n\\[2em] 0, & \text{if } m\neq n\end{cases}$$

Also,

$$\int_0^R r J_0(\lambda_m r)\, dr = \frac{R}{\lambda_m} J_1(\lambda_m R) \qquad (10.95)$$

For proof of Equations (10.94) and (10.95) see Appendices C.2 and C.3.

Multiplying Equation (10.93) by $r J_0(\lambda_m r)\, dr$ and integrating from 0 to R, gives

$$(T_0 - T_\infty)\frac{R}{\lambda_m} J_1(\lambda_m R) = A_m \frac{\lambda_m^2 R^2 + \left(\dfrac{hR}{k}\right)^2}{2\lambda_m^2}[J_0(\lambda_m R)]^2 \qquad (10.96)$$

Thus,

$$A_m = \frac{2(T_0 - T_\infty)\lambda_m R}{(\lambda_{mR})^2 + \left(\dfrac{hR}{k}\right)^2} \times \frac{J_1(\lambda_m R)}{[J_0(\lambda_m R)]^2} \qquad (10.97)$$

Finally, the temperature ratio $TR(r,\ t) = \dfrac{T(r,t) - T_\infty}{T_0 - T_\infty} = \dfrac{\vartheta(r,t)}{T_0 - T_\infty}$ is given by

$$TR(r,t) = \frac{\vartheta(r,t)}{T_0 - T_\infty} = \sum_{n=1}^\infty \frac{2\lambda_n R}{(\lambda_n R)^2 + \left(\dfrac{hR}{k}\right)^2} \times \frac{J_1(\lambda_n R) J_0(\lambda_n r)}{\left[J_0(\lambda_n R)\right]^2} e^{-a\lambda_n^2 t} \qquad (10.98)$$

10.3 Unsteady Heat Transfer in 2-D

Consider a bar having a rectangular cross-section, initially at temperature T_0, that is suddenly immersed in a huge bath at a temperature T_∞ (see Figure 10.9). The governing PDE is

$$\frac{1}{\alpha}\frac{\partial T}{\partial t} = \frac{\partial^2 T}{\partial x^2} + \frac{\partial^2 T}{\partial y^2} \qquad (10.99)$$

To obtain homogeneous boundary conditions let $\vartheta(x,y,t) = T(x,y,t) - T_\infty$; then the PDE and the initial condition and boundary conditions in variable ϑ are

$$\frac{1}{\alpha}\frac{\partial \vartheta}{\partial t} = \frac{\partial^2 \vartheta}{\partial x^2} + \frac{\partial^2 \vartheta}{\partial y^2} \qquad (10.100)$$

$$\vartheta(x,y,0) = T_0 - T_\infty \qquad (10.101)$$

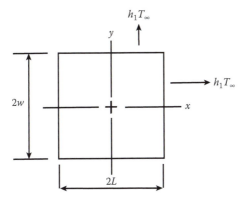

Figure 10.9   An unsteady 2-D heat transfer problem.

$$\frac{\partial \vartheta}{\partial x}(0, y, t) = 0 \quad \text{(due to symmetry)} \tag{10.102}$$

$$\frac{\partial \vartheta}{\partial y}(x, 0, t) = 0 \quad \text{(due to symmetry)} \tag{10.103}$$

$$\frac{\partial \vartheta}{\partial x}(L, y, t) + \frac{h}{k}\vartheta(L, y, t) = 0 \tag{10.104}$$

$$\frac{\partial \vartheta}{\partial y}(x, w, t) + \frac{h}{k}\vartheta(x, w, t) = 0 \tag{10.105}$$

Assume that

$$\vartheta(x, y, t) = F(t)G(x)H(y) \tag{10.106}$$

Substituting Equation (10.106) into Equation (10.100) gives

$$\frac{1}{\alpha}F'(t)G(x)H(y) = F(t)G''(x)H(y) + F(t)G(x)H''(y) \tag{10.107}$$

Dividing both sides by FGH gives

$$\frac{1}{\alpha}\frac{F'(t)}{F(t)} = \frac{G''(x)}{G(x)} + \frac{H''(y)}{H(y)} \tag{10.108}$$

The left-hand side is only a function of t and the right-hand side is only a function of x and y. Since x, y, and t are independent variables, the only way that the left-hand side can equal the right-hand side is for both to be equal to the same constant, say, $-\lambda^2$. That is,

$$\frac{1}{\alpha} \frac{F'(t)}{F(t)} = \frac{G''(x)}{G(x)} + \frac{H''(y)}{H(y)} = -\lambda^2$$

This gives the following equation for $F(t)$:

$$F' + \alpha \lambda^2 F = 0 \tag{10.109}$$

The solution of Equation (10.109) is

$$F = C e^{-\alpha \lambda^2 t} \tag{10.110}$$

The right-hand side of Equation (10.108) can be written as

$$\frac{G''(x)}{G(x)} + \lambda^2 = -\frac{H''(y)}{H(y)} \tag{10.111}$$

Again, the only way for the left-hand side to equal the right-hand side is for both to equal the same constant, say, β^2; then

$$\frac{G''(x)}{G(x)} + \lambda^2 = -\frac{H''(y)}{H(y)} = \beta^2$$

This gives

$$G'' + (\lambda^2 - \beta^2)G = 0 \tag{10.112}$$

and

$$H'' + \beta^2 H = 0 \tag{10.113}$$

The solution for G is

$$G = A_1 \cos(\sqrt{\lambda^2 - \beta^2}\ x) + A_2 \sin(\sqrt{\lambda^2 - \beta^2}\ x) \tag{10.114}$$

and

$$G' = \sqrt{\lambda^2 - \beta^2} \left\{ -A_1 \sin(\sqrt{\lambda^2 - \beta^2}\ x) + A_2 \cos(\sqrt{\lambda^2 - \beta^2}\ x) \right\} \tag{10.115}$$

Applying the boundary condition Equation (10.102), $G'(0)=0$, gives $A_2=0$.

Applying the boundary condition Equation (10.104), $(G'(L)+\dfrac{h}{k}(G(L)=0)$, gives

$$-\sqrt{\lambda^2-\beta^2}\ A_1 \sin(\sqrt{\lambda^2-\beta^2}\ L)+\frac{h}{k}A_1\cos(\sqrt{\lambda^2-\beta^2}\ L)=0$$

or

$$\tan\left(\sqrt{\lambda^2-\beta^2}\ L\right)=\frac{hL}{k}\times\frac{1}{\sqrt{\lambda^2-\beta^2}\ L} \tag{10.116}$$

Let $\gamma L=\sqrt{\lambda^2-\beta^2}\ L$; then Equation (10.116) can be written as

$$\tan(\gamma L)=\frac{hL}{k}\times\frac{1}{\gamma L} \tag{10.117}$$

As seen in Figure 10.4, there are an infinite number of $\gamma L's$ that satisfy Equation (10.117), say, $(\gamma L)_1, (\gamma L)_2, (\gamma L)_3,\dots$, then $\gamma_i=\dfrac{(\gamma L)_i}{L}$ and

$$G(x)=\sum_{i=1}^{\infty}A_i\cos(\gamma_i x) \tag{10.118}$$

Returning to Equation (10.113), and noting the similarity between the G function and the H function, we can determine that

$$H(y)=\sum_{j=1}^{\infty}B_j\cos(\beta_j y) \tag{10.119}$$

By the definition of γ, we see that

$$\lambda_{ij}^2=\gamma_i^2+\beta_j^2 \tag{10.120}$$

Combining Equations (10.110), (10.118), and (10.119) and replacing A_i and B_j by a single coefficient a_{ij}, we obtain

$$\vartheta(x,y,t)=\sum_i\sum_j a_{ij}\ \exp(\alpha\gamma_i^2 t)\cos(\gamma_i^2 x)\cos(\beta_j\ y) \tag{10.121}$$

The initial condition provides the means for determining the coefficients a_{ij}, that is,

$$\vartheta(x,y,0)=T_0-T_\infty \tag{10.122}$$

or

$$T_0 - T_\infty = \sum_i \sum_j a_{ij} \cos(\gamma_i x) \cos(\beta_j y) \tag{10.123}$$

Now multiply both sides of Equation (10.123) by $\cos(\gamma_n x) \, \cos(\beta_m y) \, dx \, dy$ and integrate for x from 0 to L and y from 0 to w.

As was shown by Equations (10.65) through (10.67), the functions $\cos(\gamma_i x)$ and $\cos(\beta_j y)$ are orthogonal. Thus,

$$\int_0^L \cos(\gamma_i x) \cos(\gamma_n x) \, dx = \begin{cases} 0, & \text{if } i \neq n \\ \dfrac{L}{2} + \dfrac{\sin(\gamma_n L) \cos(\gamma_n L)}{2\gamma_n}, & \text{if } i = n \end{cases} \tag{10.124}$$

and

$$\int_0^w \cos(\beta_j y) \cos(\beta_m y) \, dx = \begin{cases} 0, & \text{if } j \neq m \\ \dfrac{L}{2} + \dfrac{\sin(\beta_m w) \cos(\beta_m w)}{2\beta_m}, & \text{if } j = m \end{cases} \tag{10.125}$$

Also,

$$(T_0 - T_\infty) \int_0^L \int_0^w \cos(\gamma_n x) \cos(\beta_m y) \, dx \, dy = \frac{\sin(\gamma_n L)}{\gamma_n} \times \frac{\sin(\beta_m w)}{\beta_m} \tag{10.126}$$

The orthogonality of the $\cos(\gamma_i x)$ and $\cos(\beta_j y)$ functions eliminates the summation signs, giving

$$a_{nm} = \frac{(T_0 - T_\infty) \left(\dfrac{\sin(\gamma_n L)}{\gamma_n} \times \dfrac{\sin(\beta_m w)}{\beta_m} \right)}{\left(\dfrac{L}{2} + \dfrac{\sin(\gamma_n L)\cos(\gamma_n L)}{2\gamma_n} \right) \times \left(\dfrac{w}{2} + \dfrac{\sin(\beta_m w)\cos(\beta_m w)}{2\beta_m} \right)} \tag{10.127}$$

Finally,

$$\vartheta(x,y,t) = \sum_{n=1}^{\infty} \sum_{m=1}^{\infty} \frac{(T_0 - T_\infty) \left(\dfrac{\sin(\gamma_n L)}{\gamma_n} \times \dfrac{\sin(\beta_m w)}{\beta_m} \right) \exp(-\alpha \lambda_{nm}^2 t) \cos(\gamma_n x) \cos(\beta_m y)}{\left(\dfrac{L}{2} + \dfrac{\sin(\gamma_n L)\cos(\gamma_n L)}{2\gamma_n} \right) \times \left(\dfrac{w}{2} + \dfrac{\sin(\beta_m w)\cos(\beta_m w)}{2\beta_m} \right)}$$

$$\tag{10.128}$$

10.4 Perturbation Theory and Sound Waves

When there is a problem involving a small disturbance from some equilibrium condition, perturbation theory may be applied to determine the relevant governing differential or partial differential equations. Developing the governing equations that describe sound wave behavior is such a problem. A sound wave (one that is detectable by the human ear) is an oscillatory pressure disturbance of small amplitude. In air, the disturbance can be considered as taking place in an ideal, isentropic, and compressible fluid. Although sound waves are not in general one dimensional, they can be made so by placing a speaker inside a tube. The governing equations describing this phenomenon are as follows:

$$\frac{\partial \rho}{\partial t} + \frac{\partial (\rho u)}{\partial x} = 0 \tag{10.129}$$

$$\rho \left(\frac{\partial u}{\partial t} + u \frac{\partial u}{\partial x} \right) = -\frac{\partial p}{\partial x} \tag{10.130}$$

$$p = \alpha \rho^k \tag{10.131}$$

where
  ρ = fluid density.
  u = fluid velocity.
  p = fluid pressure.
  k = ratio of specific heats.
  α = a constant.

For a small disturbance, we may take

$$\rho = \rho_0 + \bar{\rho}, \quad p = p_0 + \bar{p}, \quad u = \bar{u} \tag{10.132}$$

where
  ρ_0 and p_0 = the fluid density and pressure, respectively, in the undisturbed fluid.
  $\bar{\rho}$ and $\bar{p}$ = the disturbed fluid density and pressure, respectively.
  $\frac{\bar{\rho}}{\rho_0} \ll 1, \frac{\bar{p}}{p_0} \ll 1, \frac{\bar{u}}{c_0} \ll 1$
  c_0 = the speed of sound in the undisturbed fluid.
  ρ_0, p_0, and c_0 are considered as constants.

Substituting Equation (10.132) into Equations (10.129) and (10.130) gives

$$\frac{\partial \bar{\rho}}{\partial t} + \rho_0 \frac{\partial \bar{u}}{\partial x} + \bar{\rho} \frac{\partial \bar{u}}{\partial x} + \bar{u} \frac{\partial \bar{\rho}}{\partial x} = 0 \qquad (10.133)$$

$$\rho_0 \frac{\partial \bar{u}}{\partial t} + \bar{\rho} \frac{\partial \bar{u}}{\partial t} + \bar{u} \frac{\partial \bar{u}}{\partial x} = -\frac{\partial \bar{p}}{\partial x} \qquad (10.134)$$

In perturbation theory, the product of any two perturbed variables is considered as higher-order terms and neglected. Therefore, Equations (10.133) and (10.134) reduce to

$$\frac{\partial \bar{\rho}}{\partial t} + \rho_0 \frac{\partial \bar{u}}{\partial x} = 0 \qquad (10.135)$$

$$\rho_0 \frac{\partial \bar{u}}{\partial t} = -\frac{\partial \bar{p}}{\partial x} \qquad (10.136)$$

Let $\bar{u} = \dfrac{\partial \phi}{\partial x}$; substituting this term into Equation (10.135) and dividing by ρ_0 gives

$$\frac{\partial^2 \phi}{\partial x^2} = -\frac{1}{\rho_0} \frac{\partial \bar{\rho}}{\partial t} \qquad (10.137)$$

Similarly, Equation (10.136) becomes

$$\rho_0 \frac{\partial^2 \phi}{\partial t \partial x} = \rho_0 \frac{\partial}{\partial x} \frac{\partial \phi}{\partial t} = -\frac{\partial \bar{p}}{\partial x} \qquad (10.138)$$

Equation (10.138) can be written as

$$\frac{\partial}{\partial x} \left(\frac{\partial \phi}{\partial t} + \frac{\bar{p}}{\rho_0} \right) = 0 \qquad (10.139)$$

By Equation (10.139), the term inside the brackets can only be a function of t, that is,

$$\frac{\partial \phi}{\partial t} + \frac{\bar{p}}{\rho_0} = f(t)$$

or

$$\frac{\partial \phi}{\partial t} - f(t) = -\frac{\bar{p}}{\rho_0} \tag{10.140}$$

$f(t)$ can be absorbed into ϕ by letting $\bar{\phi} = \phi - g(t)$. Then

$$\frac{\partial \bar{\phi}}{\partial x} = \frac{\partial \phi}{\partial x} = \bar{u} \tag{10.141}$$

and

$$\frac{\partial \bar{\phi}}{\partial t} = \frac{\partial \phi}{\partial t} - \frac{dg}{dt}$$

Let $\frac{dg}{dt} = f(t)$, then

$$\frac{\partial \bar{\phi}}{\partial t} = \frac{\partial \phi}{\partial t} - f(t) = -\frac{\bar{p}}{\rho_0} \tag{10.142}$$

Now define $c^2 = \frac{dp}{d\rho}$. Returning to Equation (10.131)

$$p = \alpha \rho^k \tag{10.131}$$

Then

$$c^2 = \frac{dp}{d\rho} = k\alpha\rho^{k-1} = \frac{k}{\rho}\alpha\rho^k = k\frac{p}{\rho} \tag{10.143}$$

Take

$$c_0^2 = k\frac{p_0}{\rho_0} \tag{10.144}$$

Applying perturbation concepts to Equation (10.131) gives

$$(p_0 + \bar{p}) = \alpha(\rho_0 + \bar{\rho})^k = \alpha\rho_0^k\left(1 + \frac{\bar{\rho}}{\rho_0}\right)^k \tag{10.145}$$

But

$$\left(1+x\right)^{k} = 1 + kx + \frac{k(k-1)}{2!}x^2 + \cdots \quad \text{for } x^2 < 1$$

Neglecting powers of $(\bar{\rho}/\rho_0)$ of two and higher gives

$$p_0 + \bar{p} = \alpha\rho_0^k + \alpha\rho_0^k \, k \, \frac{\bar{\rho}}{\rho_0}$$

or

$$\bar{p} = k \, p_0 \, \frac{\bar{\rho}}{\rho_0} = c_0^2 \, \bar{\rho} \tag{10.146}$$

Taking the second derivative with respect to x of both sides of Equation (10.146) gives

$$\frac{\partial^2 \bar{p}}{\partial x^2} = c_0^2 \frac{\partial^2 \bar{\rho}}{\partial x^2} \tag{10.147}$$

Returning to Equation (10.141),

$$\frac{\partial \bar{\phi}}{\partial t} = -\frac{\bar{p}}{\rho_0} \tag{10.141}$$

Now take the second derivative with respect to x of both sides of Equation (10.142) giving

$$\frac{\partial^2}{\partial x^2}\left(\frac{\partial \bar{\phi}}{\partial t}\right) = \frac{\partial}{\partial t}\left(\frac{\partial^2 \bar{\phi}}{\partial x^2}\right) = -\frac{1}{\rho_0}\left(\frac{\partial^2 \bar{p}}{\partial x^2}\right) \tag{10.148}$$

By Equations (10.147) and (10.137),

$$\frac{\partial}{\partial t}\left(\frac{\partial^2 \bar{\phi}}{\partial x^2}\right) = \frac{\partial}{\partial t}\left(-\frac{1}{\rho_0}\frac{\partial \bar{\rho}}{\partial t}\right) = -\frac{c_0^2}{\rho_0}\left(\frac{\partial^2 \bar{\rho}}{\partial x^2}\right) \tag{10.149}$$

Thus,

$$\frac{1}{c_0^2}\frac{\partial^2 \bar{\rho}}{\partial t^2} = \frac{\partial^2 \bar{\rho}}{\partial x^2} \tag{10.150}$$

and

$$\frac{1}{c_0^2}\frac{\partial^2 \bar{p}}{\partial t^2} = \frac{\partial^2 \bar{p}}{\partial x^2}$$ (10.151)

Equations (10.150) and (10.151) are wave equations.

D'Alembert's Solution

Let $\varsigma = x - c_0 t$ and $\eta = x + c_0 t$. We now wish to consider $\bar{p} = \bar{p}(\varsigma, \eta)$ and transform Equation (10.150) in terms of ς and η using the chain rule. Then

$$\frac{\partial \varsigma}{\partial x} = 1, \quad \frac{\partial \eta}{\partial x} = 1, \quad \frac{\partial \varsigma}{\partial t} = -c_0 \quad \text{and} \quad \frac{\partial \eta}{\partial t} = c_0$$ (10.152)

$$\frac{\partial \bar{p}}{\partial x} = \frac{\partial \bar{p}}{\partial \varsigma}\frac{\partial \varsigma}{\partial x} + \frac{\partial \bar{p}}{\partial \eta}\frac{\partial \eta}{\partial x} = \frac{\partial \bar{p}}{\partial \varsigma} + \frac{\partial \bar{p}}{\partial \eta}$$

$$\frac{\partial^2 \bar{p}}{\partial x^2} = \frac{\partial}{\partial x}\left(\frac{\partial \bar{p}}{\partial \varsigma} + \frac{\partial \bar{p}}{\partial \eta}\right) = \frac{\partial}{\partial \varsigma}\left(\frac{\partial \bar{p}}{\partial \varsigma} + \frac{\partial \bar{p}}{\partial \eta}\right)\frac{\partial \varsigma}{\partial x} + \frac{\partial}{\partial \eta}\left(\frac{\partial \bar{p}}{\partial \varsigma} + \frac{\partial \bar{p}}{\partial \eta}\right)\frac{\partial \eta}{\partial x}$$

or

$$\frac{\partial^2 \bar{p}}{\partial x^2} = \frac{\partial^2 \bar{p}}{\partial \varsigma^2} + 2\frac{\partial^2 \bar{p}}{\partial \varsigma \partial \eta} + \frac{\partial^2 \bar{p}}{\partial \eta^2}$$ (10.153)

Similarly,

$$\frac{\partial^2 \bar{p}}{\partial t^2} = c_0^2\left(\frac{\partial^2 \bar{p}}{\partial \varsigma^2} - 2\frac{\partial^2 \bar{p}}{\partial \varsigma \partial \eta} + \frac{\partial^2 \bar{p}}{\partial \eta^2}\right)$$ (10.154)

Substituting Equations (10.153) and (10.154) into Equation (10.151) gives

$$\left(\frac{\partial^2 \bar{p}}{\partial \varsigma^2} - 2\frac{\partial^2 \bar{p}}{\partial \varsigma \partial \eta} + \frac{\partial^2 \bar{p}}{\partial \eta^2}\right) = \left(\frac{\partial^2 \bar{p}}{\partial \varsigma^2} + 2\frac{\partial^2 \bar{p}}{\partial \varsigma \partial \eta} + \frac{\partial^2 \bar{p}}{\partial \eta^2}\right)$$

or

$$4\frac{\partial^2 \bar{p}}{\partial \varsigma \partial \eta} = 0$$ (10.155)

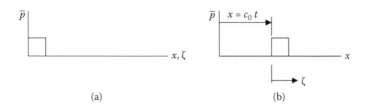

Figure 10.10   Example of a small step pressure disturbance, (a) $t = 0$, (b) $t > 0$.

Equation (10.154) can be rewritten as

$$\frac{\partial}{\partial \varsigma}\left(\frac{\partial \bar{p}}{\partial \eta}\right) = 0$$

which implies that

$$\frac{\partial \bar{p}}{\partial \eta} = F(\eta)$$

and

$$\bar{p} = \int F(\eta)\,d\eta + f_2(\varsigma) = f_1(\eta) + f_2(\varsigma) = f_1(x + c_0 t) + f_2(x - c_0 t)$$

If at $t = 0$, $f_1 = 0$ and f_2 is a step as shown in Figure 10.10a. At $t > 0$, the disturbance has moved to $x = c_0 t$ (see Figure 10.10b). Therefore, c_0 is the speed of sound. The disturbance, f_2, is a forward-moving wave and f_1 is a backward-moving wave. From Equation (10.144) and the ideal gas law the following expression for the speed of sound can be obtained:

$$c_0^2 = k\frac{p_0}{\rho_0} \tag{10.144}$$

and

$$p_0 = \rho_0\, R\, T_0$$

Therefore,

$$c_0 = \sqrt{k\, R\, T_0} \tag{10.156}$$

10.5 Review of Finite Difference Formulas

Given $y = y(x)$ and for a uniform subdivision on the x axis

$$y_i' = \frac{y_{i+1} - y_i}{\Delta x} \qquad \text{forward difference formula for } y'(x_i)$$

$$y_i' = \frac{y_i - y_{i-1}}{\Delta x} \qquad \text{backward difference formula for } y'(x_i)$$

$$y_i'' = \frac{y_{i+1} + y_{i-1} - 2y_i}{\Delta x^2} \qquad \text{central difference formula for } y''(x_i)$$

$$y_i' = \frac{3y_i - 4y_{i-1} + y_{i-2}}{2\Delta x} \qquad \text{backward difference formula for } y'(x_i) \text{ of order } (\Delta x^2)$$

$$y_i' = \frac{-y_{i+2} + 4y_{i+1} - 3y_i}{2\Delta x} \qquad \text{forward difference formula for } y'(x_i) \text{ of order } (\Delta x^2)$$

10.6 Example of Applying Finite Difference Methods to Partial Differential Equations

Consider a thin plate at initial temperature, T_0, that is suddenly immersed in a huge bath at temperature T_∞ (see Figure 10.4).

The governing PDE for the temperature field, $T(x,t)$, and initial and boundary conditions are

$$\frac{1}{\alpha}\frac{\partial T}{\partial t} = \frac{\partial^2 T}{\partial x^2} \tag{10.157}$$

$$T(x,0) = T_0 \tag{10.158}$$

$$\frac{\partial T}{\partial x}(0,t) = 0 \tag{10.159}$$

$$\frac{\partial T}{\partial x}(L,t) + \frac{h}{k}[T(L,t) - T_\infty] = 0 \tag{10.160}$$

where

α = thermal diffusivity of the plate and $2L$ is the plate thickness.

h = convective heat transfer coefficient.

k = thermal conductivity of the plate material.

To solve this heat transfer problem numerically, subdivide the x and t domains into I and J subdivisions, respectively, giving

$$x_1, x_2, x_3, \ldots, x_{I+1} \text{ and } t_1, t_2, t_3, \ldots, t_{J+1}$$

There are two finite difference numerical methods for solving this problem, the explicit method and the implicit method. The explicit method has a stability problem if the following condition is not satisfied:

$$\beta = \frac{\alpha \Delta t}{\Delta x^2} < 0.5$$

The implicit method does not have a stability problem.

10.6.1 The Explicit Method

Write the governing partial differential equation at (x_i, t_j) using the forward finite difference formula for $\dfrac{\partial T}{\partial t}(x_i, t_j)$ and the central difference formula for $\dfrac{\partial^2 T}{\partial x^2}(x_i, t_j)$ giving

$$\frac{\partial T}{\partial t}(x_i, t_j) \approx \frac{1}{\Delta t}[T(x_i, t_{j+1}) - T(x_i, t_j)] \tag{10.161}$$

and

$$\frac{\partial^2 T}{\partial x^2}(x_i, t_j) = \frac{1}{\Delta x^2}[T(x_{i+1}, t_j) + T(x_{i-1}, t_j) - 2T(x_i, t_j)] \tag{10.162}$$

To simplify the notation, use

$$T(x_i, t_j) = T_i^j$$

then

$$\frac{\partial T}{\partial t}(x_i, t_j) \approx \frac{1}{\Delta t}\left[T_i^{j+1} - T_i^j\right] \tag{10.163}$$

$$\frac{\partial^2 T}{\partial x^2}(x_i, t_j) = \frac{1}{\Delta x^2}\left[T_{i+1}^j + T_{i-1}^j - 2T_i^j\right] \tag{10.164}$$

The governing PDE becomes

$$\frac{1}{\alpha \Delta t}\left[T_i^{j+1} - T_i^{j}\right] = \frac{1}{\Delta x^2}\left[T_{i+1}^{j} + T_{i-1}^{j} - 2T_i^{j}\right] \tag{10.165}$$

Solving for T_i^{j+1} gives

$$T_i^{j+1} = T_i^{j} + \frac{\alpha \Delta t}{\Delta x^2}\left[T_{i+1}^{j} + T_{i-1}^{j} - 2T_i^{j}\right] \tag{10.166}$$

Equation (10.166) is valid for $i = 2, 3, \ldots, I$.
  Initial condition reduces to

$$T_i^1 = T_o \text{ for } i = 1, 2, 3, \ldots I + 1$$

Boundary condition $\dfrac{\partial T}{\partial x}(0,t) = 0$ is also valid at $t + \Delta t$.

Using the forward differences formula of order Δx^2 gives

$$\frac{-T_3^{j+1} + 4T_2^{j+1} - 3T_1^{j+1}}{2\Delta x} = 0$$

Solving for T_1^{j+1} gives

$$T_1^{j+1} = \frac{1}{3}\left[4T_2^{j+1} - T_3^{j+1}\right] \tag{10.167}$$

Boundary condition $\dfrac{\partial T}{\partial x}(L,t) + \dfrac{h}{k}[T(L,t) - T_\infty] = 0$ is also valid at t + Δt.

Using the backward difference formula for $\dfrac{\partial T}{\partial x}(L,t)$ of order Δx^2 gives

$$\frac{3T_{I+1}^{j+1} - 4T_I^{j+1} + T_{I-1}^{j+1}}{2\Delta x} + \frac{h}{k}\left[T_{I+1}^{j+1} - T_\infty\right] = 0$$

Solving for T_{I+1}^{j+1} gives

$$T_{I+1}^{j+1} = \frac{k}{3k + 2h\Delta x}\left[4T_I^{j+1} - T_{I-1}^{j+1} + \frac{2h\Delta x}{k}T_\infty\right] \tag{10.168}$$

The solution is obtained by marching in time. A sketch of the order of calculations is shown in Figure 10.11.
  Finally, the amount of heat transfer per unit surface area, Q, that occurs in time t_f is given by

$$Q = -2k\int_0^{t_f}\frac{\partial T}{\partial x}(L,t)dt = -\frac{k}{\Delta x}\int_0^{t_f}[3T_{I+1}(t) - 4T_1(t) + T_{I-1}(t)]dt \tag{10.169}$$

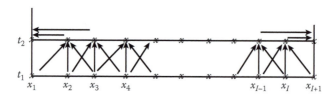

Figure 10.11   The order of calculations and marching in time.

10.6.2 The Implicit Method

Write the governing PDE using the forward finite difference formula for $\dfrac{\partial T}{\partial t}$ and the central difference formula for $\dfrac{\partial^2 T}{\partial x^2}$, but take the time position at $j+1$ giving

$$\frac{1}{\alpha}\frac{T_i^{j+1}-T_i^{j}}{\Delta t}=\frac{T_{i+1}^{j+1}+T_{i-1}^{j+1}-2T_i^{j+1}}{\Delta x^2}$$

Solving for T_i^{j+1} gives

$$T_i^{j+1}=T_i^{j}+\frac{\alpha\,\Delta t}{\Delta x^2+2\alpha\,\Delta t}\left(T_{i+1}^{j+1}+T_{i-1}^{j+1}\right)$$

$$(10.170)$$

Equation (10.170) is valid for $i=2, 3,\dots, I$. There are three unknowns in Equation (10.170): T_i^{j+1}, T_{i+1}^{j+1}, and T_{i-1}^{j+1}. The term T_i^{j} is assumed to be known. So far the set fits into a tri-diagonal system. The boundary conditions need to be checked to see if they also fit into a tri-diagonal system. The initial condition is

$$T_i^1=T_o,\text{ valid for }i=1, 2, 3,\dots, I+1 \qquad (10.171)$$

Boundary condition $\dfrac{\partial T}{\partial x}(0,t)=0$ is also valid at $t+\Delta t$.

Using the forward differences formula of order Δx gives

$$\frac{T_2^{j+1}-T_1^{j+1}}{\Delta x}=0$$

or

$$T_1^{j+1}=T_2^{j+1} \qquad (10.172)$$

The boundary condition $\dfrac{\partial T}{\partial x}(L,t)+\dfrac{h}{k}[T(L,t)-T_\infty]=0$ is also valid at t + Δt.

Using the backward difference formula for $\dfrac{\partial T}{\partial x}(L,t)$ of order Δx gives

$$\frac{T_{I+1}^{j+1}-T_I^{j+1}}{\Delta x}+\frac{h}{k}\left(T_{I+1}^{j+1}-T_\infty\right)=0$$

Solving for T_{I+1}^{j+1} gives

$$T_{I+1}^{j+1} = \frac{k}{k+h\Delta x} T_I^{j+1} + \frac{h\Delta x}{k+h\Delta x} T_\infty \tag{10.173}$$

Equations (10.170), (10.172), and (10.173) fall into a tri-diagonal system, allowing for a solution of all temperatures at t^{j+1} by the method described in Section 6.7. A complete solution can be obtained by marching in time.

Projects

Project 10.1

Solve the vibrating string problem discussed in Section 10.2 for the initial condition shown in Figure P10.1a.

Take $\rho = 8240 \text{ kg/m}^3$, $T_o = 90 \text{ N}$, d = diameter of string = 0.16 cm, L = 1 m, h = 6 cm. Plot Y vs. x at the following times: $t = 0.0001$ s, $t = 1.0$ s, $t = 10$ s, and $t = 100$ s. *Note*: $\rho(kg/m) = \rho(kg/m^3)\, A$, where A is the cross-sectional area of the string.

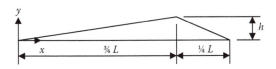

Figure P10.1a   An initial string displacement.

Project 10.2

Rearrange Equation (10.53) to read

$$TRATIO = \frac{T\left(\dfrac{x}{L}, t\right) - T_\infty}{T_0 - T_\infty} = 2\sum_{n=1}^{\infty} \frac{\sin(\delta_n)\cos\left(\delta_n \dfrac{x}{L}\right) e^{-a\delta_n^2 t/L^2}}{\cos(\delta_n)\sin(\delta_n) + \delta_n} \tag{P10.2a}$$

where $\delta_n = \lambda_n L$.

Write a computer program using 50 δ_n values and solve for *TRATIO* for the following parameters: $h = 890.0 \text{ w/m}^2\text{-°C}$, $k = 386.0 \text{ w/m-°C}$, $L = 0.5$ m, $a = 11.234$ e-05 m²/s, $T_0 = 300°C$, $T_\infty = 30°C$, $x/L = 0.0, 0.2, 0.4, 0.6, 0.8, 1.0$, and $t = 0, 20, 40, \ldots, 400$ seconds. Print out results in table form as shown in Table P10.2. Also, create a plot of *TRATIO* vs. t for $x/L = 1.0, 0.8, 0.6, 0.4, 0.2, 0.0$.

Table P10.2   Temperature Ratio (TRATIO) vs. Time (t)

Time (seconds)	X/L					
	0.0	0.2	0.4	0.6	0.8	1.0
0	1.0	1.0	1.0	1.0	1.0	1.0
20	—	—	—	—	—	—
40	—	—	—	—	—	—
400	—	—	—	—	—	—

Project 10.3

Consider the circular cylinder problem described in Section 10.4. Write a computer program that will

(a) Plot J_1 vs. λR for $0 \leq \lambda R \leq 1000$.
(b) Determine all the roots of J_1 in the λR range from 0 to 1000.
(c) Determine and print out all the $\lambda_n R$ values that satisfy Equation (10.91) in the λR range from 0 to 1000.
(d) Create a table of TR vs. time t for $r/R = [0.0, 0.2, 0.4, 0.6, 0.8, 1.0]$ and $t = [0, 20, 40, \ldots, 220, 300]$ s. The table should be similar to the table shown in P10.2.
(e) Plot TR vs. t for $0 \leq t \leq 300$ s and $r/R = [0.0, 0.2, 0.4, 0.6, 0.8, 1.0]$ all on the same graph.

Use the following values:

$$h = 890.0 \frac{\text{w}}{\text{m}^2 - \text{C}}, \quad k = 35.0 \frac{\text{w}}{\text{m} - \text{C}}, \quad R = 0.12\,\text{m}, \quad a = 0.872 \times 10^{-5} \frac{\text{m}^2}{\text{s}}$$

Hint: First determine the zeroes of J_1 by the *fzero* function; then knowing that the roots of Equation (10.91) lie between the zeros of J_1, determine $\lambda_n R$ for $n = 1\text{-}30$ by the *fzero* function. Having values for $\lambda_n R$ you can the determine TR by Equation (10.98).

Project 10.4

Write a computer program to solve numerically, by the explicit method, the problem described in Section 10.6. Use the parameters described in Project 10.2, that is, h = 890.0 w/m²-°C, k = 386.0 w/m-°C, L = 0.5 m, a = 11.234e-05 m²/s, T_0 = 300°C, T_∞ = 30°C. Take $dx = 0.005$ m and $dt = 0.1$ second. Carry the calculations

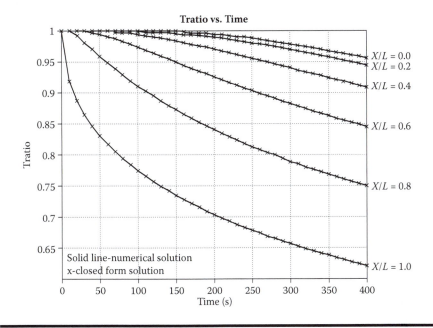

Figure P10.4   The plot of TRATIO vs. time. Numerical solution is by the explicit method.

to 400 seconds. To compare the results obtained by this numerical method with the results obtained by the closed-form solution (Project 10.2) write your answer in the form

$$TRATIO = \frac{T\left(\dfrac{x}{L}, t\right) - T_\infty}{T_0 - T_\infty} \tag{P10.4a}$$

for x/L = 1.0, 0.8, 0.6, 0.4, 0.2, 0.0, and for t = 0, 20, 40, ..., 400 seconds. Print out a table as shown in Table P10.2. Also create a plot of *TRATIO* vs. t for x/L = 1.0, 0.8, 0.6, 0.4, 0.2, 0.0. If you also did Project 10.2, superimpose the solution obtained in Project 10.2 on the plot created in Project 10.4. The resulting plot should be similar to the plot shown in Figure P10.4.

Project 10.5

Repeat Project 10.4, but this time use the implicit method. If you also did Project 10.2, superimpose the solution obtained in Project 10.2 on the plot created in Project 10.5. The resulting plot should be similar to the plot shown in Figure P10.5.

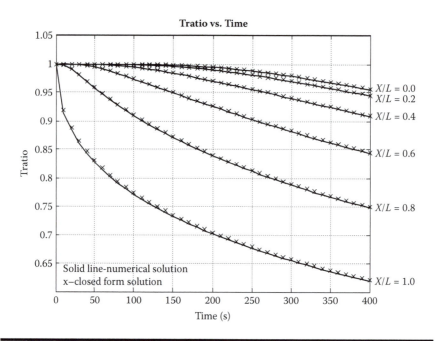

Figure P10.5 The plot of TRATIO vs. time. Numerical solution is by the implicit method.

Project 10.6

Using separation of variables, show that the steady-state temperature distribution in the slab shown in Figure P10.6 is given by

$$T(x, y) = T_\infty + 2(T_w - T_\infty) \sum_1^\infty \frac{\sin(\lambda_n w)}{\sinh(\lambda_n L)} \times \frac{\sinh(\lambda_n x)\cos(\lambda_n y)}{\lambda_n w + \sin(\lambda_n w)\cos(\lambda_n w)} \quad \text{(P10.6a)}$$

where λ_n is determined from Equation (P10.6b).

$$\tan(\lambda w) - \frac{h w}{k} \times \frac{1}{\lambda w} = 0 \quad \text{(P10.6b)}$$

Note that

$$\frac{\sinh(\lambda_n x)}{\sinh(\lambda_n L)} = \exp(-\lambda_n(L-x)) \times \frac{1 - \exp(-2\lambda_n x)}{1 - \exp(-2\lambda_n L)} \quad \text{(P10.6c)}$$

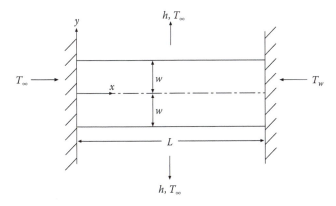

Figure P10.6 Slab geometry.

The use of Equation (P10.6c) will avoid numerical problems for large x and large λ. Print out a table for the first 50 eigenvalues (λ_n) applicable to this problem.

(a) Print out a table of $T(x, y)$ at every second x position and every second y position.

(b) Create a plot of $T(x, 0)$ and $T(x, w)$ vs. x, both on the same graph. Use the following values:

$L = 2.0$ m, $w = 0.2$ m, $T_w = 300°C$, $T_\infty = 50°C$, $k = 59$ W/m-°C, $h = 890$ W/m²-°C

Project 10.7

A safe, as a result of a fire in a nearby room, is suddenly subjected to surrounding air temperature of 800°C. The ignition temperature of paper inside the safe is 160°C. Both the inner and outer shells of the safe are constructed of 1% carbon steel. An appropriate insulating material is placed between the steel shells. The interior volume of the safe is 1 m³ (1 m x 1 m x 1 m). The insulating material is nonflammable. A finite difference numerical analysis may be used to determine the temperature distribution of the safe material and the interior temperature of the safe. The following assumptions shall be made:

1. The air inside the safe is well mixed and uniform.
2. The paper temperature is the same as the air temperature inside the safe.
3. The interior contents of the safe consist of 30% paper and 70% air by volume.
4. Radiation from the fire and the variation of the thermal properties of all materials are neglected. A description of the numerical method follows.

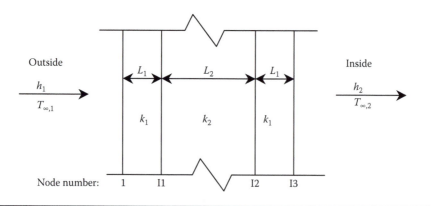

Figure P10.7a   Wall geometry.

Numerical Method

Consider the I-D unsteady heat flow through the wall shown in Figure P10.7a. Subdivide the x domains as follows: (a) subdivide the steel plate regions into $(I1-1)$ subdivisions and (b) subdivide the insulation region into $(I2-I1)$ subdivisions.

In each region the governing PDE for the temperature field is

$$\frac{1}{\alpha}\frac{\partial T}{\partial t} = \frac{\partial^2 T}{\partial x^2} \tag{P10.7a}$$

Since α and Δx differ in the two different material-type regions, the governing finite difference equation for each region is

$$T_i^{n+1} = T_i^n + \frac{\alpha_1 \Delta t}{\Delta x_1^2}\left(T_{i+1}^n + T_{i-1}^n - 2T_i^n\right) \tag{P10.7b}$$

for i = 2, 3, ... $(I1 - 1)$ and i = $(I2 + 1)$, $(I2 + 2)$,..., $(I3 - 1)$, where $T(x_i, t_n) = T_i^n$ and

$$T_i^{n+1} = T_i^n + \frac{\alpha_2 \Delta t}{\Delta x_2^2}\left(T_{i+1}^n + T_{i-1}^n - 2T_i^n\right) \tag{P10.7c}$$

for i = $(I1 + 1)$, $(I1 + 2)$, ..., $(I2 - 1)$.

To complete the problem formulation, one needs to add the initial and boundary conditions:

Initial Condition

$$T(x,0) = T_o$$

$$T(x_i, t_1) = T_i^1 = T_o, \quad \text{for } i = 1, 2, ..., I3 \tag{P10.7d}$$

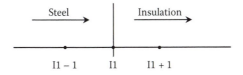

Figure P10.7b   Left steel–insulation interface.

Boundary Conditions

(a) At the left steel–insulation interface (see Figure P10.7b)
The rate that heat flows out of the steel plate per unit surface area = the rate that heat flows into the insulation per unit surface area. Expressed mathematically,

$$\vec{q}\,(I1^-,t)\cdot\vec{i} = -\vec{q}\,(I1^+,t)\cdot(-\vec{i})$$

or

$$-k_1\frac{\partial T}{\partial x}(I1^-,t) = -k_2\frac{\partial T}{\partial x}(I1^+,t)$$

The boundary condition is valid at t_n and t_{n+1}. The simplest finite difference form of the above equation is

$$-k_1\frac{T_{I1}^{n+1}-T_{I1-1}^{n+1}}{\Delta x_1} = -k_2\frac{T_{I1+1}^{n+1}-T_{I1}^{n+1}}{\Delta x_2}$$

Solving for T_{I1}^{n+1} gives

$$T_{I1}^{n+1} = \frac{\dfrac{k_1}{\Delta x_1}T_{I1-1}^{n+1} + \dfrac{k_2}{\Delta x_2}T_{I1+1}^{n+1}}{\dfrac{k_1}{\Delta x_1} + \dfrac{k_2}{\Delta x_2}} \qquad\qquad \text{(P10.7e)}$$

(b) At the right insulation–steel interface (see Figure P10.7c)

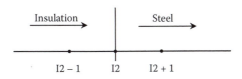

Figure P10.7c   Right insulation–steel interface.

The rate that heat flows out of insulation per unit surface area = the rate that heat flows into the steel plate per unit surface area.

Similarly to the equations developed at node I1, the equation at node I2 is

$$-k_2 \frac{T_{I2}^{n+1} - T_{I2-1}^{n+1}}{\Delta x_2} = -k_1 \frac{T_{I2+1}^{n+1} - T_{I2}^{n+1}}{\Delta x_1}$$

Solving for T_{I2}^{n+1} gives

$$T_{I2}^{n+1} = \frac{\dfrac{k_2}{\Delta x_2} T_{I2-1}^{n+1} + \dfrac{k_1}{\Delta x_1} T_{I2+1}^{n+1}}{\dfrac{k_2}{\Delta x_2} + \dfrac{k_1}{\Delta x_1}} \tag{P10.7f}$$

(c) At the left air–steel interface (see Figure P10.7d)

The rate that heat is carried to the wall by convection per unit surface area = the rate that heat enters the steel plate by conduction per unit surface area; that is,

$$h_1[T_{\infty,1} - T(x_1, t)] = -\vec{q}(x_1, t) \cdot (-\hat{i}) = -k_1 \frac{\partial T}{\partial x}(x_1, t)$$

The boundary condition is valid at t_n and t_{n+1}. The simplest finite difference form of the above equation is

$$h_1[T_{\infty,1} - T_1^{n+1}] = -k_1 \frac{T_2^{n+1} - T_1^{n+1}}{\Delta x_1}$$

Solving for T_1^{n+1} gives

$$T_1^{n+1} = \frac{h_1 T_{\infty 1} + \dfrac{k_1}{\Delta x_1} T_2^{n+1}}{h_1 + \dfrac{k_1}{\Delta x_1}} \tag{P10.7g}$$

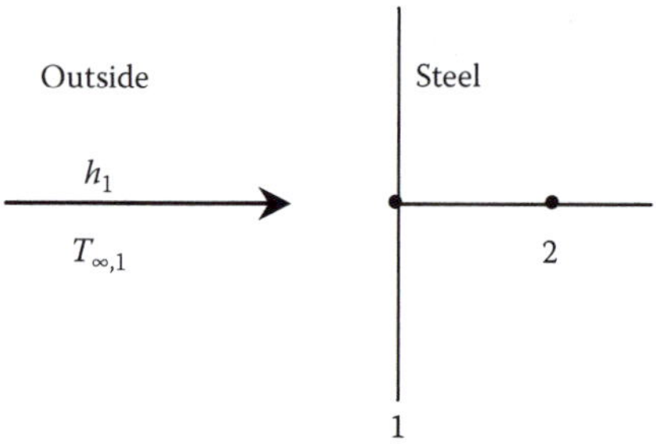

Figure P10.7d   Outside air–steel interface.

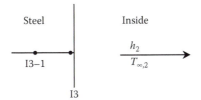

Figure P10.7e   Steel-inside air interface.

(d) At the right steel–air interface (see Figure P10.7e)
The rate that heat leaves the steel plate by conduction per unit surface = the rate that heat enters the safe by convection.

$$\vec{q}(x_{13},t) \cdot \hat{i} = -k_1 \frac{\partial T}{\partial x}(x_{13},t)\hat{i} \cdot \hat{i} = h_2[T(x_{13},t) - T_{\infty,2}(t)]$$

The boundary condition is valid at t_n and t_{n+1}. The simplest difference form of the above equation is

$$-k_1 \frac{T_{13}^{n+1} - T_{13-1}^{n+1}}{\Delta x_3} = h_2\left[T_{13}^{n+1} - T_{\infty,2}^{n+1}\right]$$

Solving for T_{13}^{n+1} gives

$$T_{13}^{n+1} = \frac{\dfrac{k_1}{\Delta x_3}T_{13-1}^{n+1} + h_2 T_{\infty2}^{n+1}}{h_2 + \dfrac{k_1}{\Delta x_3}} \tag{P10.7h}$$

Inside the Safe

Heat transfer into the safe is slow. It has been assumed that all material inside the safe is at the same uniform temperature. The safe interior consists of paper and air. The finite difference formula for the temperature inside the safe is

$$T_a^{n+1} = T_a^n + \frac{A_s h_2 \Delta t\left(T_{13}^n - T_a^n\right)}{\rho_a V_a C_{v,a} + \rho_p V_p C_{v,p}} \tag{P10.7i}$$

where
  ρ = density   subscripts
  V = volume   a ~ air
  C = specific heat   p ~ paper

This numerical method has a stability criteria; that is, in each region

$$r = \frac{\alpha \Delta t}{\Delta x^2} < \frac{1}{2}$$

Method of Solution

Since T_i^1, $i = 1, 2, \ldots, 13$ is known T_i^2 can be obtained for all i by marching as indicated in the following procedure.

Procedure

1. Solve for T_i^2 for $i = 2,\ldots, (I1 - 1)$ by Equation (P10.7b); for (I1 + 1), (I1 + 2),..., (I2 − 1) by Equation (P10.7c); and for (I2 + 1), (I2 + 2),..., (I3 − 1) by Equation (P10.7b).
2. Solve for T_{I1}^2 by Equation (P10.7e) and for T_{I2}^2 by Equation (P10.7f).
3. Solve for T_1^2 by Equation (P10.7g) (need T_2^2 first).
4. Solve for T_a^2 by Equation (P10.7i).
5. Solve for T_{I3}^2 by Equation (P10.7h) (need T_{I3-1}^2 and T_a^2 first).
6. Use a counter and an if statement to determine when to print out temperature values.
7. Use a loop to reset $T_i^1 = T_i^2$ for all i.
8. Repeat the process until $t =$ the specified time for the study.

Develop a computer program in MATLAB® to solve the wall temperature distribution and the air/paper temperature inside the safe. Use the following constants for the problem:

$$k_1 = 61.0 \ \text{W/(m-°C)}, \ k_2 = 0.166 \ \text{W/(m-°C)}, \alpha_1 = 1.665 \ e - 5 \ \text{m}^2/\text{s}, \alpha_2 = 3.5 \ e - 7,$$

$$\rho_a = 0.9980 \ \text{kg/m}^3, \rho_p = 930.0 \ \text{kg/m}^3, C_{v,a} = 0.722 e + 3 \ \text{W/(kg-°C)},$$

$$C_{v,p} = 1.340 e + 3 \ \text{W/(kg-°C)},$$

$$h_1 = 200.0 \ \text{W/(m}^2\text{-°C)}, \ h_2 = 10.0 \ \text{W/(m}^2\text{-°C)}, T_{\infty,1} = 800.0 \ \text{°C}, \ A_s = 6.0 \ \text{m}^2,$$

$$T_i^1 = 25 \ \text{°C for } i = 1,2,\ldots,13$$

$$L_1 = 0.005 \ \text{m}, \ L_2 = 0.02 \ \text{m}, \ I1 = 6, \ I2 = 86, \ I3 = 91, \ dx_1 = 0.001 \ \text{m},$$

$$dx_2 = 0.00025 \ \text{m}, \ dt = 0.025 \ \text{seconds},$$

$$V_a = 0.60 \ \text{m}^3, V_p = 0.4 \ \text{m}^3, \text{ time of study} = 3600 \ \text{seconds}.$$

Print out property values, the problem constants, and a temperature table as shown in Table P10.7.

Note: Print out temperatures every 100 seconds.

Table P10.7   Temperature Distribution Table

Time (seconds)	Exterior Surface x_1	x_{I1}	x-Position (m) $\frac{1}{2}(x_{I1}+x_{I2})$	x_{I2}	Interior Surface x_{I3}	Air Temp (°C)
0.0	25.00	25.00	25.00	25.00	25.00	25.00
100.0	–	–	–	–	–	–
200.0	–	–	–	–	–	–
3600.0						–

Project 10.8

The temperature ratio, TR (x, y, t), of the 2-D bar described in Section 10.3, "Unsteady Heat Transfer in 2-D," is given by

$$TR(x,y,t) = \frac{\vartheta(x,y,t)}{T_0 - T_\infty}$$

Then

$$TR(x,y,t) = \sum_{n=1}^{\infty}\sum_{m=1}^{\infty} \frac{\left(\dfrac{\sin(\gamma_n L)}{\gamma_n} \times \dfrac{\sin(\beta_m w)}{\beta_m}\right)\exp(-\alpha\lambda_{nm}^2 t)\cos(\gamma_n x)\cos(\beta_m y)}{\left(\dfrac{L}{2}+\dfrac{\sin(\gamma_n L)\cos(\gamma_n L)}{2\gamma_n}\right)\times\left(\dfrac{w}{2}+\dfrac{\sin(\beta_m w)\cos(\beta_m w)}{2\beta_m}\right)}$$

Develop a computer program in MATLAB to evaluate TR at $\dfrac{x}{L} = 0.0, 0.5, 1.0$ and $\dfrac{y}{w} = 0.0, 0.5, 1.0$ for $0 \le t \le 400$ seconds. Create the following plots:

Plot on the same graph, TR vs. t for $\dfrac{x}{L} = 0.0, 0.5, 1.0$ and $\dfrac{y}{w} = 0.0$.

Plot on the same graph, TR vs. t for $\dfrac{x}{L} = 0.0, 0.5, 1.0$ and $\dfrac{y}{w} = 0.5$.

Plot on the same graph, TR vs. t for $\dfrac{x}{L} = 0.0, 0.5, 1.0$ and $\dfrac{y}{w} = 1.0$.

Plot on the same graph, TR vs. t for $\dfrac{y}{w} = 0.0, 0.5, 1.0$ and $\dfrac{x}{L} = 0.0$.

Plot on the same graph, TR vs. t for $\dfrac{y}{w} = 0.0, 0.5, 1.0$ and $\dfrac{x}{L} = 0.5$.

Plot on the same graph, TR vs. t for $\dfrac{y}{w} = 0.0, 0.5, 1.0$ and $\dfrac{x}{L} = 1.0$.

Use the following parameters:

$$L = w = 0.5 \text{ m}, \quad h = 890 \ \frac{\text{W}}{\text{m}^2\text{-C}}, \quad k = 386 \frac{\text{W}}{\text{m-C}}, \quad \alpha = 11.234\,e - 5 \ \frac{\text{m}^2}{\text{s}}$$

Chapter 11

Iteration Method

11.1 Iteration in Pipe Flow Analysis

Some engineering problems are best solved by an iteration procedure. For example, the determination of the flow rate, Q, and the head loss in a pipe system is commonly solved by an iteration procedure. Consider the piping system shown in Figure 11.1. The energy equation for the system is [1]

$$\left(\frac{p}{\gamma} + \frac{V^2}{2g} + z \right)_1 + (h_p)_{sys} - \sum h_L = \left(\frac{p}{\gamma} + \frac{V^2}{2g} + z \right)_2 \qquad (11.1)$$

where

p = pressure.
V = average fluid velocity.
γ = specific weight of the fluid.
z = elevation.
g = gravitational constant.
$\sum h_L$ = sum of head losses.
$(h_p)_{sys}$ = head developed by the pump in the system.

For this system

$$p_1 = p_2 = p_{atm}, \quad V_1 \approx 0, \quad V_2 \approx 0, \quad (z_2 - z_1) \text{ are specified.}$$

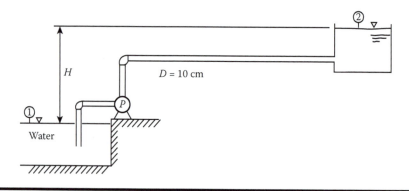

Figure 11.1   A piping system.

The sum of head losses consists of a head loss, h_f, in the pipe due to viscous or turbulent effects and minor head losses due to valves, elbows, and pipe entrance and exit losses. The head loss in the pipe is given by

$$h_f = \frac{V^2}{2g}\frac{L}{D}f \tag{11.2}$$

where
   L = pipe length (known).
   D = pipe diameter (known).
   f = friction factor.

For a smooth pipe, f can be determined by the equation [2]

$$f = (1.82\log_{10} \text{Re}_d - 1.64)^{-2} \tag{11.3}$$

where Re_d is the Reynolds number, which is given by

$$\text{Re} = \frac{VD}{\upsilon} = \frac{4Q}{\pi D \upsilon} \tag{11.4}$$

and υ is the kinematic viscosity. The expression for V that was used in Equation (11.4) is

$$V = \frac{4Q}{\pi D^2} \tag{11.5}$$

The minor head losses are expressed by the equation

$$h_{L,i} = K_i \frac{V^2}{2g} \tag{11.6}$$

The K values for the minor head losses are $K_{entrance} = 0.5$, $K_{exit} = 1.0$, $K_{elbow} = 1.5$.

Substituting these relationships into Equation (11.1) and rearranging gives

$$(h_p)_{sys} = z_2 - z_1 + \frac{V^2}{2g}\left(\frac{L}{D}f + 4.5\right) \tag{11.7}$$

The pump manufacturer provides a performance curve of $(h_p)_{pc}$ vs. Q for the pump. Suppose the $(h_p)_{pc}$ vs. Q is approximated by the following quadratic equation:

$$(h_p)_{pc} = 120 - 500\, Q^2 \tag{11.8}$$

But

$$(h_p)_{sys} = (h_p)_{pc} \tag{11.9}$$

or

$$120 - 500Q^2 = z_2 - z_1 + \frac{V^2}{2g}\left(\frac{L}{D}f + 4.5\right) \tag{11.10}$$

To express the above equation in terms of Q, substitute for $V = \dfrac{4Q}{\pi D^2}$. Then Equation (11.10) becomes

$$120 - 500Q^2 = z_2 - z_1 + \frac{8Q^2}{g\pi^2 D^4}\left(\frac{L}{D}f + 4.5\right) \tag{11.11}$$

Since f is a function of Re, which is a function of Q, Equation (11.11) is best solved by iteration. The iterative procedure for determining Q is as follows:

1. Assume a value for f, say, $f_1 = 0.03$.
2. Solve Equation (11.11) for Q.
3. Solve Equation (11.4) for Re.
4. Solve Equation (11.3) for f, say, f_2.
5. If $|f_2 - f_1| < \varepsilon$, say, $\varepsilon = 1.0 \times 10^{-5}$, then Q is the correct value, otherwise set $f_1 = f_2$ and repeat process until condition of item 5 is satisfied.

11.2 The Gauss–Seidel Method

The Gauss–Seidel iteration method may be used to solve Laplace's Equation. Consider the steady-state heat conduction problem of the slab shown in Figure P10.6. For the derivation of the heat conduction equation, see B.1 and B.2 in Appendix B.

The governing partial differential equation for the temperature distribution is

$$\frac{\partial^2 T}{\partial x^2} + \frac{\partial^2 T}{\partial x^2} = 0 \tag{11.12}$$

The boundary conditions are

$$T(0, y) = T_\infty \tag{11.13}$$

$$T(L, y) = T_w \tag{11.14}$$

$$\frac{\partial T}{\partial y}(x,0) = 0 \text{ (by symmetry)} \tag{11.15}$$

$$\frac{\partial T}{\partial y}(x,w) + \frac{h}{k}(T(x,w) - T_\infty) = 0 \tag{11.16}$$

The finite difference form of the partial differential equation is

$$\frac{1}{(\Delta x)^2}\{T(x + \Delta x, y) - 2T(x, y) + T(x - \Delta x, y)\}$$

$$+ \frac{1}{(\Delta y)^2}\{T(x, y + \Delta y) - 2T(x, y) + (T(x, y - \Delta y)\} = 0 \tag{11.17}$$

Now subdivide the x domain into N subdivisions and the y domain into M subdivisions, giving positions (x_n, y_m), where $n = 1,2,\ldots,N+1$ and $m = 1,2,\ldots,M+1$ and $\Delta x = \dfrac{L}{N}$ and $\Delta y = \dfrac{w}{M}$.

Let $T(x_n, y_m) = T_{n,m}$, then the finite difference form of the partial differential equation is

$$T_{n,m} = \frac{1}{2(1+\beta^2)}\left(T_{n,m+1} + T_{n,m-1} + \beta^2 T_{n+1,m} + \beta^2 T_{n-1,m}\right) \tag{11.18}$$

The above equation is valid at all interior points. Thus, it is valid for $n = 2, 3,\ldots, N$ and $m = 2, 3,\ldots, M$. There are $(N-1)(M-1)$ such equations.

The finite difference form for the boundary conditions are

$$T_{1,m} = T_\infty \qquad \text{for } m = 1, 2, 3,\ldots, M+1 \tag{11.19}$$

$$T_{N+1,m} = T_w \qquad \text{for } m = 1, 2, 3,\ldots, M+1 \tag{11.20}$$

Using the forward difference formula for $\dfrac{\partial T}{\partial y}(x,0)$ of order $(\Delta y)^2$, the boundary condition $\dfrac{\partial T}{\partial y}(x,0) = 0$ becomes

$$T_{n,1} = \frac{1}{3}(4T_{n,2} - T_{n,3}) \tag{11.21}$$

Using the backward difference formula for $\dfrac{\partial T}{\partial y}(x,w)$ of order $(\Delta y)^2$, the boundary condition $\dfrac{\partial T}{\partial y}(x,w) + \dfrac{h}{k}(T(x,w) - T_\infty) = 0$ becomes

$$T_{n,M+1} = \dfrac{1}{3 + \dfrac{2h\,\Delta y}{k}} \left\{ 4T_{n,M} - T_{n,M-1} + \dfrac{2h\,\Delta y\,T_\infty}{k} \right\} \qquad (11.22)$$

Equations (11.18) through (11.22) represent the finite difference equations describing the temperature distribution in the slab.

Method of Solution

1. Assume a set of values for $T_{n,m}$, say, $T_{n,m}^1$ for $n = 2, 3,\dots$, N and $m = 2, 3,\dots$, $M+1$.
2. Successively substitute into Equations (11.18) through (11.22), obtaining a new set of values for $T_{n,m}$, say, $T_{n,m}^2$, using the updated values in the equations when available.
3. Repeat this process until $|T_{n,m}^2 - T_{n,m}^1| < \varepsilon$ for all n,m.

Faster convergence may be obtained by overrelaxing the set of equations. This is done by adding and subtracting $T_{n,m}^1$ from Equation (11.18) and introducing a relaxation parameter, ω, giving

$$T_{n,m}^2 = T_{n,m}^1 + \dfrac{\omega}{2(1+\beta^2)} \left(T_{n,m+1}^1 + T_{n,m-1}^2 + \beta^2 T_{n+1,m}^1 + \beta^2 T_{n-1,m}^2 - \dfrac{2(1+\beta^2)}{\omega} T_{n,m}^1 \right) \qquad (11.23)$$

where $1 < \omega < 2$. A similar procedure is carried out for Equations (11.21) and (11.22), giving

$$T_{n,1}^2 = T_{n,1}^1 + \dfrac{\omega}{3}\left(4T_{n,2}^1 - T_{n,3}^1 - \dfrac{3}{\omega} T_{n,1}^1 \right) \qquad (11.24)$$

and

$$T_{n,M+1}^2 = T_{n,M+1}^1 + \dfrac{\omega k}{3k + 2h\,\Delta y}\left\{ 4T_{n,M}^2 - T_{n,M-1}^2 + \dfrac{2h\,\Delta y\,T_\infty}{k} - \dfrac{3k+2h\,\Delta y}{\omega k} T_{n,M+1}^1 \right\} \qquad (11.25)$$

The method of solution described earlier is still valid, except Equations (11.23), (11.24), and (11.25) are substituted for Equations (11.18), (11.21), and (11.22), respectively. Sometimes in order to get convergence, one might have to underrelax; that is, $0 < \omega < 1$.

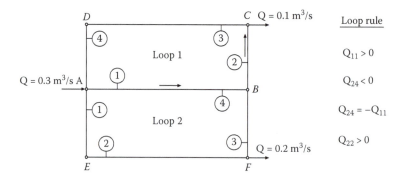

Figure 11.2 Loop rule for pipe network.

11.3 The Hardy Cross Method

The Hardy Cross method, which is an iterative method, provides the means for determining the flow rates and head losses throughout a pipe network, if the pipe diameters, lengths, and pipe roughnesses are known. The description of the method is taken from References [1,3].

The following two definitions are used in describing the method:

1. A *loop* is a series of pipes forming a closed path (see Figure 11.2). A sign convention is used in describing the loop rules. The flow rate, Q, and the head loss, h_f, are considered positive if the flow is in the counterclockwise direction around the loop. It should be realized that two loops with a common pipe may have a positive Q in one loop and a negative Q in the other loop.
2. A *node* is a point where two or more lines are joined. A sign convention is also used for node rules (see Figure 11.3). A flow is considered as positive if the flow direction is toward the node.

It should be realized that Q may be positive when the loop rule is applied and negative when the node rule is applied.

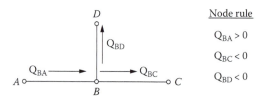

Figure 11.3 Node rule for pipe network.

The Hardy Cross method is based on two concepts:

1. The law of mass conservation
2. The fact that the total head at a node is single valued

Concept (1) leads to the node rule, which is applied at each node in the network. The node rule is

$$\sum_{\beta} Q_{\alpha\beta} = 0 \tag{11.26}$$

where α indicates the node under consideration and β indicates the connecting node. The sign convention gives the direction of flow. Concept (2) leads to the loop rule, which may be stated as follows: For loop i,

$$\sum h_{ij} = 0 \tag{11.27}$$

where h_{ij} is the head loss in the jth line in the ith loop. For the loop rule, Q_{ij} is the flow rate in the jth line in the ith loop. In Equation (11.27) the subscript f, which is usually written with h to indicate a head loss due to viscous or turbulent effects, has been omitted to reduce the number of subscripts. Minor head losses are usually neglected in network analysis. Elevation changes along a loop cancel and therefore need not be included. Finally, it should be noted that these rules are analogous to Kirchoff's rules for electrical circuits involving resistances. In the analogy, Q corresponds to electrical current and pressure drop corresponds to voltage drop. A description of the method follows:

1. Subdivide the network into a number of loops, making sure that all lines are included in at least one loop.
2. Determine the zeroth estimate for the flow rates, $Q_{\alpha\beta}^{(0)}$, for each line according to the following procedure. Let s equal the total number of nodes in the network and r the total number of lines. Invariably, r will be greater than s. Writing the law of mass conservation at each node gives s equations in r unknowns. Therefore one needs to assume $(r - s + 1)$ $Q_{\alpha\beta}^{(0)}$ values, which are consistent with the mass conservation rule. The remaining $Q_{\alpha\beta}^{(0)}$ are to be determined by applying the law of mass conservation at each node. This should give a set of linear equations in s unknowns that can readily be solved for the remaining $Q_{\alpha\beta}^{(0)}$ unknowns.
3. This initial guess will not satisfy Equation (11.27); as a result one needs to apply a correction to each $Q_{\alpha\beta}^{(0)}$ value. This is done by applying a Taylor Series expansion (using only two terms in the expansion) to the $h(Q)$ equation.

$$h(Q + \Delta Q) = h(Q) + \left(\frac{dh}{dQ}\right)_Q \Delta Q \tag{11.28}$$

Taking $h(Q + \Delta Q) = h^{(1)}$ and $h(Q) = h^{(0)}$, Equation (11.28) becomes

$$h^{(1)} = h^{(0)} + \left(\frac{dh}{dQ}\right)_{Q^{(0)}} \Delta Q \tag{11.29}$$

Applying Equation (11.29) to Equation (11.27) gives

$$\sum_j h_{ij}^{(1)} = \sum_j \left[h_{ij}^{(0)} + \left(\frac{dh}{dQ}\right)_{ij} \Delta Q_i \right] = 0 \tag{11.30}$$

For each loop, the ΔQ_i can be factored out, thus giving a correction factor equation for each loop; that is,

$$\Delta Q_i = -\frac{\sum_j h_{ij}^{(0)}}{\sum_j \left(\frac{dh}{dQ}\right)_{ij}} \tag{11.31}$$

where $\left(\dfrac{dh}{dQ}\right)_{ij}$ is evaluated at $Q_{ij}^{(0)}$.

The Darcy–Weisbach equation relates h to the friction factor f, which is

$$h = \frac{V^2}{2g} \frac{L}{D} f = \frac{8LQ^2}{\pi^2 g D^5} f = K Q^2 f \tag{11.32}$$

The Swamee–Jain formula [3] gives an explicit formula for f, which is

$$f = \frac{1.325}{\left[\ln\left(\dfrac{e}{3.7\,D} + \dfrac{5.74}{Re^{0.9}} \right) \right]^2} \tag{11.33}$$

where
Re = Reynolds number = $\dfrac{4Q}{\pi D \upsilon}$
υ = the kinamatic viscosity (m²/s)
D = the pipe diameter (m)
e = pipe roughness

Equation (11.33) is valid for $10^{-6} \le e/D \le 10^{-2}$ and $5 \times 10^3 \le Re \le 10^8$.

In applying the loop rule, some of the lines will experience a head gain and not a head loss. This occurs when the flow direction is opposite to the positive loop direction. To account for this, take

$$h = \begin{cases} KQ^2 f & \text{if } Q \geq 0 \\ -KQ^2 f & \text{if } Q < 0 \end{cases} \tag{11.34}$$

and

$$\frac{dh}{dQ} = \pm \left(2KfQ + KQ^2 \frac{df}{dQ} \right) \tag{11.35}$$

where the (+) sign is used if $Q > 0$ and the (−) sign is used if $Q < 0$.

The formula for $\dfrac{df}{dQ}$ is

$$\frac{df}{dQ} = \frac{13.69 \left(\dfrac{e}{3.7D} + \dfrac{5.74}{|Re|^{0.9}} \right)^{-1}}{|Re|^{0.9} Q \left[\ln \left(\dfrac{e}{3.7D} + \dfrac{5.74}{|Re|^{0.9}} \right) \right]^3} \tag{11.36}$$

Lines that are in common in two loops need to be treated as follows. If line j in loop i is in common with line m in loop k, then $Q_{km} = -Q_{ij}$ and for the first iteration

$$Q_{ij}^{(1)} = Q_{ij}^{(0)} + \Delta Q_i - \Delta Q_k$$

For example, referring to Figure 11.2, for the first iteration,

$$Q_{11}^{(1)} = Q_{11}^{(0)} + \Delta Q_1 - \Delta Q_2 \tag{11.37}$$

The formulation for the Hardy Cross method is now complete. Project 11.3 involves the Hardy Cross method to determine the flow rate distribution in a three-loop network.

Projects

Project 11.1

Determine the flow rate, Q, and the friction factor, f, for the problem described in Section 11.1 using the following values: $L = 3000$ m, $D = 10$ cm, $\gamma = 9790$ N/m³, $\upsilon = 1.005 \times 10^{-6}$ m²/s, $(z_2 - z_1) = 50$ m, $g = 9.81$ m/s².

Print out values for Q, f, the sum of the head losses, Σh_L, and the head, $(h_p)_{sys}$, developed by the pump in the system.

Project 11.2

Determine the temperature distribution in the slab shown in Figure P10.6 by the Gauss–Seidel method described in Section 11.2. Use the following values: $L = 2.0$ m, $w = 0.2$ m, $T_w = 300°C$, $T_\infty = 50.0°C$, $h = 890$ W/m²-C, $k = 59.0$ W/m-C, $N = 200$, $M = 50$, $\varepsilon = 0.001$. As an initial guess take the centerline temperature to be the solution of heat flow through a wall; that is,

$$T_{n,1}^1 = T_\infty + (T_w - T_\infty) \times x_n / L, \quad \text{for } n = 2, 3, \dots, N$$

and for positions other than the centerline, assume a temperature of a flow through a wall with a convective boundary condition; that is,

$$T_{n,M+1}^1 = \frac{w}{k - hw}\left(\frac{k}{w}T_{n,1}^1 - hT_\infty\right)$$

and

$$T_{n,m}^1 = T_{n,1}^1 + \left(T_{n,M+1}^1 - T_{n,1}^1\right) \times \frac{y_m}{w}$$

for $n = 2, 3, \dots, N$, $m = 2, 3, \dots, M$.

Once convergence has been achieved, define $T1_n = T_{n,1}^2$, $T2_n = T_{n,26}^2$, and $T3_n = T_{n,M+1}^2$ for $n = 1, 2, 3, \dots, N+1$. Construct a table for temperatures T1, T2, and T3 for $n = 1, 6, 11, \dots, N + 1$. Also create plots of T1, T2, and T3 all on the same page.

Project 11.3

Use the Hardy Cross method to determine the flow rate distribution (Q's) in the network shown in Figure P11.3. Print out a table indicating the loop number, the line number, and the flow rate in that line. The network parameters are described in Table P11.3. *Hint:* To apply the ΔQ corrections to lines that are common in two loops, use an ID matrix for every line identifying the loop number of the common line. Set the ID element to zero if the line is not a common line. Example: Suppose line (1,1) is in common with line (2,4), then ID(1,1) = 2, and suppose line (1,3) is not a common line, then ID(1,3) = 0.

Take $v = 1.308 \times 10^{-6}$ m²/s, $e = 0.026$ cm, and $g = 9.81$ m/s².

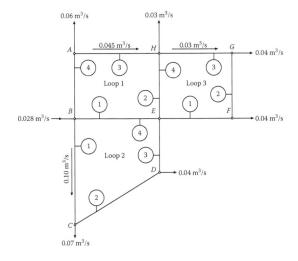

Figure P11.3   A pipe network.

Table P11.3   Network Parameters

Loop Number	Line Number	Length (m)	Diameter (cm)	Initial Guess Q (m³/s)
1	1	3220	40	
	2	4830	30	
	3	3200	35	–0.45
	4	4830	40	
2	1	5630	40	0.1
	2	4020	35	
	3	3200	30	
	4	3200	40	
3	1	4830	30	
	2	4830	25	
	3	4830	30	–0.03
	4	4830	30	

Project 11.4

This project involves determining the volume flow rate a pump will deliver to a closed tank as a function of time [5]. The pump characteristic curve (H vs. Q) was taken from a pump manufacturer's catalog. The configuration for this project is shown in Figure P11.4. Assume that the tank receiving water is closed. Thus, as water fills up in the tank, the air in the tank is compressed. Isothermal compression is to be assumed. The problem is to determine the time it takes to raise the water level in the tank by a specified amount. Data points of the (H vs. Q) curve provided by the pump manufacturer (units changed to SI units) is shown in Table P11.4.

Determine the coefficients of the third degree polynomial by the method of least squares using MATLAB's polyfit function that represents the approximating function for the data in Table P11.4. The polyfit function returns the coefficients a_1, a_2, a_3, a_4 for the third degree polynomial that best fits the data. The approximating function as described by Equation (P11.4a) is used in the analysis.

$$H = a_1 Q^3 + a_2 Q^2 + a_3 Q + a_4 \tag{P11.4a}$$

Application of the energy equation [4] to the system shown in Figure P11.4 gives

$$(h_p)_{sys} = h_f + (z_2 - z_1 + \ell) + \frac{p_a - p_{atm}}{\gamma} + h_{minor\ losses} \tag{P11.4b}$$

where

$(h_p)_{sys}$ = the head the pump delivers to the water flowing through the pipe.
h_f = viscous head loss in the system.
z = elevation.

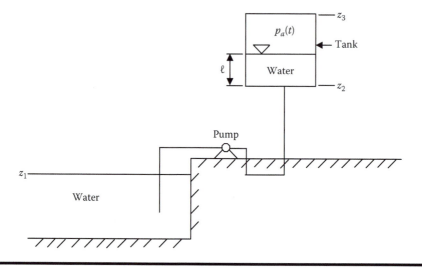

Figure P11.4 A sketch of a reservoir, pump, tank system. (From Reference [5]. With permission of Manchester University Press.)

Table P11.4   H vs. Q Data from Pump Manufacturer

Q (m³/h)	H (m)	Q (m³/h)	H (m)
3.3	43.3	61.6	40.8
6.9	43.4	68.5	39.6
13.7	43.6	75.3	38.7
20.5	43.6	82.2	37.2
27.4	43.3	89	36.3
34.2	43	95.8	34.4
41.1	42.7	102.7	32.6
47.9	42.4	109.6	30.5
54.8	41.8		

P_a = absolute air pressure in the tank.
γ = specific weight of water.
P_{atm} = surrounding atmospheric pressure.
ℓ = water level above the bottom of the tank.

The flow rate developed by the pump must satisfy

$$(h_p)_{sys} = H \qquad\qquad (P11.4c)$$

Viscous head loss, h_f, in a pipe is given by [1,4]

$$h_f = \frac{8Q^2 L}{\pi^2 g D^5} f \qquad\qquad (P11.4d)$$

where
   Q = volume flow rate through pipe.
   L = pipe length.
   g = gravitational constant.
   f = friction factor.
   D = diameter of pipe.

Minor losses due to elbows, entrance losses, etc. are expressed by [1,4]

$$h_{\text{minor losses}} = K \frac{8Q^2}{\pi^2 g D^4} \qquad\qquad (P11.4e)$$

where K = minor head loss coefficient.

Substituting Equations (P11.4a), (P11.4b), (P11.4d), and (P11.4e) into Equation (P11.4c) gives

$$a_1 Q^3 + a_2 Q^2 + a_3 Q + a_4 = \frac{8Q^2}{\pi^2 g D^4}\left(\frac{L}{D}f + K\right) + \frac{p_a - p_{atm}}{\gamma} + (z_2 - z_1 + \ell) \quad \text{(P11.4f)}$$

Rearranging terms gives the following cubic equation:

$$a_1 Q^3 + \left\{a_2 - \frac{8}{\pi^2 g D^4}\left(\frac{L}{D}f + K\right)\right\}Q^2 + a_3 Q$$

$$+ \left\{a_4 - \frac{p_a - p_{atm}}{\gamma} - (z_2 - z_1 + \ell)\right\} = 0 \qquad \text{(P11.4g)}$$

For a smooth pipe line, the friction factor, f, can be approximated by [2]

$$f = (1.82\log_{10} \mathrm{Re} - 1.64)^{-2} \qquad \text{(P11.4h)}$$

where

$$\mathrm{Re} = \text{Reynolds number} = \frac{4Q}{\pi D v} \qquad \text{(P11.4i)}$$

where

v = kinematic viscosity of water.

The air pressure, p_a, at time t, is determined from the ideal gas law for an isothermal process; that is,

$$p_a(t) = \frac{p_{a,i} \forall_i}{\forall(t)} = p_{a,i}\frac{A_T(z_3 - z_2 - \ell_i)}{A_T(z_3 - z_2 - \ell(t))} = p_{a,i}\frac{(z_3 - z_2 - \ell_i)}{(z_3 - z_2 - \ell(t))} \qquad \text{(P11.4j)}$$

where

$p_{a,i}$ = initial air pressure.
A_T = cross-sectional area of tank = $\dfrac{\pi D_T^{\,2}}{4}$.
$\forall$ = air volume in tank.
D_T = diameter of tank.

An iterative method of solution for determining Q is as follows:

1. Assume a value for f, say, f_1; start with $f_1 = 0.03$.
2. Solve Equation (P11.4g) for Q by the MATLAB function ROOTS, selecting the maximum positive root.

3. Solve for Re by Equation (P11.4i).
4. Solve for f, say, f_2, by Equation (P11.4h).
5. If $|f_2 - f_1| < 1.0 \times 10^{-6}$ then $f = f_2$ and Q = the value obtained in step 2.
6. If $|f_2 - f_1| > 1.0 \times 10^{-6}$ then set $f_1 = f_2$ and repeat the process (steps 1 through 5).
7. Continue iteration until $|f_2 - f_1| < 1.0 \times 10^{-6}$.

The governing equation for the time it takes to raise the water level in the tank is as follows:

$$\frac{d(A_T \ell)}{dt} = Q(\ell, p_a) \quad \text{or} \quad \frac{d\ell}{dt} = \frac{Q(\ell, p_a)}{A_T} \quad \text{(P11.4k)}$$

Separating the variables and integrating from $\ell = \ell_i$ to $\ell = \ell_f$ gives

$$t_f = A_T \int_{\ell_i}^{\ell_f} \frac{d\ell}{Q(\ell, p_a)} = A_T \int_{\ell_i}^{\ell_f} F(\ell, p_a)\, d\ell \quad \text{(P11.4l)}$$

Use Simpson's rule to obtain t_f. A review of Simpson's rule follows:

Given:

$$I = \int_a^b F(x)\, dx$$

Subdivide the x domain into N even subdivisions giving $x_1, x_2, x_3, \ldots x_{N+1}$. Let the function values at $x_1, x_2, x_3, \ldots x_{N+1}$ be $F_1, F_2, F_3, \ldots F_{N+1}$, then

$$I = \frac{\Delta x}{3}(F_1 + 4F_2 + 2F_3 + 4F_4 + 2F_5 + \cdots + 2F_N + F_{N+1}) \quad \text{(P11.4m)}$$

Then $\ell \sim x$ and $\dfrac{1}{Q} \sim F$.

Procedure:

1. At each ℓ_j determine p_{aj}, by Equation (P11.4j).
2. For each ℓ_j, iterate for Q_j by the iteration procedure described above, obtaining all the Q_j' s. Then determine all the F_j's.
3. Apply Equation (P11.4m) to obtain the time t_f.

Use the following values for the variables:

$(z_2 - z_1) = 30$ m, $\ell_i = 1.2$ m, $\ell_f = 4.2$ m, $(z_3 - z_2) = 7.6$ m

$v = 1.0 \times 10^{-6}$ m^2/s, $p_{atm} = 1.0132 \times 10^5$ N/m^2,

$\gamma = 9790$ N/m^3, $K = 4.5$, $D = 15$ cm, $L = 60$ m, $p_a(0) = p_{atm}$, $N = 100$, $D_T = 1.5$ m

1. Create a plot of the approximating curve of H (m) vs. Q (m^3/h) (solid line) and on the same plot include the data points as circles. *Note*: Except for the plots, Q needs to be in (m^3/s).
2. Create plots of Q (m^3/h) vs. ℓ(m), $t(s)$ vs. ℓ(m), p_a(N/m^{2}gage) vs. ℓ(m), and f vs. ℓ(m).
3. Print out the time, t_f(s), it takes to raise the water level in the tank by 3 m.
4. Print out the air pressure (N/m^2 gage) in the tank at time t_f.
5. Print out the initial flow rate, Q_i, and final flow rate, Q_f, in (m^3/h).

References

1. Bober, W. and Kenyon, R. A., *Fluid Mechanics*, John Wiley & Sons, New York, 1980.
2. Holman, J. P., *Heat Transfer*, 9th ed., McGraw-Hill, New York, 2002.
3. Bober, W., The use of the Swamee-Jain formula in pipe network problems, *Journal of Pipelines*, 4, 315–317, 1984.
4. White, F., *Fluid Mechanics*, 6th ed., McGraw-Hill, New York, 2008.
5. Bober, W., Fluid mechanics computer project for mechanical engineering students, *IJMEE*, 36, 3, July 2008.

Chapter 12

Laplace Transforms

12.1 Laplace Transform and Inverse Transform

Laplace Transforms [1,2] can be used to solve ordinary and partial differential equations (PDEs). The method reduces an ordinary differential equation to an algebraic equation that can be manipulated to a form such that the inverse transform can be obtained from tables. The inverse transform is the solution to the differential equation. The inverse transform can also be obtained by residue theory in complex variables. The method is applicable to problems where the independent variable domain is from (0 to ∞). The method is particularly useful for linear, nonhomogeneous differential equations, such as vibration problems where the forcing function is piecewise continuous.

Let $f(t)$ be a function defined for all $t \geq 0$; then

$$\mathscr{L}(f(t)) = F(s) = \int_0^\infty e^{-st} f(t) \, dt \tag{12.1}$$

$F(s)$ is called the *Laplace Transform* of $f(t)$.

The *inverse transform* of $F(s)$ is defined to be the function $f(t)$, that is,

$$\mathscr{L}^{-1}(F(s)) = f(t) \tag{12.2}$$

We can create a table that contains both $f(t)$ and the corresponding $F(s)$.

Example 12.1

Let $f(t) = 1; t \geq 0$. Then

$$\mathcal{L}(1) = \int_0^\infty e^{-st}\, dt = -\left[\frac{e^{-st}}{s}\right]_0^\infty = \frac{1}{s} \tag{12.3}$$

Example 12.2

Let $f(t) = e^{at}$.

$$\mathcal{L}(e^{at}) = \int_0^\infty e^{at} e^{-st}\, dt = \int_0^\infty e^{-(s-a)t}\, dt = \left[-\frac{e^{-(s-a)t}}{(s-a)}\right]_0^\infty = \frac{1}{s-a} \tag{12.4}$$

Linearity of Laplace Transforms:

$$\mathcal{L}(a f(t) + b g(t)) = a\,\mathcal{L}(f(t) + b\,\mathcal{L}(g(t)) \tag{12.5}$$

Example 12.3

Let $f(t) = e^{i\omega t}$

$$\mathcal{L}(e^{i\omega t}) = \frac{1}{s - i\omega} = \frac{1}{s - i\omega} \times \frac{s + i\omega}{s + i\omega} = \frac{s + i\omega}{s^2 + \omega^2}$$

$$\mathcal{L}(e^{i\omega t}) = \frac{s}{s^2 + \omega^2} + i\frac{\omega}{s^2 + \omega^2} \tag{12.6}$$

$$\mathcal{L}(e^{i\omega t}) = \mathcal{L}(\cos \omega t) + i\,\mathcal{L}(\sin \omega t) \tag{12.7}$$

Equating the real and imaginary components of Equations (12.6) and (12.7) gives

$$\mathcal{L}(\cos \omega t) = \frac{s}{s^2 + \omega^2} \tag{12.8}$$

$$\mathcal{L}(\sin \omega t) = \frac{\omega}{s^2 + \omega^2} \tag{12.9}$$

If $\mathcal{L}(f(t)) = F(s)$, then

$$F(s-a) = \int_0^\infty f(t) e^{-(s-a)t}\, dt = \int_0^\infty f(t) e^{at} e^{-st}\, dt \tag{12.10}$$

or

$$F(s-a) = \mathcal{L}\left(f(t)\,e^{at}\right) \tag{12.11}$$

Combining Equation (12.8) and Equation (12.11) we obtain

$$\mathcal{L}(e^{at}\cos\omega t) = \frac{s-a}{(s-a)^2 + \omega^2} \tag{12.12}$$

Similarly,

$$\mathcal{L}(e^{at}\sin\omega t) = \frac{\omega}{(s-a)^2 + \omega^2} \tag{12.13}$$

Example 12.4

$$\mathcal{L}\left(t^{n+1}\right) = \int_0^\infty t^{n+1}\,e^{-st}\,dt$$

Let $dv = e^{-st}\,dt$, then $v = -\dfrac{e^{-st}}{s}$ and let $u = t^{n+1}$; then $du = (n+1)\,t^n\,dt$.
But

$$\int_0^\infty u\,dv = [u\,v]_0^\infty - \int_0^\infty v\,du \tag{12.14}$$

Substituting the above values into Equation (12.14) gives

$$\int_0^\infty t^{n+1}\,e^{-st}\,dt = -\left[t^{n+1}\,\frac{e^{-st}}{s}\right]_0^\infty + \frac{n+1}{s}\int_0^\infty t^n\,e^{-st}\,dt$$

or

$$\int_0^\infty t^{n+1}\,e^{-st}\,dt = \frac{n+1}{s}\,\mathcal{L}\left(t^n\right) \tag{12.15}$$

From Equation (12.15), we can see that

$$\mathcal{L}\left(t^n\right) = \frac{n}{s}\,\mathcal{L}\left(t^{n-1}\right)$$

$$\mathcal{L}\left(t^{n-1}\right) = \frac{n}{s}\,\mathcal{L}\left(t^{n-2}\right)$$

$$\mathcal{L}(t) = \frac{1}{s}\,\mathcal{L}\left(t^0\right) = \frac{1}{s}\,\mathcal{L}(1) = \frac{1}{s^2}$$

See Table 12.1, "Table of Laplace Transforms."

Table 12.1   Table of Laplace Transforms

	$f(s)$	$F(t)$
1	$\dfrac{1}{s}$	1
2	$\dfrac{1}{s^2}$	t
3	$\dfrac{1}{s^n}$ $(n = 1, 2, \ldots)$	$\dfrac{t^{n-1}}{(n-1)!}$
4	$\dfrac{1}{\sqrt{s}}$	$\dfrac{1}{\sqrt{\pi t}}$
5	$s^{-\frac{3}{2}}$	$2\sqrt{\dfrac{t}{\pi}}$
6	$s^{-\left(n+\frac{1}{2}\right)}$ $(n = 1, 2, \ldots)$	$\dfrac{2^n \, t^{n-1/2}}{1 \times 3 \times 5 \ldots (2n-1)\sqrt{\pi}}$
7	$\dfrac{1}{s-a}$	e^{at}
8	$\dfrac{1}{(s-a)^2}$	te^{at}
9	$\dfrac{1}{(s-a)^n}$ $(n = 1, 2, \ldots)$	$\dfrac{1}{(n-1)!}t^{n-1}e^{at}$
10	$\dfrac{\Gamma(k)}{(s-a)^k}$ $(k > 0)$	$t^{k-1}e^{at}$
11	$\dfrac{1}{(s-a)(s-b)}$ $(a \neq b)$	$\dfrac{1}{(a-b)}(e^{at} - e^{bt})$
12	$\dfrac{s}{(s-a)(s-b)}$ $(a \neq b)$	$\dfrac{1}{(a-b)}(ae^{at} - be^{bt})$
13	$\dfrac{1}{(s-a)(s-b)(s-c)}$	$-\dfrac{(b-a)e^{at} + (c-a)e^{bt} + (a-b)e^{ct}}{(a-b)(b-c)(c-a)}$

Table 12.1   Table of Laplace Transforms (Continued)

	$f(s)$	$F(t)$
14	$\dfrac{1}{s^2 + a^2}$	$\dfrac{1}{a}\sin at$
15	$\dfrac{s}{s^2 + a^2}$	$\cos at$
16	$\dfrac{1}{s^2 - a^2}$	$\dfrac{1}{a}\sinh at$
17	$\dfrac{s}{s^2 - a^2}$	$\cosh at$
18	$\dfrac{1}{s(s^2 + a^2)}$	$\dfrac{1}{a^2}(1 - \cos at)$
19	$\dfrac{1}{s^2(s^2 + a^2)}$	$\dfrac{1}{a^3}(at - \sin at)$
20	$\dfrac{1}{(s^2 + a^2)^2}$	$\dfrac{1}{2a^3}(\sin at - at\cos at)$
21	$\dfrac{s}{(s^2 + a^2)^2}$	$\dfrac{t}{2a}\sin at$
22	$\dfrac{s^2}{(s^2 + a^2)^2}$	$\dfrac{1}{2a}(\sin at + at\cos at)$
23	$\dfrac{s^2 - a^2}{(s^2 + a^2)^2}$	$t\cos at$
24	$\dfrac{s}{(s^2 + a^2)(s^2 + b^2)}\ (a^2 \neq b^2)$	$\dfrac{\cos at - \cos bt}{b^2 - a^2}$
25	$\dfrac{1}{(s - a)^2 + b^2}$	$\dfrac{1}{b}e^{at}\sin bt$
26	$\dfrac{s - a}{(s - a)^2 + b^2}$	$e^{at}\cos bt$

(Continued)

Table 12.1   Table of Laplace Transforms (Continued)

	$f(s)$	$F(t)$
27	$\dfrac{3a^2}{s^3+a^3}$	$e^{-at}-e^{\frac{at}{2}}\left(\cos\dfrac{at\sqrt{3}}{2}-\sqrt{3}\sin\dfrac{at\sqrt{3}}{2}\right)$
28	$\dfrac{4a^3}{s^4+4a^4}$	$\sin at\cosh at-\cos at\sinh at$
29	$\dfrac{s}{s^4+4a^4}$	$\dfrac{1}{2a^2}\sin at\sinh at$
30	$\dfrac{1}{s^4-a^4}$	$\dfrac{1}{2a^3}(\sinh at-\sin at)$
31	$\dfrac{s}{s^4-a^4}$	$\dfrac{1}{2a^2}(\cosh at-\cos at)$
32	$\dfrac{8a^3s^2}{(s^2+a^2)^3}$	$(1+a^2t^2)\sin at-at\cos at$
33	$\dfrac{1}{s}\left(\dfrac{s-1}{s}\right)^n$	$L_n(t)=\dfrac{e^t}{n!}\dfrac{d^n}{dt^n}(t^n e^{-t})$
34	$\dfrac{s}{(s-a)^{3/2}}$	$\dfrac{1}{\sqrt{\pi t}}e^{at}(1+2at)$
35	$\sqrt{s-a}-\sqrt{s-b}$	$\dfrac{1}{2\sqrt{\pi t^3}}(e^{bt}-e^{at})$
36	$\dfrac{1}{\sqrt{s}+a}$	$\dfrac{1}{\sqrt{\pi t}}-ae^{a^2t}\operatorname{erfc}(a\sqrt{t})$
37	$\dfrac{\sqrt{s}}{s-a^2}$	$\dfrac{1}{\sqrt{\pi t}}+ae^{a^2t}\operatorname{erfc}(a\sqrt{t})$
38	$\dfrac{1}{\sqrt{s}(s-a^2)}$	$\dfrac{1}{a}e^{a^2t}\operatorname{erf}(a\sqrt{t})$
39	$\dfrac{1}{\sqrt{s}(\sqrt{s}+a)}$	$e^{a^2t}\operatorname{erfc}(a\sqrt{t})$

Table 12.1   Table of Laplace Transforms (Continued)

	$f(s)$	$F(t)$
40	$\dfrac{1}{(s+a)\sqrt{s+b}}$	$\dfrac{1}{\sqrt{b-a}}e^{-at}\operatorname{erf}(\sqrt{b-a}\sqrt{t})$
41	$\dfrac{b^2-a^2}{\sqrt{s}(s-a^2)(\sqrt{s}+b)}$	$e^{a^2t}\left[\dfrac{b}{a}\operatorname{erf}(a\sqrt{t})-1\right]$
42	$\dfrac{1}{\sqrt{s^2+a^2}}$	$J_0(at)$
43	$\dfrac{1}{s}e^{-k/s}$	$J_0(2\sqrt{kt})$
44	$\dfrac{1}{\sqrt{s}}e^{-k/s}$	$\dfrac{1}{\sqrt{\pi t}}\cos 2\sqrt{kt}$
45	$\dfrac{1}{\sqrt{s}}e^{k/s}$	$\dfrac{1}{\sqrt{\pi t}}\cosh 2\sqrt{kt}$
46	$\dfrac{1}{s^{3/2}}e^{-k/s}$	$\dfrac{1}{\sqrt{\pi k}}\sin 2\sqrt{kt}$
47	$\dfrac{1}{s^{3/2}}e^{k/s}$	$\dfrac{1}{\sqrt{\pi k}}\sinh 2\sqrt{kt}$
48	$e^{-k\sqrt{s}}\ (k>0)$	$\dfrac{k}{2\sqrt{\pi t^3}}\exp\left(-\dfrac{k^2}{4t}\right)$
49	$\dfrac{1}{s}e^{-k\sqrt{s}}\ (k\geq 0)$	$\operatorname{erfc}\left(\dfrac{k}{2\sqrt{t}}\right)$
50	$\dfrac{1}{\sqrt{s}}e^{-k\sqrt{s}}\ (k\geq 0)$	$\dfrac{1}{\sqrt{\pi t}}\exp\left(-\dfrac{k^2}{4t}\right)$
51	$s^{-3/2}e^{-k\sqrt{s}}\ (k\geq 0)$	$2\sqrt{\dfrac{t}{\pi}}\exp\left(-\dfrac{k^2}{4t}\right)-k\operatorname{erfc}\left(\dfrac{k}{2\sqrt{t}}\right)$

(Continued)

Table 12.1   Table of Laplace Transforms (Continued)

	$f(s)$	$F(t)$
52	$\dfrac{ae^{-k\sqrt{s}}}{s(a+\sqrt{s})}$ $(k \geq 0)$	$-e^{ak}e^{a^2t}\,\text{erfc}\left(a\sqrt{t}+\dfrac{k}{2\sqrt{t}}\right)+\text{erfc}\left(\dfrac{k}{2\sqrt{t}}\right)$
53	$\dfrac{e^{-k\sqrt{s}}}{\sqrt{s}(a+\sqrt{s})}$ $(k \geq 0)$	$e^{ak}e^{a^2t}\,\text{erfc}\left(a\sqrt{t}+\dfrac{k}{2\sqrt{t}}\right)$
54	$\log\dfrac{s-a}{s-b}$	$\dfrac{1}{t}(e^{bt}-e^{at})$
55	$\log\dfrac{s^2+a^2}{s^2}$	$\dfrac{2}{t}(1-\cos at)$
56	$\log\dfrac{s^2-a^2}{s^2}$	$\dfrac{2}{t}(1-\cosh at)$
57	$\arctan\dfrac{k}{s}$	$\dfrac{1}{t}\sin kt$

12.2 Transforms of Derivatives

$$\mathcal{L}(f') = \int_0^\infty f'(t)e^{-st}\,dt$$

Let $dv = \dfrac{d f}{dt}\,dt$, then $v = f$

Let $u = e^{-st}$, then $du = -s\,e^{-st}\,dt$

$$\int_0^\infty f'(t)e^{-st}\,dt = [f\,e^{-st}]_0^\infty + s\int_0^\infty f\,e^{-st}\,dt = -f(0)+s\mathcal{L}(f) \qquad (12.16)$$

$$\mathcal{L}(f'') = \int_0^\infty f''(t)e^{-st}\,dt$$

Let $dv = \dfrac{d f'}{dt}\,dt$, then $v = f'$

Let $u = e^{-st}$, then $du = -s\, e^{-st}\, dt$

$$\int_0^\infty f''(t)e^{-st}\, dt = [\, f'e^{-st}\,]_0^\infty + s\int_0^\infty f'e^{-st}\, dt = -f'(0) - s\, f(0) + s^2\, \mathscr{L}(f) \qquad (12.17)$$

By Equations (12.16) and (12.17) we can see the pattern for the Laplace Transform of the nth derivative, that is,

$$\mathscr{L}(f^{(n)}) = s^n\, \mathscr{L}(f) - s^{n-1} f(0) - s^{n-2}\, f'(0) - \cdots - f^{(n-1)}(0) \qquad (12.18)$$

12.3 Ordinary Differential Equations, Initial Value Problem

Consider the differential equation arising from a spring-dashpot-mass system with a driving force (see Projects P2.10 and P2.11). The governing differential equation for the motion of the mass, $y(t)$, is

$$y'' + \frac{c}{m}y' + \frac{k}{m}y = \frac{F_0(t)}{m} \qquad (12.19)$$

where

   $m =$ mass.
   $k =$ the spring constant.
   $c =$ the damping coefficient.
   $F_0 =$ driving force.

Take the initial conditions to be

$$y(0) = \alpha, \quad y'(0) = \beta$$

Let $p = \dfrac{c}{m}$, $q = \dfrac{k}{m}$ and $r = \dfrac{F_0(t)}{m}$, then Equation (12.19) becomes

$$y'' + p\, y' + q\, y = r(t) \qquad (12.20)$$

The Laplace Transform of each of the terms in Equation (12.20) follows:

$$\mathscr{L}(y'') = s^2\, \mathscr{L}(y) - s\, y(0) - y'(0)$$

$$\mathscr{L}(y') = s\, \mathscr{L}(y) - y(0)$$

Let $\mathcal{L}(y) = Y$, and $R = \mathcal{L}(r)$; then the Laplace Transform of Equation (12.20) becomes

$$(s^2 Y - s\alpha - \beta) + p(sY - \alpha) + qY = R$$

or

$$(s^2 + ps + q)Y = (s + p)\alpha + \beta + R$$

Let $H(s) = \dfrac{1}{s^2 + ps + q}$
Then

$$Y(s) = [(s + p)\alpha + \beta]H(s) + R(s)H(s) \qquad (12.21)$$

By the use of partial fractions and the Laplace Transform tables, we can frequently obtain the inverse transform, $\mathcal{L}^{-1}(Y(s)) = y(t)$.

Example 12.5

Given the following differential equation (no damping and an exponentially decaying driving force), determine the solution.

$$y'' + y = 5e^{-t}$$
$$y(0) = 2, \qquad y'(0) = 0 \qquad\qquad (12.22)$$

This problem fits the general form of Equation (12.20), with $p = 0, q = 1, r = 5e^{-t}$, $\alpha = 2$, and $\beta = 0$. Thus,

$$Y(s) = [2s]H(s) + R(s)H(s)$$

where

$$H(s) = \frac{1}{s^2 + 1} \quad \text{and} \quad R(s) = \mathcal{L}(5e^{-t}) = \frac{5}{s+1}$$

Then

$$Y(s) = \frac{2s}{s^2 + 1} + \frac{5}{(s+1)(s^2 + 1)} \qquad (12.23)$$

From Table 12.1,

$$\mathcal{L}^{-1}\left(\frac{s}{s^2+1}\right) = \cos t$$

$$\mathcal{L}^{-1}\left(\frac{1}{s^2+1}\right) = \sin t$$

For the second term on the right-hand side of Equation (12.23), we need to decompose the term by the method of partial fractions, that is,

$$\frac{5}{(s+1)(s^2+1)} = \frac{A}{s+1} + \frac{Bs+C}{s^2+1} = \frac{A(s^2+1)+(Bs+C)(s+1)}{(s+1)(s^2+1)}$$

Then

$$A+B=0, \quad C+B=0, \quad A+C=5 \rightarrow A=\frac{5}{2}, \quad B=-\frac{5}{2}, \quad C=\frac{5}{2}$$

Therefore,

$$\frac{5}{(s+1)(s^2-1)} = \frac{5}{2(s+1)} - \frac{5}{2}\frac{(s-1)}{(s^2+1)}$$

Thus,

$$Y(s) = \frac{2s}{(s^2+1)} + \frac{5}{2(s+1)} - \frac{5s}{2(s^2+1)} + \frac{5}{2(s^2+1)}$$

From Table 12.1,

$$\mathcal{L}^{-1}\left(\frac{1}{s+1}\right) = e^{-t}$$

$$\mathcal{L}^{-1}\left(\frac{s}{s^2+1}\right) = \cos t$$

$$\mathcal{L}^{-1}\left(\frac{1}{s^2+1}\right) = \sin t$$

Therefore, the solution to Equation (12.22) is

$$y = -\frac{1}{2}\cos t + \frac{5}{2}\sin t + \frac{5}{2}e^{-t}$$

A plot of y vs. t is shown in Figure 12.1.

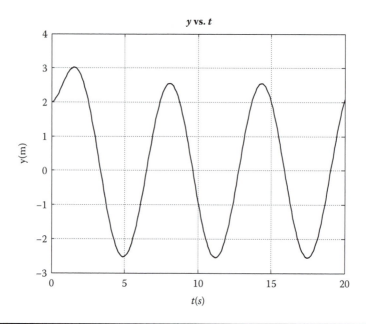

Figure 12.1   The plot of *y* vs. *t*.

12.4 A Shifting Theorem

If $\mathscr{L}(f(t)) = F(s)) = \int_0^\infty f(t)e^{-st}\,dt$, then

$$F(s-a) = \int_0^\infty f(t)e^{-(s-a)t}\,dt = \int_0^\infty f(t)e^{at}\,e^{-st}\,dt = \mathscr{L}(f(t)e^{at}) \qquad (12.24)$$

For damped vibrations,

$$\mathscr{L}(e^{at}\cos\omega t) = \frac{s-a}{(s-a)^2 + \omega^2} \qquad (12.25)$$

$$\mathscr{L}(e^{at}\sin\omega t) = \frac{\omega}{(s-a)^2 + \omega^2} \qquad (12.26)$$

Example 12.6

Determine the solution of the following differential equation:

$$y'' + 3y' + 2y = 5\sin 2t$$

$$y(0) = 1, \ y'(0) = -4 \qquad (12.27)$$

$$p = 3, q = 2, r = 5 \sin 2t = 10 \frac{\sin 2t}{2}$$

$$H(s) = \frac{1}{s^2 + 3s + 2} = \frac{1}{(s+2)(s+1)}$$

$$Y(s) = [(s+3)(1) - 4] H(s) + R(s) H(s)$$

$$R(s) = \mathcal{L}\left(\frac{\sin 2t}{2}\right) = \frac{1}{s^2 + 4}$$

$$Y(s) = \frac{s-1}{(s+2)(s+1)} + \frac{10}{(s^2 + 4)(s+2)(s+1)}$$

It is left as a student exercise to show that

$$\frac{1}{(s+2)(s+1)} = -\frac{1}{(s+2)} + \frac{1}{(s+1)}$$

$$\frac{s}{(s+2)(s+1)} = \frac{2}{(s+2)} - \frac{1}{(s+1)}$$

$$\frac{1}{(s^2+4)(s+2)(s+1)} = \frac{A+Bs}{s^2+4} + \frac{C}{s+2} + \frac{D}{s+1}$$

$$\frac{1}{(s^2+4)(s+2)(s+1)} = \frac{(A+Bs)(s^2+3s+2) + C(s^2+4)(s+1) + D(s^2+4)(s+2)}{(s^2+4)(s+2)(s+1)}$$

To make the numerator of the above equation equal to 1, we require

$$B + C + D = 0, \quad A + 3B + C + 2D = 0, \quad 3A + 2B + 4C + 4D = 0, \quad 2A + 4C + 8D = 1$$

Letting $A = x_1, B = x_2, C = x_3,$ and $D = x_4,$ we may write the system of equations as a matrix equation, that is,

$$\begin{bmatrix} 0 & 1 & 1 & 1 \\ 1 & 3 & 1 & 2 \\ 3 & 2 & 4 & 4 \\ 2 & 0 & 4 & 8 \end{bmatrix} \begin{bmatrix} x_1 \\ x_2 \\ x_3 \\ x_4 \end{bmatrix} = \begin{bmatrix} 0 \\ 0 \\ 0 \\ 1 \end{bmatrix}$$

Using the *inv* matrix function in MATLAB®, we can readily obtain the solution for the x values; these are

$$x_1 = A = -0.05, \quad x_2 = B = -0.075, \quad x_3 = C = -0.125, \quad x_4 = D = 0.20$$

$$\mathcal{L}^{-1}\left(\frac{1}{s+2}\right) = e^{-2t}$$

$$\mathcal{L}^{-1}\left(\frac{1}{s+1}\right) = e^{-t}$$

$$\mathcal{L}^{-1}\left(\frac{1}{s^2+4}\right) = \frac{1}{2}\sin 2t$$

$$\mathcal{L}^{-1}\left(\frac{s}{s^2+4}\right) = \cos 2t$$

Collecting all the terms in $Y(s)$ and applying the above inverse transforms gives

$$y(t) = \mathcal{L}^{-1}(Y(s)) = 1.75 e^{-2t} - 0.75 \cos 2t - 0.25 \sin 2t$$

Although we were able to solve this problem by the use of Laplace Transforms, the solution could be obtained with fewer steps by either MATLAB's ode45 function or by the method involving complementary and particular solutions (see Figure 12.2). The program using the ode45 function in MATLAB follows.

```
% ODE_laplace.m
% In this example a single second order ordinary differential equation
% is reduced to two first order differential equations. This program
% solves the two equation system using MATLAB's ode45 function.
% The problem is to determine the y(t) position of the mass in a
% spring-dashpot system.
% Y1=y, Y2=v, Y1'=Y2, Y2'=5*sin 2t-3Y2-2Y1
% y(0)=1.0, y'(0)= -4.0
clear; clc;
initial=[1.0 -4.0];
tspan=0.0:0.1:20;
tol1=1.0e-6;
tol2=[1.0e-6 1.0e-6];
options=odeset('RelTol',tol1,'AbsTol',tol2);
[t,Y]=ode45('dYdt_laplace',tspan,initial,options);
P=[t Y];
dt=0.1;
for i=1:201
    t1=(i-1)*dt;
    ylp(i)=1.75*exp(-2*t1) -0.75*cos(2*t1) -0.25*sin(2*t1);
end
```

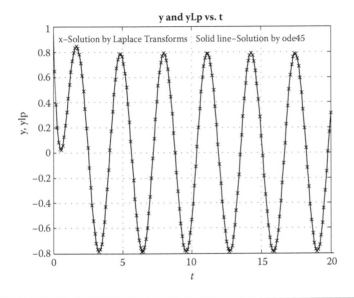

Figure 12.2   A comparison between a solution obtained by Laplace Transforms and ode45.

```
plot(t,Y(:,1),t,ylp,'x'),
xlabel('t'),ylabel('y,ylp'),title('y and ylp vs. t'), grid;
text(1.0,0.92,'x-solution by Laplace transforms'),
text(10.0,0.92,'solid line-solution by ode45');
```

```
% dYdt_laplace
% This function works with ODE_laplace
% Y(1)=y, Y(2)=v
% Y1'=Y(2), Y2'=5*sin 2t-3Y2-2Y1
function Yprime=dYdt_laplace(t,Y)
Yprime=zeros(2,1);
Yprime(1)=Y(2);
Yprime(2)=5*sin(2*t)-3.0*Y(2)-2.0*Y(1);
```

12.5 The Unit Step Function

Let (see Figure 12.3)

$$u(t-a) = \begin{cases} 0, & \text{if } t < a \\ 1, & \text{if } t \geq a \end{cases} \qquad (12.28)$$

The unit step function is useful in analyzing beams and electrical circuits.

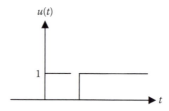

Figure 12.3   Unit step function.

The function

$$f(t-a)u(t-a) = \begin{cases} 0, & \text{if } t < a \\ f(t-a), & \text{if } t > a \end{cases}$$

has the Laplace Transform $e^{-as} F(s)$, where $\mathscr{L}(f(t)) = F(s)$.

Proof:

$$e^{-as} F(s) = e^{-as} \int_0^\infty f(\tau) e^{-s\tau} d\tau = \int_0^\infty f(\tau) e^{-s(\tau+a)} d\tau$$

Let $t = \tau + a$, then $dt = d\tau$. When $\tau = 0$, $t = a$ and when $\tau = \infty$, $t = \infty$. Therefore,

$$e^{-as} F(s) = \int_a^\infty f(t-a) e^{-st} dt = \int_0^a 0 e^{-st} dt + \int_a^\infty f(t-a) e^{-st} dt$$

$$e^{-as} F(s) = \int_0^\infty u(t-a) f(t-a) e^{-st} dt$$

$$e^{-as} F(s) = \mathscr{L}(u(t-a) f(t-a)) \tag{12.29}$$

12.6 Laplace Transform of the Unit Step Function

$$\mathscr{L}(u(t-a)) = \int_0^\infty u(t-a) e^{-st} dt = \int_0^a 0 \cdot e^{-st} dt + \int_a^\infty 1 \cdot e^{-st} dt$$

$$\mathscr{L}(u(t-a)) = -\frac{1}{s} [e^{-st}]_a^\infty = -\frac{1}{s} [0 - e^{-as}] = \frac{1}{s} e^{-as} \tag{12.30}$$

Example 12.7

Determine the solution of the following differential equation:

$$y'' + 3y' + 2y = \begin{cases} 5t, & \text{for } t < 2 \\ 0, & \text{for } t \geq 2 \end{cases}$$

(12.31)

$$y(0) = 1, \quad y'(0) = 0$$

$$p = 3, q = 2,$$

$$r(t) = \begin{cases} 5t, & \text{for } t < 2 \\ 0, & \text{for } t \geq 2 \end{cases}$$

$$R(s) = \mathcal{L}(r(t)) = \mathcal{L}\{5t\mu(t) - 5t\mu(t-2) = 5t\mu(t) - 5(t-2)\mu(t-2) - 10\mu(t-2)\}$$

$$(s^2 + ps + q)Y = (s+p)\alpha + \beta + R$$

$$Y(s) = [(s+3)(1)]H(s) + R(s)H(s)$$

$$H(s) = \frac{1}{s^2 + 3s + 2} = \frac{1}{(s+2)(s+1)}$$

$$Y(s) = \frac{s}{(s+2)(s+1)} + \frac{3}{(s+2)(s+1)} + \frac{5}{s^2(s+2)(s+1)}$$

$$- \frac{5e^{-2s}}{s^2(s+2)(s+1)} - \frac{10e^{-2s}}{s(s+2)(s+1)}$$

Let $Y(s) = Y_1 + Y_2 + Y_3 + Y_4 + Y_5$, where

$$Y_1 = \frac{s}{(s+2)(s+1)}$$

$$Y_2 = \frac{3}{(s+2)(s+1)}$$

$$Y_3 = \frac{5}{s^2(s+2)(s+1)}$$

$$Y_4 = -\frac{5e^{-2s}}{s^2(s+2)(s+1)}$$

$$Y_5 = -\frac{10e^{-2s}}{s(s+2)(s+1)}$$

By the use of partial fractions, we can determine that

$$\mathcal{L}^{-1}(Y_1) = 2e^{-2t} - e^{-t}$$

$$\mathcal{L}^{-1}(Y_2) = -3e^{-2t} + 3e^{-t}$$

$$\mathcal{L}^{-1}(Y_3) = \frac{5}{2}t - \frac{15}{4} - \frac{5}{4}e^{-2t} + 5e^{-t}$$

$$\mathcal{L}^{-1}(Y_4) = -\left\{\frac{5}{2}t - 5 - \frac{15}{4} - \frac{5}{4}e^4 e^{-2t} + 5e^2 e^{-t}\right\}\mu(t-2)$$

$$\mathcal{L}^{-1}(Y_5) = -10\left\{\frac{1}{2} + \frac{1}{2}e^4 e^{-2t} - e^2 e^{-t}\right\}\mu(t-2)$$

Summing the above five terms gives

$$y(t) = \begin{cases} \dfrac{5}{2}t - \dfrac{15}{4} - \dfrac{9}{4}e^{-2t} + 7e^{-t}, & \text{for } t < 2 \\[2ex] -\left(\dfrac{9}{4} + \dfrac{15}{4}e^4\right)e^{-2t} + (7 + 5e^2)e^{-t}, & \text{for } t \geq 2 \end{cases}$$

This problem can also be solved numerically by the use of the ode45 function in MATLAB. The MATLAB program used to solve the above problem follows.

```
% ODE_laplace2.m
% This program solves a system of 2 ordinary differential equations
% by using ode45 function. The problem is to determine the
% y(t)positions of a mass in a mass-spring-dashpot system
% Y1=y, Y2=v, Y1'=Y2, Y2'=5*t-3Y2-2Y1, for t < 2,
% Y2'=-3Y2-2Y1, for t >= 2,
% y(0)=1.0, y'(0)=0
clear; clc;
initial=[1.0 0.0];
tspan=0.0:0.1:4;
tol1=1.0e-6;
tol2=[1.0e-6 1.0e-6];
options=odeset('RelTol',tol1,'AbsTol',tol2);
[t,Y]=ode45('dYdt_laplace2',tspan,initial,options);
P=[t Y];
dt=0.1;
for i=1:41
    t1(i)=(i-1)*dt;
    t2=(i-1)*dt;
    if t2 < 2
```

```
        ylp(i)=2.5*t2-15/4-9/4*exp(-2*t2)+7*exp(-t2);
    end
    if t2 >= 2
    ylp(i)= - (9/4+15/4*exp(4))*exp(-2*t2)+(7+5*exp(2))*exp(-t2);
    end
end
plot(t,Y(:,1),t1,ylp,'x'),
xlabel('t'),ylabel('y,ylp'),title('y & ylp vs. t'), grid;
```

```
% dYdt_laplace2
% This function works with ODE_laplace
% Y(1)=y, Y(2)=v
% Y1'=Y(2), Y2'=5*t-3Y2-2Y1, if t >2, Y2'=-3Y2-2Y1, if t >= 2
function Yprime=dYdt_laplace(t,Y)
Yprime=zeros(2,1);
Yprime(1)=Y(2);
if t < 2
    Yprime(2)=5*t-3.0*Y(2) -2.0*Y(1);
else
    Yprime(2)= -3.0*Y(2) -2.0*Y(1);
end
```

A comparison of the ode45 and the Laplace Transform solution is shown in Figure 12.4.

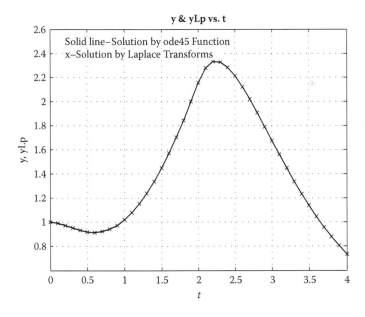

Figure 12.4 A comparison between a solution obtained by Laplace Transforms and ode45.

12.7 Convolution

Given two transforms $F(s)$ and $G(s)$ whose transforms are $f(t)$ and $g(t)$ are known, then

$$\mathcal{L}^{-1}(F(s)G(s)) = \int_0^t f(\tau)g(t-\tau)d\tau = \int_0^t g(\tau)f(t-\tau)d\tau \qquad (12.32)$$

Proof:

$$F(s)G(s) = \left(\int_0^\infty f(p)e^{-ps}\,dp\right)\left(\int_0^\infty g(\tau)e^{-\tau s}\,d\tau\right) \qquad (12.33)$$

$$F(s)G(s) = \int_0^\infty g(\tau)\left\{\int_0^\infty f(p)e^{-(p+\tau)s}\,dp\right\}d\tau \qquad (12.34)$$

Let $p+\tau = t$, then when $p = 0, t = \tau$ and when $p = \infty, t = \infty$. Also $dp = dt$.

$$F(s)G(s) = \int_0^\infty g(\tau)\left\{\int_\tau^\infty f(t-\tau)e^{-ts}\,dt\right\}d\tau = \iint_R g(\tau)f(t-\tau)\,e^{-ts}dt\,d\tau \qquad (12.35)$$

where R is the region below the line $t = \tau$ as shown in Figure 12.5.

The integration order of the multiple integral in Equation (12.35) is to integrate with respect to t first, then with respect to τ second. In this case, the order of integration does not matter. So we can integrate with respect to τ first, then with

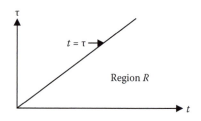

Figure 12.5   Integration in the (τ, t) plane.

respect to t second. This gives

$$F(s)G(s) = \iint\limits_{R} g(\tau) f(t-\tau) e^{-st} dt\, d\tau$$

$$= \int_{0}^{\infty} e^{-st} \left\{ \int_{0}^{t} g(\tau) f(t-\tau)\, d\tau \right\} dt = \mathscr{L} \left(\int_{0}^{t} g(\tau) f(t-\tau)\, d\tau \right)$$

Thus,

$$\mathscr{L}^{-1}(F(s)G(s)) = \int_{0}^{t} g(\tau) f(t-\tau)\, d\tau \qquad (12.36)$$

We can reverse the roles of f and g, giving

$$\mathscr{L}^{-1}(F(s)G(s)) = \int_{0}^{t} g(t-\tau) f(\tau)\, d\tau$$

Example 12.8

Let's apply the convolution formula to the second term in the $Y(s)$ function in Example 12.6. In that example the second term in the $Y(s)$ function is the Y_3 term in Equation (12.37).

$$Y(s) = \frac{s}{(s+2)(s+1)} - \frac{1}{(s+2)(s+1)} + \frac{10}{(s^2+4)(s+2)(s+1)} = Y_1 + Y_2 + Y_3 \qquad (12.37)$$

$$\mathscr{L}^{-1}(Y_1) = \mathscr{L}^{-1}\left(\frac{s}{(s+2)(s+1)} \right) = \mathscr{L}^{-1}\left(\frac{2}{s+2} - \frac{1}{s+1} \right) = 2e^{-2t} - e^{-t}$$

$$\mathscr{L}^{-1}(Y_2) = -\mathscr{L}^{-1}\left(\frac{1}{(s+2)(s+1)} \right) = -\left(-\frac{1}{s+2} + \frac{1}{s+1} \right) = e^{-2t} - e^{-t}$$

$$\mathscr{L}^{-1}(Y_1 + Y_2) = 3e^{-2t} - 2e^{-t} \qquad (12.38)$$

$$\mathscr{L}^{-1}(Y_3) = \left(\frac{10}{(s^2+4)(s+2)(s+1)} \right) = 10\,\mathscr{L}^{-1}(F(s)G(s)) \qquad (12.39)$$

where

$$F(s) = \frac{1}{s^2 + 4} \text{ and } G(s) = \frac{1}{(s+2)(s+1)}$$

$$\mathscr{L}^{-1}(F(s)) = \frac{1}{2}\sin 2t = f(t) \text{ and } \mathscr{L}^{-1}(G(s)) = -e^{-2t} + e^{-t} = g(t)$$

$$\mathscr{L}^{-1}(F(s)G(s)) = \int_0^t g(t-\tau) f(\tau) d\tau = \int_0^t \left(-e^{-2(t-\tau)} + e^{-(t-\tau)}\right)\frac{1}{2}\sin 2\tau\, d\tau$$

$$\mathscr{L}^{-1}(F(s)G(s)) = -\frac{1}{2}e^{-2t}\int_0^t e^{2\tau}\sin 2\tau\, d\tau + \frac{1}{2}e^{-t}\int_0^t e^{\tau}\sin 2\tau\, d\tau \quad (12.40)$$

From integral tables

$$\int e^{ax}\sin px\, dx = \frac{1}{a^2 + p^2}e^{ax}(a\sin px - p\cos px)$$

Applying the above equation to Equation (12.40) gives

$$\mathscr{L}^{-1}(F(s)G(s)) = -\frac{1}{2}e^{-2t}\left\{\frac{e^{2\tau}(2\sin 2\tau - 2\cos 2\tau)}{4+4}\right\}_0^t + \frac{1}{2}e^{-t}\left\{\frac{e^{\tau}(\sin 2\tau - 2\cos 2\tau)}{1+4}\right\}_0^t$$

After collecting like terms, the above equation reduces to

$$\mathscr{L}^{-1}(F(s)G(s)) = -\frac{4}{160}\sin 2t - \frac{12}{160}\cos 2t - \frac{1}{8}e^{-2t} + \frac{1}{5}e^{-t} \quad (12.41)$$

Combining Equations (12.38), (12.39), and (12.41), we obtain

$$\mathscr{L}^{-1}(Y(s)) = 1.75\,e^{-2t} - 0.25\sin 2t - 0.75\cos 2t \quad (12.42)$$

which is the same as we obtained earlier.

12.8 Laplace Transforms Applied to Partial Differential Equations

Let us obtain the Laplace Transform of the following PDE:

$$A\frac{\partial^2 \theta}{\partial x^2} + B\frac{\partial^2 \theta}{\partial t^2} + C\frac{\partial \theta}{\partial x} + D\frac{\partial \theta}{\partial t} + E\theta = F(x,t) \qquad (12.43)$$

where $\theta = \theta(x,t)$ and $0 \leq t \leq \infty$.

Multiply both sides of the above equation by e^{-st} and integrate from 0 to ∞.

$$\underset{t \to s}{\mathscr{L}}\left(A\frac{\partial^2 \theta}{\partial x^2}\right) = \int_0^\infty A\frac{\partial^2 \theta}{\partial x^2}e^{-st}dt = A\frac{d^2}{dx^2}\int_0^\infty \theta e^{-st}dt = A\frac{d^2}{dx^2}\bar{\theta}(x;s)$$

where $\bar{\theta}(x;s) = $ the Laplace Transform of $\theta(x,t)$.

$$\underset{t \to s}{\mathscr{L}}\left(B\frac{\partial^2 \theta}{\partial t^2}\right) = \int_0^\infty B\frac{\partial^2 \theta}{\partial t^2}e^{-st}dt = B\left(s^2\bar{\theta} - s\theta(x,0) - \frac{d\theta}{dt}(x,0)\right)$$

$$\underset{t \to s}{\mathscr{L}}\left(C\frac{\partial \theta}{\partial x}\right) = \int_0^\infty C\frac{\partial \theta}{\partial x}e^{-st}dt = C\frac{d}{dx}\int_0^\infty \theta e^{-st}dt = C\frac{d}{dx}\bar{\theta}(x;s)$$

$$\underset{t \to s}{\mathscr{L}}\left(D\frac{\partial \theta}{\partial t}\right) = \int_0^\infty D\frac{\partial \theta}{\partial t}e^{-st}dt = D(s\bar{\theta}(x;s) - \theta(x,0))$$

$$\underset{t \to s}{\mathscr{L}}(E\theta) = \int_0^\infty E\theta e^{-st}dt = E(\bar{\theta}(x;s))$$

$$\underset{t \to s}{\mathscr{L}}(F) = \int_0^\infty F\,e^{-st}dt = \bar{F}(x;s)$$

From the above relations, it can be seen that Equation (12.43) becomes an ordinary differential equation with respect to x. In this equation s is considered a constant. That is,

$$A\frac{d^2\bar{\theta}}{dx^2} + C\frac{d\bar{\theta}}{dx} + (Bs^2 + Ds + E)\bar{\theta} = \bar{F}(x;s) + (Bs+D)\theta(x,0) + B\frac{\partial \theta}{\partial t}(x,0) \quad (12.44)$$

One also needs to take the Laplace Transforms of the boundary conditions. Suppose that

$$\theta(0,t) = g_1(t) \text{ and } \frac{\partial \theta}{\partial x}(L,t) = g_2(t)$$

Then

$$\overline{\theta}(0;s) = \mathcal{L}(g_1) \tag{12.45}$$

$$\mathcal{L}\left(\frac{\partial \theta}{\partial x}(L,t)\right) = \lim_{x \to L} \frac{d}{dx} \int_0^\infty \theta(x,t) e^{-st} \, dt = \lim_{x \to L} \frac{d}{dx} \overline{\theta}(x;s)$$

Thus,

$$\frac{d\overline{\theta}}{dx}(L;s) = \mathcal{L}(g_2) \tag{12.46}$$

The initial conditions $\theta(x,0)$ and $\frac{\partial \theta}{\partial t}(x,0)$ are directly entered into Equation (12.44).

Example 12.9

Consider a semi-infinite slab, initially at a uniform temperature, T_i, that is suddenly subjected to temperature T_0 at its free surface (see Figure 12.6). The governing PDE is

$$\frac{1}{a}\frac{\partial T}{\partial t} = \frac{\partial^2 T}{\partial x^2}$$

where a is the thermal diffusivity of the slab material.

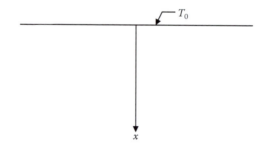

Figure 12.6 Semi-infinite slab.

Initial Condition

$$T(x,0) = T_i$$

Boundary Conditions

$$T(0, t) = T_0$$

$$T(\infty, t) = T_i$$

To simplify the method for obtaining the solution, let $\theta(x,t) = T(x,t) - T_i$, then the above equations reduce to

$$\frac{1}{a}\frac{\partial \theta}{\partial t} = \frac{\partial^2 \theta}{\partial x^2} \tag{12.47}$$

$$\theta(x,0) = 0 \tag{12.48}$$

$$\theta(0,t) = T_0 - T_i \tag{12.49}$$

$$\theta(\infty, t) = 0 \tag{12.50}$$

Taking the Laplace Transform of both sides of Equations (12.47), (12.49), and (12.50) gives

$$\frac{1}{a}(s\bar{\theta} - \theta(x,0)) = \frac{d^2\bar{\theta}}{dx^2} \tag{12.51}$$

$$\bar{\theta}(\infty ; s) = 0 \tag{12.52}$$

$$\bar{\theta}(0;s) = \frac{T_0 - T_i}{s} \tag{12.53}$$

where $\bar{\theta} = \underset{t \to s}{\mathcal{L}}(\theta(x,t))$. Equation (12.51) becomes

$$\frac{d^2\bar{\theta}}{dx^2} - \frac{s}{a}\bar{\theta} = 0 \tag{12.54}$$

The solution is

$$\bar{\theta}(x;s) = c_1 e^{\sqrt{\frac{s}{a}}x} + c_2 e^{-\sqrt{\frac{s}{a}}x} \tag{12.55}$$

Applying a boundary condition, Equation (12.52) gives

$$\bar{\theta}(\infty; s) = 0 \rightarrow c_1 = 0$$

Then

$$\bar{\theta} = c_2 e^{-\sqrt{\frac{s}{a}}\, x}$$

Applying a boundary condition, Equation (12.53) gives

$$\bar{\theta}(0; s) = \frac{T_0 - T_i}{s} = c_2 \quad (1)$$

Thus,

$$\bar{\theta} = \frac{T_0 - T_i}{s} e^{-\sqrt{\frac{s}{a}}\, x} \qquad (12.56)$$

From Table 12.1

$$\mathcal{L}^{-1}\left(\frac{1}{s} e^{-k\sqrt{s}} \right) = erfc\left(\frac{k}{2\sqrt{t}} \right)$$

Therefore,

$$\theta(x,t) = (T_0 - T_i)\, erfc\left(\frac{x}{2\sqrt{a t}} \right)$$

or

$$T(x,t) - T_i = T_0 - T_i - (T_0 - T_i)\, erf\left(\frac{x}{2\sqrt{a t}} \right)$$

$$\frac{T(x,t) - T_0}{T_i - T_0} = erf\left(\frac{x}{2\sqrt{a t}} \right) \qquad (12.57)$$

Example 12.10

Suppose we consider the same problem as Example 12.9 except we will replace the boundary condition at $x = 0$ with the convection boundary condition

$$\frac{\partial T}{\partial x}(0,t) - \frac{h}{k}(T(0,t) - T_\infty) = 0 \qquad (12.58)$$

Again, let $\theta(x,t) = T(x,t) - T_i$, then the PDE, the initial condition, and the boundary condition at $x = \infty$ are the same as in Example 12.9, that is,

$$\frac{1}{a}\frac{\partial \theta}{\partial t} = \frac{\partial^2 \theta}{\partial x^2} \tag{12.59}$$

$$\theta(x,0) = 0 \tag{12.60}$$

$$\theta(\infty, t) = 0 \tag{12.61}$$

By adding and subtracting T_i to the terms inside the parentheses of the second term in Equation (12.58), the boundary condition becomes

$$\frac{\partial \theta}{\partial x}(0,t) - \frac{h}{k}[\theta(0,t) + T_i - T_\infty] = 0 \tag{12.62}$$

Taking the Laplace Transform of both sides of Equations (12.59) and (12.61) and applying the Laplace Transform boundary condition to the Laplace Transform of the PDE, we obtain as in Example 12.9

$$\overline{\theta} = c_2 e^{-\sqrt{\frac{s}{a}}\, x}$$

Then

$$\frac{d\overline{\theta}}{dx} = -c_2 \sqrt{\frac{s}{a}}\, e^{-\sqrt{\frac{s}{a}}\, x}$$

Applying a boundary condition, Equation (12.62) gives

$$-c_2 \sqrt{\frac{s}{a}} - \frac{h}{k}\left(c_2 + \frac{T_i - T_\infty}{s}\right) = 0 \tag{12.63}$$

Solving for c_2 gives

$$c_2 = \frac{h}{k}\frac{T_\infty - T_i}{s} \times \frac{1}{\dfrac{h}{k} + \sqrt{\dfrac{s}{a}}} \tag{12.64}$$

Therefore,

$$\overline{\theta} = \frac{h}{k}\frac{T_\infty - T_i}{s} \times \frac{1}{\dfrac{h}{k} + \sqrt{\dfrac{s}{a}}}\, e^{-x\sqrt{\frac{s}{a}}} = (T_\infty - T_i) \times \frac{\dfrac{h\sqrt{a}}{k}\, e^{-x\sqrt{\frac{s}{a}}}}{s\left(\dfrac{h\sqrt{a}}{k} + \sqrt{s}\right)} \tag{12.65}$$

From Table 12.1

$$\mathcal{L}^{-1}\left\{\frac{be^{-\beta\sqrt{s}}}{s\left(b+\sqrt{s}\right)}\right\} = erfc\left(\frac{\beta}{2\sqrt{t}}\right) - e^{(b\beta+b^2 t)} erfc\left(b\sqrt{t}+\frac{\beta}{2\sqrt{t}}\right) \quad (12.66)$$

Comparing Equation (12.66) to Equation (12.65),

$$b = \frac{h\sqrt{a}}{k} \text{ and } \beta = \frac{x}{\sqrt{a}} \quad (12.67)$$

Applying Equations (12.66) and (12.67) to Equation (12.65) gives

$$\theta(x,t) = (T_\infty - T_i)\left\{erfc\left(\frac{x}{2\sqrt{at}}\right) - e^{\left(\frac{hx}{k}+\frac{h^2 a}{k^2}t\right)} erfc\left(\frac{h\sqrt{at}}{k}+\frac{x}{2\sqrt{at}}\right)\right\}$$

or

$$\frac{T(x,t)-T_i}{T_\infty - T_i} = \left\{erfc\left(\frac{x}{2\sqrt{at}}\right) - e^{\left(\frac{hx}{k}+\frac{h^2 a}{k^2}t\right)} erfc\left(\frac{h\sqrt{at}}{k}+\frac{x}{2\sqrt{at}}\right)\right\} \quad (12.68)$$

12.9 Laplace Transforms and Complex Variables

Up to now, we only considered s to be a real number. However, one may also consider s to be a complex number; that is, $s = x + iy$. Then

$$F(s) = \int_0^\infty f(t)e^{-st}\, dt = u(x, y) + iv(x, y) \quad (12.69)$$

where $f(t)$ is a real-valued function that is piecewise continuous (see Reference [1] for additional conditions for $f(t)$) and $F(s)$ is analytic in the half plane. A function $F(s)$ is said to be analytic in a domain, D, if it is defined and differentiable at all points of D. It can be shown [1] that

$$f(t) = \frac{1}{2\pi i}\lim_{\beta \to \infty}\int_{\gamma-i\beta}^{\gamma+i\beta} F(s)e^{st}\, ds \quad (12.70)$$

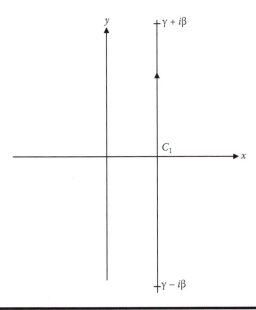

Figure 12.7 Integration of $F(S)$ on line C_1 in the complex plane gives $f(t)$.

where the line integral $(\gamma - i\beta \rightarrow \gamma + i\beta)$ is the curve C_1 in the complex plane shown in Figure 12.7. It is convenient, for reasons that will become clear later, to combine curve C_1 with curve C_2 forming the closed path as shown in Figure 12.8. Residue theory (discussed in the following section) provides the means for determining the value of the line integrals around a closed path without actually carrying out the

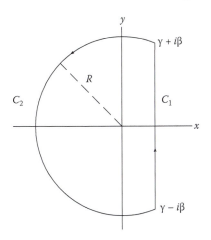

Figure 12.8 Region in the complex plane for determining $f(t)$.

line integrations. Suppose that the value of the line integrals around the closed path of curves C_1 and C_2 by residue theory is $g(t)$, then

$$2\pi i\, g(t) = \int_{\gamma-i\beta}^{\gamma+i\beta} F(s)e^{st}\, ds + \int_{C_2} F(s)e^{st}\, ds \tag{12.71}$$

It can be shown [1] that $\lim_{\beta\to\infty}\int_{C_2} F(s)e^{st}\, ds \to 0$. Thus,

$$g(t) = f(t) = \frac{1}{2\pi i}\lim_{\beta\to\infty}\oint_{C_1+C_2} F(s)e^{st}\, ds \tag{12.72}$$

12.9.1 Residues and Poles [3]

If s_0 is an isolated singular point of $G(s)$ it can be expressed as a Laurent series

$$G(s) = \sum_{n=1}^{\infty} a_n\,(s-s_0)^n + \frac{b_1}{s-s_0} + \frac{b_2}{(s-s_0)^2} + \cdots + \frac{b_m}{(s-s_0)^m} + \cdots \tag{12.73}$$

The term b_1 has particular significance, that is,

$$b_1 = \frac{1}{2\pi i}\oint_C G(s)\, ds \tag{12.74}$$

where C is a closed curve enclosing the singular point.

If in Equation (12.73), $b_{m+1}, b_{m+2}, \ldots$ are all zero and $b_m \neq 0$, then $G(s_0)$ has a *pole* of order m. If $G(s)$ is singular at s_0 and $\phi(s) = (s-s_0)^m\, G(s)$ removes the singularity, then

$$b_1 = \frac{\phi^{(m-1)}(s_0)}{(m-1)!} \tag{12.75}$$

where

$$\phi^{(m-1)}(s_0) = \frac{d^{m-1}\phi}{ds^{(m-1)}}(s_0)$$

If $m = 1$,

$$b_1 = \phi(s_0) = \lim_{s\to s_0}(s-s_0)G(s) \tag{12.76}$$

If

$$G(s) = \frac{p(s)}{q(s)}$$

where both $p(s)$ and $q(s)$ are analytic at s_0 and $p\,(s_0) \neq 0$, then both p and q can be expanded in a Taylor Series about s_0 giving

$$G(s) = \frac{p(s_0) + p'(s_0)(s-s_0) + \dfrac{1}{2!}p''(s_0)(s-s_0)^2 + \cdots}{q(s_0) + q'(s_0)(s-s_0) + \dfrac{1}{2!}q''(s_0)(s-s_0)^2 + \cdots}$$

If $G\,(s)$ has a simple pole at s_0, then $q_0(s_0) = 0, q'(s_0) \neq 0$, and

$$(s-s_0)G(s) = \frac{p(s_0) + p'(s_0)(s-s_0) + \dfrac{1}{2!}p''(s_0)(s-s_0)^2 + \cdots}{q'(s_0) + \dfrac{1}{2!}q''(s_0)(s-s_0) + \cdots}$$

By Equation (12.76)

$$b_1 = \frac{p(s_0)}{q'(s_0)}$$

The term b_1 is called the *residue* of $G\,(s)$ at the isolated singular point s_0. If curve C encloses n isolated singular points (see Figure 12.9) of $G\,(s)$, then

$$\oint_C G(s)\,ds = 2\pi i(K_1 + K_2 + \cdots + K_n) \tag{12.77}$$

where K_j = the residue at the jth isolated singular point.

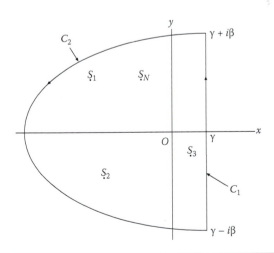

Figure 12.9   Region enclosing several poles.

Defining $G(s) = F(s)e^{st}$ and comparing Equation (12.72) with Equation (12.77), we see that

$$g(t) = f(t) = (K_1 + K_2 + \cdots + K_n) \tag{12.78}$$

Example 12.11

Suppose we take the $Y(s)$ from Example 12.5 and obtain $y(t)$ by residue theory. In that example

$$Y(s) = \frac{2s}{s^2+1} + \frac{5}{(s+1)(s^2+1)} = \frac{2s^2 + 2s + 5}{(s+1)(s^2+1)} \tag{12.79}$$

The functions $Y(s)$ corresponds to $F(s)$ and $y(t)$ corresponds to $f(t)$ in Equation (12.71). Thus,

$$y(t) = \sum_n K_n$$

where K_n is the nth residue of $G(s) = Y(s)e^{st}$, which has simple poles at $s = -1$ and $s = \pm i$.

Pole at $s = -1$:

$$K_1 = \lim_{s \to -1} (s+1)G(s) = \frac{2-2+5}{2}e^{-t} = \frac{5}{2}e^{-t} \tag{12.80}$$

Pole at $s = +i$:

$$K_2 = \lim_{s \to i} (s-i)G(s) = \frac{-2+2i+5}{(1+i)(2i)}e^{it} = \frac{3+2i}{-2+2i}(\cos t + i \sin t)$$

$$K_2 = \frac{1}{8}(-2\cos t + 10\sin t - 10i\cos t - 2i\sin t) \tag{12.81}$$

Pole at $s = -i$:

$$K_3 = \lim_{s \to -i} (s+i)G(s) = \frac{-2-2i+5}{(1-i)(-2i)}(\cos t - i\sin t)$$

$$K_3 = -\frac{1}{8}(2\cos t - 10\sin t - 10i\cos t - 2i\sin t) \tag{12.82}$$

Adding K_1, K_2, and K_3 gives

$$y(t) = \frac{5}{2}e^{-t} - \frac{1}{2}\cos t + \frac{5}{2}\sin t \qquad (12.83)$$

which is the same answer obtained in Example 12.5.

Exercises

Exercise 12.1

Determine $\mathcal{L}(\sin^2 \omega t)$.

Exercise 12.2

Determine $\mathcal{L}(e^{2t} \cos t)$.

Exercise 12.3

Determine $\mathcal{L}(e^{-3t} \sin 2t)$.

Exercise 12.4

Determine the Laplace Transform of the function shown graphically in Figure E12.4.

Exercise 12.5

Determine $\mathcal{L}^{-1}\left(\dfrac{2s+1}{(s^2 + 9)(s-2)}\right)$.

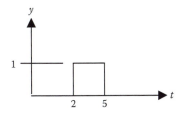

Figure E12.4 Function y (t).

Exercise 12.6

Determine $\mathscr{L}^{-1}\left(\dfrac{1}{(s^2+3s+2)}\right)$.

Exercise 12.7

Determine $\mathscr{L}^{-1}\left(\dfrac{1}{(s+1)(s^2+3s+2)}\right)$ by convolution.

Exercise 12.8

Solve the following differential equation by Laplace Transforms:

$$y''+3y'+2y=3e^{-t},\quad y(0)=1,\quad y'(0)=2$$

Exercise 12.9

Determine $y\,(t)$ of Example 12.6 by theory of residues in complex variables.

References

1. Churchill, R. V., *Operational Mathematics*, 2nd ed., McGraw-Hill, New York, 1958.
2. Kreyszig, E., *Advanced Engineering Mathematics*, 8th ed., John Wiley & Sons, New York, 1999.
3. Churchill, R. V., *Introduction to Complex Variables and Applications*, McGraw-Hill, New York, 1948.

Chapter 13

An Introduction to the Finite Element Method

C. T. Tsai

13.1 Finite Element Method for Stress Analysis

The finite element method is a powerful mathematical tool to numerically calculate the stresses and deformation in structures with complicated geometry, boundary conditions, and material properties. For a two-dimensional stress analysis problem, a solution can be obtained by solving a partial differential equation (PDE) in MATLAB®, where a complicated geometry can be built and meshed. Then the PDE can be discretized on the mesh and solved, providing an approximate solution. The pdetool graphical user interface in MATLAB provides easy-to-use graphical tools to describe complicated domains and to generate triangular meshes. It also discretizes PDEs, finds discrete solutions, and plots results. To overview the fundamentals of the finite element method (FEM) and understand how finite element formulations are developed for various PDEs, a structural mechanics plane stress problem is described below.

13.2 Structural Mechanics Plane Stress Analysis

In structural mechanics, the equations relating stress and strain arise from the applied forces and constraints in the material medium. *Plane stress* is a condition that prevails in a flat plate in the (x, y) plane, loaded only in its own plane and without z-direction restraints; that is, all the stress components in the z direction vanish.

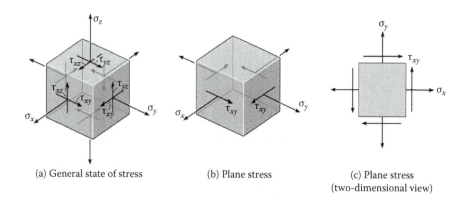

(a) General state of stress   (b) Plane stress   (c) Plane stress
(two-dimensional view)

Figure 13.1   Illustration of stress at a point in a body.

A general description of plane stress at any point of a solid is shown in Figure 13.1, where all the stress components in the z direction (σ_z, τ_{yz}, and τ_{xz}) are zero.

The three-dimensional generalized Hooke's law for an isotropic material is [1,2]

$$\varepsilon_x = \frac{1}{E}[\sigma_x - \upsilon(\sigma_y + \sigma_z)] \tag{13.1}$$

$$\varepsilon_y = \frac{1}{E}[\sigma_y - \upsilon(\sigma_x + \sigma_z)] \tag{13.2}$$

$$\varepsilon_z = \frac{1}{E}[\sigma_z - \upsilon(\sigma_x + \sigma_y)] \tag{13.3}$$

$$\gamma_{xy} = \frac{\tau_{xy}}{G}, \qquad \gamma_{yz} = \frac{\tau_{yz}}{G}, \qquad \gamma_{xz} = \frac{\tau_{xz}}{G} \tag{13.4}$$

where E is the *elastic modulus* or *Young's modulus*, υ is the Poisson's ratio, and G is the shear modulus, which is defined by

$$G = \frac{E}{2(1+\upsilon)}$$

For plane stress, Equations (13.1) through (13.4) become

$$\varepsilon_x = \frac{1}{E}[\sigma_x - \upsilon(\sigma_y)] \tag{13.5}$$

$$\varepsilon_y = \frac{1}{E}[\sigma_y - \upsilon(\sigma_x)] \tag{13.6}$$

$$\varepsilon_z = \frac{1}{E}[-\upsilon(\sigma_x + \sigma_y)] \tag{13.7}$$

$$\gamma_{xy} = \frac{\tau_{xy}}{G} = \frac{2(1+\upsilon)\tau_{xy}}{E} \tag{13.8}$$

Equation (13.7) indicates that without z-direction restraints, the strain εz may not be zero, even though the stress in the z direction, σ_z, is zero. Equations (13.5) through (13.8) can be written in matrix form as shown by Equation (13.9).

$$\begin{bmatrix} \varepsilon_x \\ \varepsilon_y \\ \gamma_{xy} \end{bmatrix} = \frac{1}{E} \begin{bmatrix} 1 & -\upsilon & 0 \\ -\upsilon & 1 & 0 \\ 0 & 0 & 2(1+\upsilon) \end{bmatrix} \begin{bmatrix} \sigma_x \\ \sigma_y \\ \tau_{xy} \end{bmatrix} \tag{13.9}$$

The inverse of Equation (13.9) gives the stress–strain relationship for the plane stress problem:

$$[\sigma] = \begin{bmatrix} \sigma_x \\ \sigma_y \\ \tau_{xy} \end{bmatrix} = \frac{E}{1-\upsilon^2} \begin{bmatrix} 1 & \upsilon & 0 \\ \upsilon & 1 & 0 \\ 0 & 0 & \frac{1-\upsilon}{2} \end{bmatrix} \begin{bmatrix} \varepsilon_x \\ \varepsilon_y \\ \gamma_{xy} \end{bmatrix} = [D][\varepsilon] \tag{13.10}$$

Matrix $[D]$ in Equation (13.10) is denoted as the material matrix for the plane stress problem.

The strain components of the material are described by the displacement components u and v in the x and y directions, respectively, and are defined by

$$\varepsilon_x = \frac{\partial u}{\partial x}, \quad \varepsilon_y = \frac{\partial v}{\partial y}, \quad \text{and} \quad \gamma_{xy} = \frac{\partial u}{\partial y} + \frac{\partial v}{\partial x} \tag{13.11}$$

The equilibrium equations are

$$\frac{\partial \sigma_x}{\partial x} + \frac{\partial \tau_{xy}}{\partial y} + f_x = 0 \tag{13.12}$$

$$\frac{\partial \tau_{xy}}{\partial x} + \frac{\partial \sigma_y}{\partial y} + f_y = 0 \tag{13.13}$$

where f_x and f_y are body forces in x and y directions, respectively. Combining Equations (13.10) and (13.11), and substituting the combination into Equations

(13.12) and (13.13) gives the PDE in terms of displacement components, u and v, as shown by Equations (13.14) and (13.15):

$$\frac{\partial^2 u}{\partial x^2} + \frac{\partial^2 u}{\partial y^2} + \frac{1+\upsilon}{1-\upsilon}\frac{\partial}{\partial x}\left(\frac{\partial u}{\partial x} + \frac{\partial v}{\partial y}\right) + \frac{2(1+\upsilon)}{E}f_x = 0 \qquad (13.14)$$

$$\frac{\partial^2 v}{\partial x^2} + \frac{\partial^2 v}{\partial y^2} + \frac{1+\upsilon}{1-\upsilon}\frac{\partial}{\partial y}\left(\frac{\partial u}{\partial x} + \frac{\partial v}{\partial y}\right) + \frac{2(1+\upsilon)}{E}f_y = 0 \qquad (13.15)$$

Equations (13.14) and (13.15) can be rearranged into the following:

$$G\left(\frac{\partial^2 u}{\partial x^2} + \frac{\partial^2 u}{\partial y^2}\right) + (G+\mu)\frac{\partial}{\partial x}\left(\frac{\partial u}{\partial x} + \frac{\partial v}{\partial y}\right) + f_x = 0 \qquad (13.16)$$

$$G\left(\frac{\partial^2 v}{\partial x^2} + \frac{\partial^2 v}{\partial y^2}\right) + (G+\mu)\frac{\partial}{\partial y}\left(\frac{\partial u}{\partial x} + \frac{\partial v}{\partial y}\right) + f_y = 0 \qquad (13.17)$$

where μ is defined as

$$\mu = \frac{E\upsilon}{(1+\upsilon)(1-2\upsilon)}$$

Equations (13.16) and (13.17) are elliptic PDEs. In MATLAB [3] $\boldsymbol{u}$ is a vector of two dimensions, $\begin{bmatrix} u \\ v \end{bmatrix}$, and is written in as

$$-\nabla \cdot (c\nabla u) = f \qquad (13.18)$$

where c is a rank-four tensor, which can be represented by four 2×2 matrices, c_{11}, c_{12}, c_{21}, and c_{22}. For homogeneous isotropic elastic materials, they are defined as

$$c_{11} = \begin{bmatrix} 2G+\mu & 0 \\ 0 & G \end{bmatrix} \qquad (13.19)$$

$$c_{12} = \begin{bmatrix} 0 & \mu \\ G & 0 \end{bmatrix} \qquad (13.20)$$

$$c_{21} = \begin{bmatrix} 0 & G \\ \mu & 0 \end{bmatrix} \qquad (13.21)$$

$$c_{22} = \begin{bmatrix} G & 0 \\ 0 & 2G+\mu \end{bmatrix} \qquad (13.22)$$

The body force, f, is defined by

$$f = \begin{bmatrix} f_x \\ f_y \end{bmatrix} \tag{13.23}$$

Substituting Equations (13.19) through (13.23) into Equation (13.18) gives

$$-\begin{bmatrix} \dfrac{\partial}{\partial x} & \dfrac{\partial}{\partial y} \end{bmatrix}(c_{11}\nabla u + c_{12}\nabla v) = -\begin{bmatrix} \dfrac{\partial}{\partial x} & \dfrac{\partial}{\partial y} \end{bmatrix}$$

$$\left(\begin{bmatrix} 2G+\mu & 0 \\ 0 & G \end{bmatrix} \begin{bmatrix} \dfrac{\partial u}{\partial x} \\ \dfrac{\partial u}{\partial y} \end{bmatrix} + \begin{bmatrix} 0 & \mu \\ G & 0 \end{bmatrix} \begin{bmatrix} \dfrac{\partial v}{\partial x} \\ \dfrac{\partial v}{\partial y} \end{bmatrix} \right) \tag{13.24}$$

$$= -(2G+\mu)\dfrac{\partial^2 u}{\partial x^2} - \mu\dfrac{\partial^2 v}{\partial x\partial y} - G\dfrac{\partial^2 u}{\partial y^2} - G\dfrac{\partial^2 v}{\partial x\partial y} = f_x$$

$$-\begin{bmatrix} \dfrac{\partial}{\partial x} & \dfrac{\partial}{\partial y} \end{bmatrix}\left(c_{21}\nabla u + c_{22}\nabla v\right) = -\begin{bmatrix} \dfrac{\partial}{\partial x} & \dfrac{\partial}{\partial y} \end{bmatrix}$$

$$\left(\begin{bmatrix} 0 & G \\ \mu & 0 \end{bmatrix} \begin{bmatrix} \dfrac{\partial u}{\partial x} \\ \dfrac{\partial u}{\partial y} \end{bmatrix} + \begin{bmatrix} G & 0 \\ 0 & 2G+\mu \end{bmatrix} \begin{bmatrix} \dfrac{\partial v}{\partial x} \\ \dfrac{\partial v}{\partial y} \end{bmatrix} \right) \tag{13.25}$$

$$= -(2G+\mu)\dfrac{\partial^2 v}{\partial y^2} - \mu\dfrac{\partial^2 u}{\partial x\partial y} - G\dfrac{\partial^2 v}{\partial x^2} - G\dfrac{\partial^2 u}{\partial x\partial y} = f_y$$

Equations (13.24) and (13.25) are identical with Equations (13.16) and (13.17). Therefore, the same finite element equations can be formulated by using either Equations (13.12) and (13.13) or Equation (13.18).

In the FEM, the domain Ω is divided into a finite number of simple geometric objects such as triangles or quadrilaterals. In MATLAB's pdftool, only 3-node triangles are used to approximate the computational domain Ω. The triangles form a mesh and each vertex is called a node. Each triangle is counted as one element. Since each edge of the triangle is a straight line, the curved boundaries of a domain cannot be modeled accurately using just a few elements. Therefore, in regions near

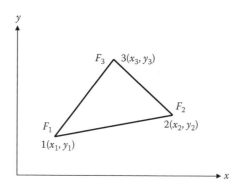

Figure 13.2   Linear triangular scalar field element.

curved boundaries, more triangular elements are needed to improve the accuracy of the solution.

In the FEM, a triangular element (see Figure 13.2) is created with known polynomials for the field variables, such as u and v, within the element. The polynomials interpolate the field variables at each node by the concept of a shape function. This concept is described in the next section.

13.3 The Shape Function for a Linear Triangle Element

For a 3-node triangle element with three prescribed values of field variables F_1, F_2, and F_3 at nodes 1, 2, and 3, as shown in Figure 13.2, one can assume a linear interpolation function for the field variable $F(x, y)$ as [4]

$$F(x, y) = a + bx + cy$$

$$= \lfloor 1 \quad x \quad y \rfloor \begin{Bmatrix} a \\ b \\ c \end{Bmatrix} \tag{13.26}$$

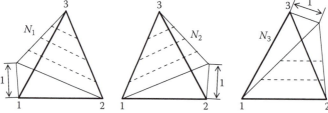

Figure 13.3   Illustration of three shape functions.

The values of F at the nodes shown in Figure 13.2 are given by

$$
\begin{Bmatrix} F_1 \\ F_2 \\ F_3 \end{Bmatrix} = \begin{bmatrix} 1 & x_1 & y_1 \\ 1 & x_2 & y_2 \\ 1 & x_3 & y_3 \end{bmatrix} \begin{Bmatrix} a \\ b \\ c \end{Bmatrix} = [A] \begin{Bmatrix} a \\ b \\ c \end{Bmatrix} \tag{13.27}
$$

The inverse of Equation (13.27) is

$$
\begin{Bmatrix} a \\ b \\ c \end{Bmatrix} = \frac{1}{|A|} \begin{bmatrix} x_2 y_3 - x_3 y_2 & x_3 y_1 - x_1 y_3 & x_1 y_2 - x_2 y_1 \\ y_2 - y_3 & y_3 - y_1 & y_1 - y_2 \\ x_3 - x_2 & x_1 - x_3 & x_2 - x_1 \end{bmatrix} \begin{Bmatrix} F_1 \\ F_2 \\ F_3 \end{Bmatrix} \tag{13.28}
$$

where

$$
|A| = x_2 y_3 - x_3 y_2 + x_3 y_1 - x_1 y_3 + x_1 y_2 - x_2 y_1 \tag{13.29}
$$

Substituting Equation (13.28) into Equation (13.26) gives

$$
F(x, y) = \frac{\lfloor 1 \quad x \quad y \rfloor}{|A|} \begin{bmatrix} x_2 y_3 - x_3 y_2 & x_3 y_1 - x_1 y_3 & x_1 y_2 - x_2 y_1 \\ y_2 - y_3 & y_3 - y_1 & y_1 - y_2 \\ x_3 - x_2 & x_1 - x_3 & x_2 - x_1 \end{bmatrix} \begin{Bmatrix} F_1 \\ F_2 \\ F_3 \end{Bmatrix}
$$

$$
= \lfloor N_1(x, y) \quad N_2(x, y) \quad N_3(x, y) \rfloor \begin{Bmatrix} F_1 \\ F_2 \\ F_3 \end{Bmatrix} \tag{13.30}
$$

where

$$
N_1(x, y) = \frac{x_2 y_3 - x_3 y_2 + x(y_2 - y_3) + y(x_3 - x_2)}{|A|} \tag{13.31}
$$

$$
N_2(x, y) = \frac{x_3 y_1 - x_1 y_3 + x(y_3 - y_1) + y(x_1 - x_3)}{|A|} \tag{13.32}
$$

$$
N_3(x, y) = \frac{x_1 y_2 - x_2 y_1 + x(y_1 - y_2) + y(x_2 - x_1)}{|A|} \tag{13.33}
$$

The illustration of three shape functions is shown in Figure 13.3. The summation of three shape functions equals unity.

13.3.1 3-Node Triangular Element for 2-D Stress Analysis

For two-dimensional (2-D) stress analysis, there are two displacement components, u and v, at each node as shown in Figure 13.4. The interpolation function of each component, u(x, y) and v(x, y), is given as [4–6]

$$u(x, y) = \lfloor 1 \quad x \quad y \rfloor \begin{Bmatrix} a_1 \\ b_1 \\ c_1 \end{Bmatrix} \quad v(x, y) = \lfloor 1 \quad x \quad y \rfloor \begin{Bmatrix} a_2 \\ b_2 \\ c_2 \end{Bmatrix} \quad (13.34)$$

Since the displacement components within the element are a linear function of x and y, the strains are constants as given below:

$$\varepsilon_x = \frac{\partial u}{\partial x} = b_1 \qquad \varepsilon_y = \frac{\partial v}{\partial y} = c_2 \qquad \gamma_{xy} = \frac{\partial u}{\partial y} + \frac{\partial v}{\partial x} = c_1 + b_2 \quad (13.35)$$

Conclusion

The strains and stresses are constant within the 3-node triangular element. Therefore, this element is also called constant-strain triangle (CST).

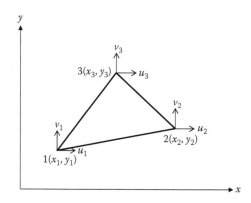

Figure 13.4   Linear triangular element for 2-D stress analysis.

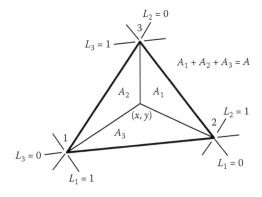

Figure 13.5   Area coordinates in a triangular element.

13.3.2 Shape Function in Area Coordinates

The shape functions shown in Equations (13.31) through (13.33) are not suitable for finite element formulation. For finite element formulation, it is essential to map physical elements of various sizes and shapes into the natural or intrinsic coordinates. For the triangular element, an area coordinate is used as natural or intrinsic coordinates. Figure 13.5 shows the definition of area coordinates, where a point (x, y) in the triangle can divide the area A into three areas, A_1, A_2, and A_3.

From Figure 13.5 it can be seen that

$$\frac{A_1}{A} + \frac{A_2}{A} + \frac{A_3}{A} = 1 \tag{13.36}$$

Define

$$L_1 = \frac{A_1}{A} \qquad L_2 = \frac{A_2}{A} \qquad L_3 = \frac{A_3}{A} \tag{13.37}$$

Then every Cartesian coordinate (x, y) can be also represented by area coordinate (L_1, L_2, L_3). From Equation (13.37), it can be seen that $L_1 = 1$ at node 1 and 0 at nodes 2 and 3; $L_2 = 1$ at node 2 and 0 at nodes 1 and 3; $L_3 = 1$ at node 3 and 0 at nodes 1 and 2 as shown in Figure 13.5. Also by Equation (13.36), $L_1 + L_2 + L_3 = 1$. Therefore, the shape function N_1, N_2, and N_3 can be defined as

$$N_1 = L_1 \qquad N_2 = L_2 \qquad N_3 = L_3 = 1 - L_1 - L_2 \tag{13.38}$$

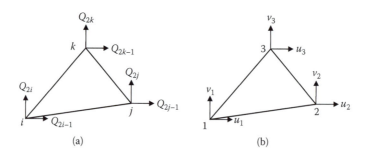

Figure 13.6 **(a) Global displacement components for an element with node numbers i, j, and k. (b) Corresponding displacement components in a local element, where local node numbers 1, 2, and 3 are corresponding to global node numbers i, j, and k, respectively.**

Now, Equation (13.30) can be also written as an interpolation of displacement components at the three nodes as given in Equation (13.39):

$$u = \lfloor N_1 \quad N_2 \quad N_3 \rfloor \begin{Bmatrix} u_1 \\ u_2 \\ u_3 \end{Bmatrix}, \quad v = \lfloor N_1 \quad N_2 \quad N_3 \rfloor \begin{Bmatrix} v_1 \\ v_2 \\ v_3 \end{Bmatrix} \quad (13.39)$$

The illustration of these shape functions is also shown in Figure 13.3.

Now the displacement solution is a simple linear function of x and y coordinates, on each triangle. If the exact displacement solutions in the whole domain are also linear functions of x and y, one could obtain an exact solution by using the triangular elements in finite element analysis. If the exact displacement solutions in the whole domain are not linear functions, one can still obtain an approximate solution by using more triangular elements in the whole domain.

13.3.3 Finite Element Formulation Using Triangular Elements

Same finite element equations can be derived by using either Equations (13.12) and (13.13) or (13.18). Equations (13.12) and (13.13) will be used here to derive the finite element equation. The same equation can be derived in the MATLAB, which uses Equation (13.18). Equations (13.12) and (13.13), which are in the form of stress components, will be converted to displacement components $\boldsymbol{u}$. What we are looking for is the best approximation of $\boldsymbol{u}$ in the class of continuous piecewise polynomials, and we need to test the equation for $\boldsymbol{u}$ against all possible functions $\boldsymbol{v}$

of that class. Testing means formally to multiply the residual against any function and then integrate, that is, determine *u* such that [4–6]

$$\int_\Omega \left[\left(\frac{\partial \sigma_x}{\partial x} + \frac{\partial \tau_{xy}}{\partial y} + f_x \right) \delta u + \left(\frac{\partial \tau_{xy}}{\partial x} + \frac{\partial \sigma_y}{\partial y} + f_y \right) \delta v \right] dA = 0 \qquad (13.40)$$

for all possible δu and δv in the area domain Ω. The functions δu and δv are usually called weighting functions or *test functions*. In stress analysis these functions are defined as virtual displacements.

Integrating Equation (13.40) by parts (Green's formula or divergence theorem) gives

$$\int_\Omega \frac{\partial \sigma_x}{\partial x} \delta u \, dA = -\int_\Omega \sigma_x \frac{\partial \delta u}{\partial x} dA + \int_S n_x \sigma_x \, \delta u \, dS \qquad (13.41)$$

$$\int_\Omega \frac{\partial \tau_{xy}}{\partial y} \delta u \, dA = -\int_\Omega \tau_{xy} \frac{\partial \delta u}{\partial y} dA + \int_S n_y \tau_{xy} \, \delta u \, dS \qquad (13.42)$$

$$\int_\Omega \frac{\partial \tau_{xy}}{\partial x} \delta v \, dA = -\int_\Omega \tau_{xy} \frac{\partial \delta v}{\partial x} dA + \int_S n_x \tau_{xy} \, \delta v \, dS \qquad (13.43)$$

$$\int_\Omega \frac{\partial \sigma_y}{\partial y} \delta v \, dA = -\int_\Omega \sigma_y \frac{\partial \delta v}{\partial y} dA + \int_S n_y \sigma_y \, \delta v \, dS \qquad (13.44)$$

where S is the boundary of the area domain Ω and $\vec{n} = n_x \vec{i} + n_y \vec{j}$ is the unit vector normal to the boundary S. Substituting Equations (13.41) through (13.44) into Equation (13.40) gives

$$-\int_\Omega \begin{bmatrix} \dfrac{\partial \delta u}{\partial x} \\[2mm] \dfrac{\partial \delta v}{\partial y} \\[2mm] \dfrac{\partial \delta v}{\partial x} + \dfrac{\partial \delta u}{\partial y} \end{bmatrix}^T \begin{bmatrix} \sigma_x \\ \sigma_y \\ \tau_{xy} \end{bmatrix} dA + \int_\Omega \begin{bmatrix} \delta u \\ \delta v \end{bmatrix}^T \begin{bmatrix} f_x \\ f_y \end{bmatrix} dA$$

$$+ \int_S [(n_x \sigma_x + n_y \tau_{xy}) \delta u + (n_x \tau_{xy} + n_y \sigma_y) \delta v] dS = 0 \qquad (13.45)$$

where

$$T_x = n_x \sigma_x + n_y \tau_{xy} = n_x \frac{E}{1-\upsilon^2}(\varepsilon_x + \upsilon\varepsilon_y) + n_y \frac{E}{2(1+\upsilon)}\gamma_{xy} \qquad (13.46)$$

$$T_y = n_x \tau_{xy} + n_y \sigma_y = n_x \frac{E}{2(1+\upsilon)}\gamma_{xy} + n_y \frac{E}{1-\upsilon^2}(\varepsilon_y + \upsilon\varepsilon_x) \qquad (13.47)$$

Substituting Equation (13.11) into Equations (13.46) and (13.47) gives

$$T_x = n_x \frac{E}{1-\upsilon^2}\left(\frac{\partial u}{\partial x} + \upsilon\frac{\partial v}{\partial y}\right) + n_y \frac{E}{2(1+\upsilon)}\left(\frac{\partial v}{\partial x} + \frac{\partial u}{\partial y}\right) \qquad (13.48)$$

$$T_y = n_x \frac{E}{2(1+\upsilon)}\left(\frac{\partial v}{\partial x} + \frac{\partial u}{\partial y}\right) + n_y \frac{E}{1-\upsilon^2}\left(\frac{\partial v}{\partial y} + \upsilon\frac{\partial u}{\partial x}\right) \qquad (13.49)$$

Equations (13.48) and (13.49) are called Neumann boundary conditions. Substituting Equations (13.48) and (13.49) into Equation (13.45) gives

$$-\int_\Omega \begin{bmatrix} \dfrac{\partial \delta u}{\partial x} \\[2mm] \dfrac{\partial \delta v}{\partial y} \\[2mm] \dfrac{\partial \delta v}{\partial x} + \dfrac{\partial \delta u}{\partial y} \end{bmatrix}^T \begin{bmatrix} \sigma_x \\ \sigma_y \\ \tau_{xy} \end{bmatrix} dA + \int_\Omega \begin{bmatrix} \delta u \\ \delta v \end{bmatrix}\begin{bmatrix} f_x \\ f_y \end{bmatrix} dA + \int_S \begin{bmatrix} \delta u \\ \delta v \end{bmatrix}^T \begin{bmatrix} T_x \\ T_y \end{bmatrix} dS = 0$$

$$(13.50)$$

By using Equations (13.10) and (13.11), Equation (13.50) can be expressed in terms of 2-D displacement field u and v as given by Equation (13.51):

$$-\int_\Omega \begin{bmatrix} \dfrac{\partial \delta u}{\partial x} \\[2mm] \dfrac{\partial \delta v}{\partial y} \\[2mm] \dfrac{\partial \delta v}{\partial x} + \dfrac{\partial \delta u}{\partial y} \end{bmatrix}^T [D] \begin{bmatrix} \dfrac{\partial u}{\partial x} \\[2mm] \dfrac{\partial v}{\partial y} \\[2mm] \dfrac{\partial v}{\partial x} + \dfrac{\partial u}{\partial y} \end{bmatrix} dA + \int_\Omega \begin{bmatrix} \delta u \\ \delta v \end{bmatrix}^T \begin{bmatrix} f_x \\ f_y \end{bmatrix} dA$$

$$+ \int_S \begin{bmatrix} \delta u \\ \delta v \end{bmatrix}^T \begin{bmatrix} T_x \\ T_y \end{bmatrix} dS = 0 \qquad (13.51)$$

If the domain is divided into a finite number of triangular elements, Equation (13.51) becomes

$$
-\sum_{i=1}^{n}\int_{e_i}
\begin{bmatrix}
\dfrac{\partial \delta u}{\partial x} \\[8pt]
\dfrac{\partial \delta v}{\partial y} \\[8pt]
\dfrac{\partial \delta v}{\partial x} + \dfrac{\partial \delta u}{\partial y}
\end{bmatrix}^T
[D]
\begin{bmatrix}
\dfrac{\partial u}{\partial x} \\[8pt]
\dfrac{\partial v}{\partial y} \\[8pt]
\dfrac{\partial v}{\partial x} + \dfrac{\partial u}{\partial y}
\end{bmatrix}
dA
+ \sum_{i=1}^{n}\int_{e_i}
\begin{bmatrix}
\delta u \\
\delta v
\end{bmatrix}^T
\begin{bmatrix}
f_x \\
f_y
\end{bmatrix}
dA
$$

(13.52)

$$
+ \sum_{j=1}^{m}\int_{s_j}
\begin{bmatrix}
\delta u \\
\delta v
\end{bmatrix}^T
\begin{bmatrix}
T_x \\
T_y
\end{bmatrix}
ds = 0
$$

where n is the total number of elements in the domain Ω, and m is the total number of element edges subjected to Neumann boundary conditions. For a 3-node triangular element, both displacement fields (u and v) and weighting functions (δu and δv) have the same interpolation function as shown in Equation (13.39). Therefore, any triangular element with global node numbers i, j, and k can be replaced by a triangular element with local node numbers 1, 2, and 3 as shown in Figure 13.6. In addition, the element stiffness matrix and load vector can also be expressed in terms of local node number. The strains in an element can now be expressed as

$$
\begin{bmatrix}
\varepsilon_x \\
\varepsilon_y \\
\gamma_{xy}
\end{bmatrix}
=
\begin{bmatrix}
\dfrac{\partial u}{\partial x} \\[8pt]
\dfrac{\partial v}{\partial y} \\[8pt]
\dfrac{\partial v}{\partial x} + \dfrac{\partial u}{\partial y}
\end{bmatrix}
=
\begin{bmatrix}
\dfrac{\partial N_1}{\partial x} & 0 & \dfrac{\partial N_2}{\partial x} & 0 & \dfrac{\partial N_3}{\partial x} & 0 \\[8pt]
0 & \dfrac{\partial N_1}{\partial y} & 0 & \dfrac{\partial N_2}{\partial y} & 0 & \dfrac{\partial N_3}{\partial y} \\[8pt]
\dfrac{\partial N_1}{\partial y} & \dfrac{\partial N_1}{\partial x} & \dfrac{\partial N_2}{\partial y} & \dfrac{\partial N_2}{\partial x} & \dfrac{\partial N_3}{\partial y} & \dfrac{\partial N_3}{\partial x}
\end{bmatrix}
\begin{bmatrix}
u_1 \\
v_1 \\
u_2 \\
v_2 \\
u_3 \\
v_3
\end{bmatrix}
$$

$$
= [B]_{3\times6}[q]_{6\times1}
$$

(13.53)

The following term can be defined as virtual strains as

$$
\begin{bmatrix} \delta\varepsilon_x \\ \delta\varepsilon_y \\ \delta\gamma_{xy} \end{bmatrix} = \begin{bmatrix} \dfrac{\partial\delta u}{\partial x} \\ \dfrac{\partial\delta v}{\partial y} \\ \dfrac{\partial\delta v}{\partial x} + \dfrac{\partial\delta u}{\partial y} \end{bmatrix} = \begin{bmatrix} \dfrac{\partial N_1}{\partial x} & 0 & \dfrac{\partial N_2}{\partial x} & 0 & \dfrac{\partial N_3}{\partial x} & 0 \\ 0 & \dfrac{\partial N_1}{\partial y} & 0 & \dfrac{\partial N_2}{\partial y} & 0 & \dfrac{\partial N_3}{\partial y} \\ \dfrac{\partial N_1}{\partial y} & \dfrac{\partial N_1}{\partial x} & \dfrac{\partial N_2}{\partial y} & \dfrac{\partial N_2}{\partial x} & \dfrac{\partial N_3}{\partial y} & \dfrac{\partial N_3}{\partial x} \end{bmatrix} \begin{bmatrix} \delta u_1 \\ \delta v_1 \\ \delta u_2 \\ \delta v_2 \\ \delta u_3 \\ \delta v_3 \end{bmatrix}
$$

$$
= [B]_{3\times6}[\delta q]_{6\times1}
$$

(13.54)

where [B] relates the displacement with the strain and is defined as the strain-displacement matrix. Therefore, the first term in Equation (13.52) can be expressed as follows:

$$
\int_{e_i} \begin{bmatrix} \dfrac{\partial\delta u}{\partial x} \\ \dfrac{\partial\delta v}{\partial y} \\ \dfrac{\partial\delta v}{\partial x} + \dfrac{\partial\delta u}{\partial y} \end{bmatrix}^T [D] \begin{bmatrix} \dfrac{\partial u}{\partial x} \\ \dfrac{\partial v}{\partial y} \\ \dfrac{\partial v}{\partial x} + \dfrac{\partial u}{\partial y} \end{bmatrix} dA = [\delta q]^T \int_{e_i} [B]^T[D][B]\,dA[q] = [k]^e[q]
$$

(13.55)

where $[k]^e$ is the stiffness matrix of the triangular element, defined as

$$
[k]^e = \int_{e_i} [B]^T[D][B]\,dA
$$

The second term in Equation (13.52) is related to the body forces, which can be expressed as

$$
\int_{e_i} \begin{bmatrix} \delta u \\ \delta v \end{bmatrix}^T \begin{bmatrix} f_x \\ f_y \end{bmatrix} dA = \begin{bmatrix} \delta u_1 \\ \delta v_1 \\ \delta u_2 \\ \delta v_2 \\ \delta u_3 \\ \delta v_3 \end{bmatrix}^T \int_{e_i} \begin{bmatrix} N_1 & 0 \\ 0 & N_1 \\ N_2 & 0 \\ 0 & N_2 \\ N_3 & 0 \\ 0 & N_3 \end{bmatrix} \begin{bmatrix} f_x \\ f_y \end{bmatrix} dA = [\delta q]^T[F]^e_b
$$

(13.56)

where $[F]_b^e$ is the equivalent load vector of body forces in this element, which is

$$[F]_b^e = \int_{e_i} \begin{bmatrix} N_1 & 0 \\ 0 & N_1 \\ N_2 & 0 \\ 0 & N_2 \\ N_3 & 0 \\ 0 & N_3 \end{bmatrix} \begin{bmatrix} f_x \\ f_y \end{bmatrix} dA = \int_{e_i} \begin{bmatrix} N_1 f_x \\ N_1 f_y \\ N_2 f_x \\ N_2 f_y \\ N_3 f_x \\ N_3 f_y \end{bmatrix} dA = \begin{bmatrix} F_1 \\ F_2 \\ F_3 \\ F_4 \\ F_5 \\ F_6 \end{bmatrix} \qquad (13.57)$$

The equivalent forces are acting at each node of the triangle as shown in Figure 13.7. The third term in Equation (13.52) relates the surface traction boundary condition (that is, the Neumann boundary condition). If the surface traction acts on edge 1–2 of the triangles as shown in Figure 13.8a, this term can be expressed as follows:

$$\int_{s_j} \begin{bmatrix} \delta u \\ \delta v \end{bmatrix}^T \begin{bmatrix} T_x \\ T_y \end{bmatrix} ds = \begin{bmatrix} \delta u_1 \\ \delta v_1 \\ \delta u_2 \\ \delta v_2 \\ \delta u_3 \\ \delta v_3 \end{bmatrix}^T \int_{l_{1-2}} \begin{bmatrix} N_1 & 0 \\ 0 & N_1 \\ N_2 & 0 \\ 0 & N_2 \\ 0 & 0 \\ 0 & 0 \end{bmatrix} \begin{bmatrix} T_x \\ T_y \end{bmatrix} dl = [\delta q]^T [F]_T^s$$

$$(13.58)$$

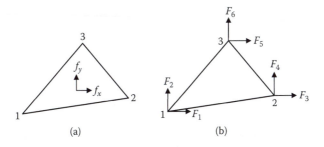

(a)   (b)

Figure 13.7 (a) Body forces acting on an element. (b) Equivalent nodal force at each node due to the body forces.

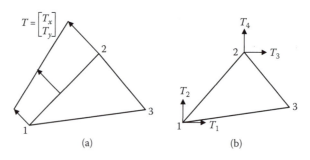

Figure 13.8 **(a) Surface tractions T acting on the edge of an element. (b) Equivalent nodal force at each node due to the surface tractions.**

where $[F]_T^s$ is the equivalent load vector due to surface traction acting on the boundary, which is

$$[F]_T^s = \int_{l_{1-2}} \begin{bmatrix} N_1 & 0 \\ 0 & N_1 \\ N_2 & 0 \\ 0 & N_2 \\ 0 & 0 \\ 0 & 0 \end{bmatrix} \begin{bmatrix} T_x \\ T_y \end{bmatrix} dl = \begin{bmatrix} T_1 \\ T_2 \\ T_3 \\ T_4 \\ 0 \\ 0 \end{bmatrix} \qquad (13.59)$$

The equivalent loads are only acting at nodes 1 and 2 since the surface traction is applied on edge 1–2 as shown in Figure 13.8.

Finally, Equation (13.52) can be rewritten as

$$\sum_{i=1}^{n} [\delta q]_i^T [k]_i^e [q]_i = \sum_{i=1}^{n} [\delta q]_i^T [F]_{bi}^e + \sum_{j=1}^{m} [\delta q]^T [F]_{Tj}^s \qquad (13.60)$$

Now all the triangle elements, with node number i, j, and k as shown in Figure 13.6, are assembled together and expressed in terms of the global displacement components [Q] as shown in Figure 13.6a; then, Equation (13.60) becomes

$$[\delta Q]^T [K][Q] = [\delta Q]^T ([F]_b + [F]_T + [P]) = [\delta Q]^T [F] \qquad (13.61)$$

where [K] is the global stiffness matrix for the whole domain, $[F]_b$ and $[F]_T$ are the global load vector due to body force and surface traction, and [P] is the concentrated force acting at the nodes. [F] is the assembly of all forces acting on the domain. For an arbitrary value of the test function [δQ], Equation (13.61) becomes

$$[K][Q] = [F] \qquad (13.62)$$

Equation (13.62) can be solved by applying the known displacement value at nodes on the boundary (that is, Dirichlet boundary conditions). Once the unknown displacements at nodes are obtained, they will be employed in Equation (13.53) to calculate the strains in each element. Finally, the strains are substituted into Equation (13.10) to calculate the stress component in each element.

13.4 Finite Element Analysis Using MATLAB's PDE Toolbox

MATLAB's PDE Toolbox™ software [3] can be used to numerically solve a 2-D PDE problem by defining the 2-D region, the boundary conditions, the PDE coefficients, and generating meshes for the region. After the problem is solved, the results can be visualized. Advanced applications are also possible by downloading the domain geometry, boundary conditions, and mesh description to the MATLAB workspace. From the command line (or M-files) you can call functions to do the hard work, for example, generate meshes, discretize your problem, perform interpolation, plot data on unstructured grids, etc., while you retain full control over the global numerical algorithm.

Before you start the PDE toolbox, it is better to formulate a PDE problem on paper (draw the domain, write the boundary conditions, and the PDE). To start the PDE toolbox, at the MATLAB command line, type

    pdetool

and press the enter button as shown in Figure 13.9. This invokes the graphical user interface (GUI) as shown in Figure 13.10, which is a self-contained graphical environment for PDE solving.

You can click the Option button as shown in Figure 13.11 to create the grids, define grid spacing, select axes limits, select the type of problem you want to solve, etc. For example, if you are solving a plane stress problem, from the drop-down menu select the "Application Option," then select the "Structural Mechanics, Plane Stress" option.

To draw the geometry of the model, click the Draw button and start to draw solid objects (rectangle, circle, ellipse, and polygon) as shown in Figure 13.12. You can combine these objects by adding or subtracting them to create the geometry you want. Similarly, to apply the boundary conditions, click the Boundary button and you can specify different types of boundary conditions on different edges.

In the PDE mode, you interactively specify the type of PDE and the coefficients c, a, f, and d. You can specify the coefficients for each subdomain independently. For structural mechanics, heat transfer, or other specific problems, you just need to specify the material and the thermal and physical properties in the model.

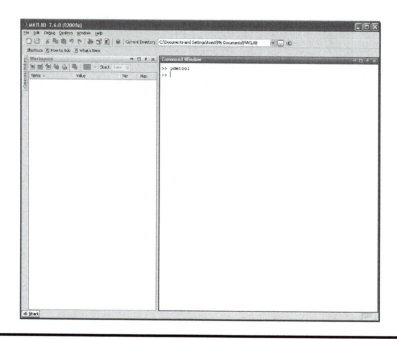

Figure 13.9 MATLAB command window. (From MATLAB. With permission.)

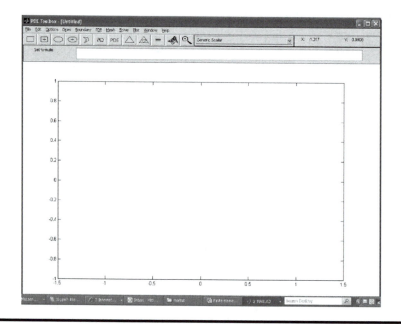

Figure 13.10 MATLAB graphical user interface (GUI) for PDE toolbox. (From MATLAB. With permission.)

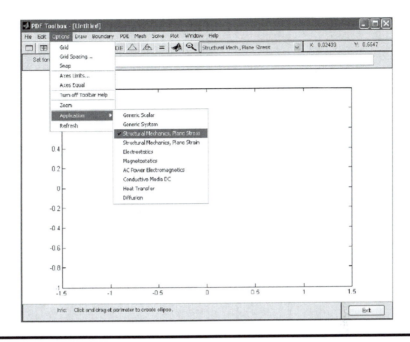

Figure 13.11   Options button. (From MATLAB. With permission.)

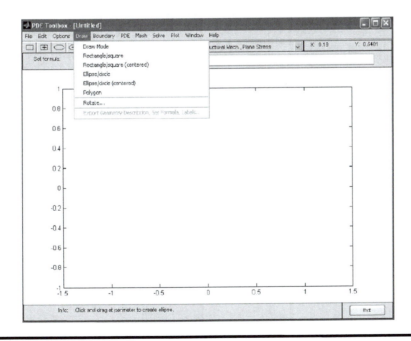

Figure 13.12   Draw button. (From MATLAB. With permission.)

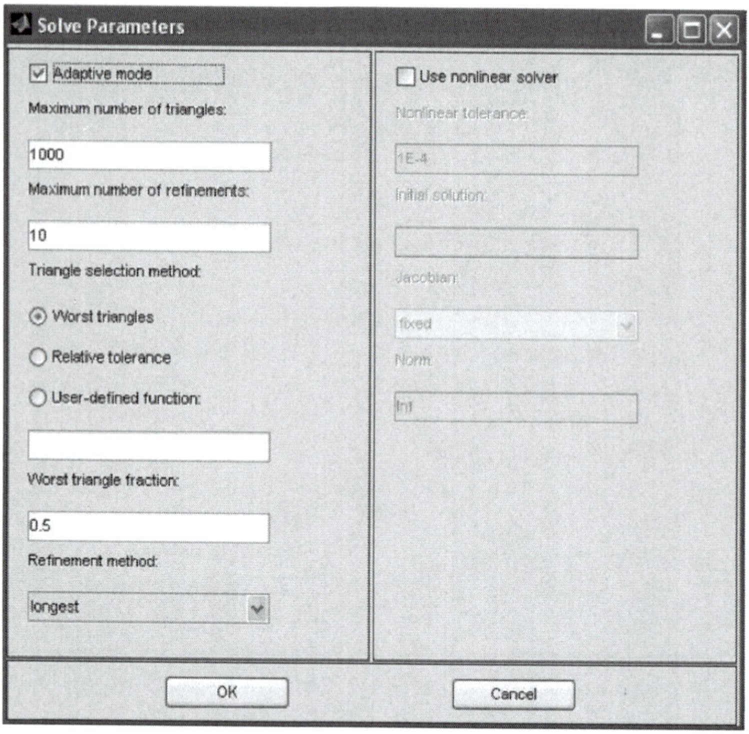

Figure 13.13   Solve parameters window. (From MATLAB. With permission.)

To generate the mesh, click the mesh button. You can control the parameters of the automated mesh generator. After solving a problem, you can return to the mesh mode to further refine your mesh and then solve again. If solutions from both meshes are sufficiently close, it indicates that the solution has converged to an accurate solution and further mesh refinement is not needed. Otherwise, keep refining the mesh until the solutions converge to an accurate solution. You can also employ the adaptive mesh refiner and solver by clicking on the "Solve" menu and selecting the "Parameters" drop-down menu item; a window will pop up as shown in Figure 13.13. Then you select the adaptive mode item from the menu bar and input the parameters in the window. In this mode, the mesh is automatically refined until a solution has converged to accurate solution.

In the solve mode, you can click the "Solve" menu item and select the "Solve PDE" button to solve the model you created or you can select the "parameters..." button to control the nonlinear solvers for elliptic problems as shown in Figure 13.13. For parabolic and hyperbolic problems, you can specify the initial values and the times for which the output should be generated. For the eigenvalue solver, you can specify the interval in which to search for eigenvalues.

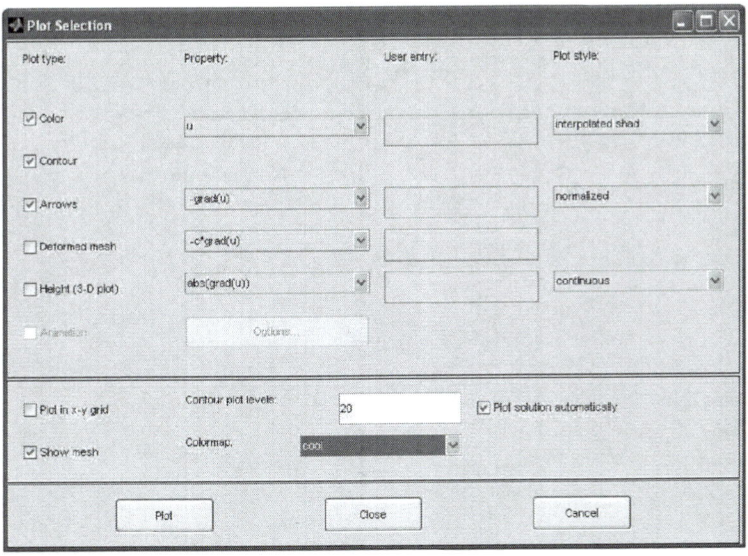

Figure 13.14   Plot solution window. (From MATLAB. With permission.)

To visualize the results, you can click the "plot" button and select the "Plot Solution" to plot the results. You can also select the "parameters...", then a window will pop up as shown in Figure 13.14 and you can select the options in the window to control your plots with a wide range of visualization possibilities. You can visualize both inside the pdetool GUI and in separate figures. You can plot three different solution properties at the same time, using color, height, and vector field plots. Surface, mesh, contour, and arrow (quiver) plots are available. For surface plots, you can choose between interpolated and flat rendering schemes. The mesh may be hidden or exposed in all plot types. For parabolic and hyperbolic equations, you can even produce an animated movie of the solution's time dependence. All visualization functions are also accessible from the command line.

The PDE Toolbox software is easy to use in the most common areas due to the application interfaces. Eight application interfaces are available [3], in addition to the generic scalar and system (vector valued *u*) cases:

- Structural Mechanics—Plane Stress
- Structural Mechanics—Plane Strain
- Electrostatics
- Magnetostatics
- AC Power Electromagnetics
- Conductive Media DC
- Heat Transfer
- Diffusion

These interfaces have dialog boxes where the PDE coefficients, boundary conditions, and solution are explained in terms of physical entities. The application interfaces enable you to enter specific parameters, such as Young's modulus in the structural mechanics problems. In this book, we will use structural mechanics and heat transfer problems as examples to explain how to use the PDE toolbox.

The typical processes of using the PDE toolbox to create the finite element model, obtain solutions, and display the results are listed in the following three major steps:

1. Preprocessor: Create a finite element model.
   (a) Specify the appropriate PDE-type problem.
   (b) Set up the drawing area and create a solid model.
   (c) Specify the boundary conditions.
   (d) Input the physical constants.
   (e) Generate the mesh.
2. Solution: Obtain the solution.
3. Postprocessor: Display the results.

Example 13.1 Plane Stress Analysis of a Plate with a Hole

A plate with a hole at the center is subjected to a uniform load p at its free end, as shown in Figure 13.15. $E = 200 \times 10^9$ N/m², v = 0.3, thickness = 0.01 m, and p = 10^6 N/m². If L = 0.20 m, w = 0.04 m, and r = 0.01 m, find the stress concentration factor (K) from finite element analysis and compare the result with the curve from strength of materials.

1. Preprocessor
   (a) Specify the appropriate PDE type of problem.
   Click the Option button as shown in Figure 13.11, select the "Application," then select the "Structural Mechanics, Plane Stress" option. Now you are in plane stress mode.
   (b) Set up the drawing area and create a solid model.
   To set up the drawing area, you need to know the dimension of the plate, which is 0.2 × 0.04 m. You can set up a drawing area a little bigger

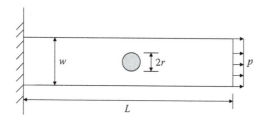

Figure 13.15 Geometry of a plate with a hole at the center. (From MATLAB. With permission.)

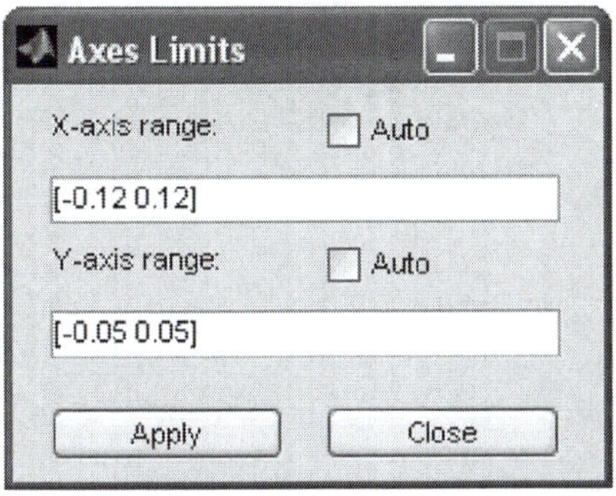

Figure 13.16   Axes limits window. (From MATLAB. With permission.)

than 0.2×0.04 m, say, 0.24×0.1 m. Click the "Option" button, select "Axes Limit…," and an Axes Limits dialog box will pop up as shown in Figure 13.16. Input the range of the *x* axis and the *y* axis. The axis range should be entered as a 1×2 MATLAB vector such as [−0.12 0.12] for x-axis range and [−0.05 0.05] for y-axis range as shown in Figure 13.16. If you select the Auto check box, automatic scaling of the axis is used. Clicking the "Apply" button applies the entered axis ranges and the drawing area is set up as shown in Figure 13.17. Clicking the "Close" button will end the Axes Limits dialog box.

You can also set up grid spacing by clicking on the "Option" button, and select "Grid Spacing…"; a grid-spacing dialog box will pop up as shown in Figure 13.18a, where −0.1:0.05:0.1 and −0.05:0.01:0.05 is the default spacing. −0.1:0.05:0.1 means the grid starts at x = −0.1 and ends at x = 0.1 with an increment of 0.05 grid units each. You can adjust the *x*-axis and *y*-axis grid spacing if you do not like the default spacing. By default, the MATLAB automatic linear grid spacing is used. If you turn off the Auto check box, the edit fields for linear spacing and extra ticks are enabled. For example, unchecking the "Auto" box, the default linear spacing −0.1:0.05:0.1 can be changed to −0.12:0.02:012, as shown in Figure 13.18b. Clicking the "Apply" button applies the entered grid spacing. Clicking on the "Done" button closes the Grid Spacing dialog box. Click the "Option" button and then click the "Grid" button to turn on or off the grid. A grid is shown in Figure 13.19.

Now you can draw the model. To draw a circle from its center, click the "Option" button, and select the "ellipse/circle (centered)" button, then click-and-drag from the center [0, 0], using the *right* mouse button, to drag a point to the circle's perimeter (radius = 0.01) to create a circle, but use the *left* mouse button to drag an ellipse. If the circle does not look like a circle, it means your x- and y-axes spacing is not equal; you can click on the "Option" button and select "Axes Equal"; a circle will be shown as in Figure 13.20 and a name C1 is assigned to this circle. You can also click the "ellipse with + sign

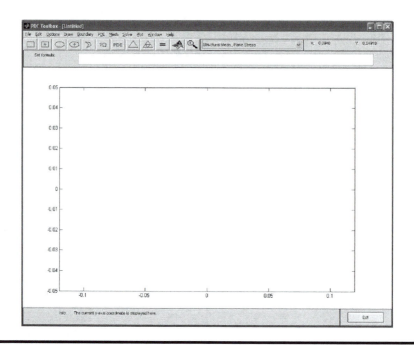

Figure 13.17 Drawing area of [–12 12] for x-axis range and [–8 8] for y-axis range is created. (From MATLAB. With permission.)

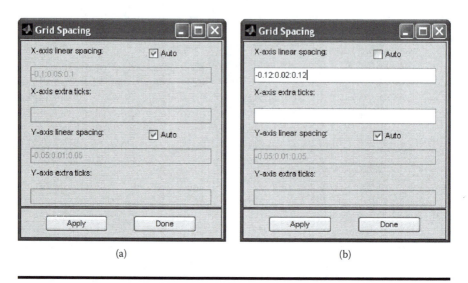

(a)   (b)

Figure 13.18 (a) Default (auto) grid spacing. (b) Manual grid spacing. (From MATLAB. With permission.)

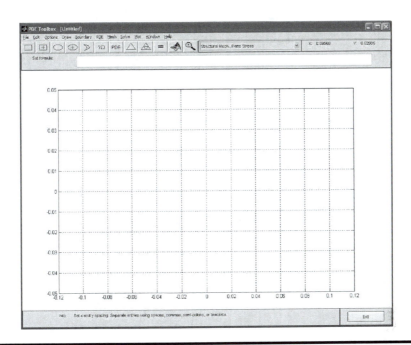

Figure 13.19 Drawing area with grids. (From MATLAB. With permission.)

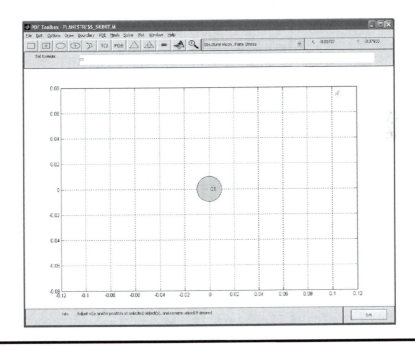

Figure 13.20 Draw a circle. (From MATLAB. With permission.)

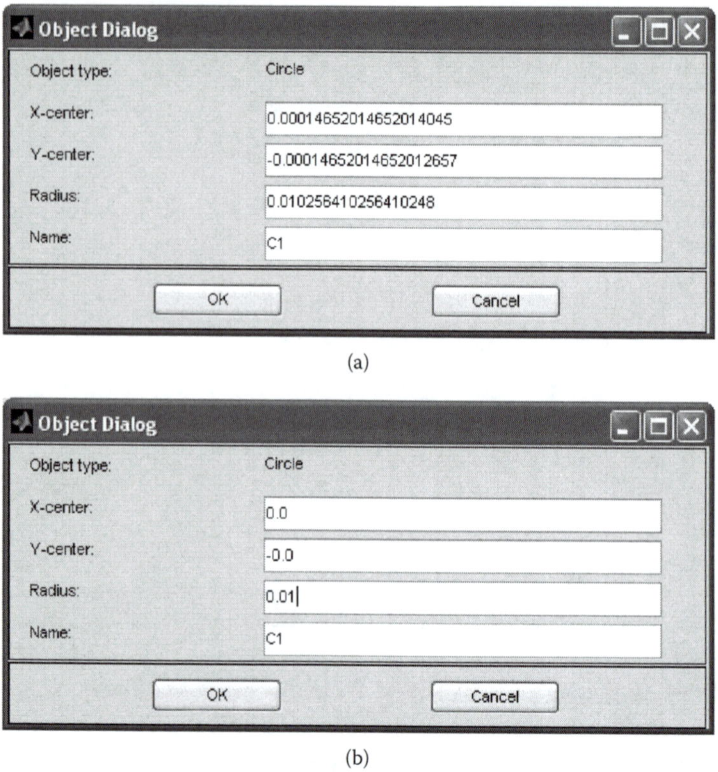

(a)

(b)

Figure 13.21 (a) Object dialog box shows the existing center location and radius. (b) Entering the exact center location and radius if Figure 13.21a is not the exact number. (From MATLAB. With permission.)

in the center" button to draw the ellipses or circles. The button with the + sign is used when you want to draw starting at the center. If you want to move or resize the circle, you can easily do so. Click-and-drag an object to move it, or double-click an object to open a dialog box as shown in Figure 13.21a, where you can enter the exact center location to (0,0) and radius to 2 as shown in Figure 13.21b. From the dialog box, you can also alter the name of the circle. You can also turn on the "snap-to-grid" feature by clicking the "Option" button and selecting "Snap" to force the circle to line up with the grid. If you are not satisfied and want to restart, you can delete the rectangle by clicking the Delete key or by selecting Clear from the Edit menu. You can also select "ellipse/circle" to draw ellipses and circles in a similar manner.

To draw a rectangle or a square starting at a corner, click the "Rectangle/Square" button. Then put the cursor at the desired corner (here is [−0.1, 0.02]), and click-and-drag using the *left* mouse button to create a rectangle

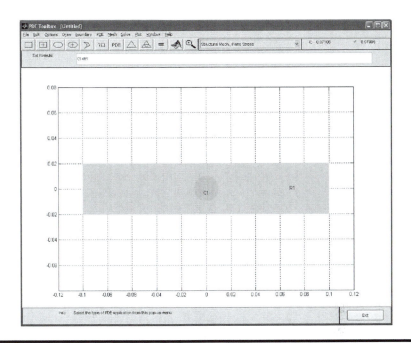

Figure 13.22   Draw a rectangle. (From MATLAB. With permission.)

with the desired side lengths (here width is 0.2 m, height is 0.04 m). (Use the right mouse button to create a square.) A rectangle is created and assigned the name R1. You can move or resize the rectangle by clicking-and-dragging an object to move it, and double-clicking an object to open a dialog box, where you can enter exact location coordinates. The resulting model is now the union of the rectangle R1 and the circle C1 as shown in Figure 13.22, described as C1 + R1. The area where the two objects overlap is clearly visible, as the overlap area is drawn with a darker shade of gray than the surrounding area. The object that you are selecting has a black border. A selected object can be moved, resized, copied, and deleted. You can select more than one object by Shift + clicking the objects that you want to select. Also, a Select All option is available from the Edit menu.

The desired model is formed by subtracting the circle C1 from the rectangle R1. You do this by editing the set formula that by default is the union of all objects; that is, R1 + C1. You can type any other valid set formula into the Set formula edit field. Click in the edit field and use the keyboard to change the set formula to R1 – C1. If you want, you can save this model as an M-file. Use the Save As option from the File menu, and enter a file name of your choice. It is good practice to continue to save your model at regular intervals using Save. All the additional steps in the process of modeling and solving your PDE are then saved to the same M-file.

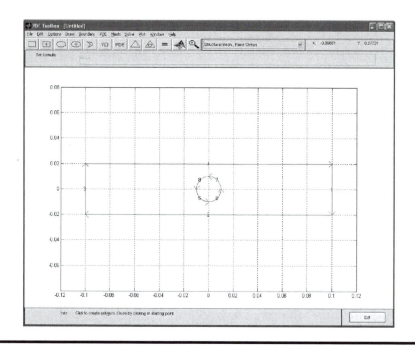

Figure 13.23   Boundaries of the model; there are eight boundary edges in this model. (From MATLAB. With permission.)

(c) Specifying the boundary conditions.

Click the "Boundary" button and select the "Boundary Mode" to turn on the boundary mode. Also click the "Show Edge Labels" to turn on the edge number and the boundary lines will be displayed as shown in Figure 13.23. Click on the line whose boundary conditions (edge 3) want to be set, and then select the "Specify Boundary Conditions" button. Alternatively, you can double-click the selected edge that brings up the dialog box shown in Figure 13.24 with the Neumann boundary condition being checked. Since edge 3 is a Dirichlet boundary condition, check the "Dirichlet" button below the condition type as shown in Figure 13.25. Since edge 3 is fixed, both u and v displacement components are zero. From the boundary condition equations: [3]

$$h_{11}u + h_{12}v = r_1 \qquad h_{21}u + h_{22}v = r_2$$

Enter $h_{11} = 1$, $h_{12} = 0$, and $r_1 = 0$, which implies that $u = 0$. Similarly, enter $h_{21} = 0$, $h_{22} = 1$, and $r_2 = 0$, which implies that $v = 0$. After these numbers are entered into Figure 13.25, click the "OK" button and this boundary condition is applied to edge 3 (edge 3 color changes to red). Next click on edge 1 and then select the "Specify Boundary Conditions" box again; a dialog

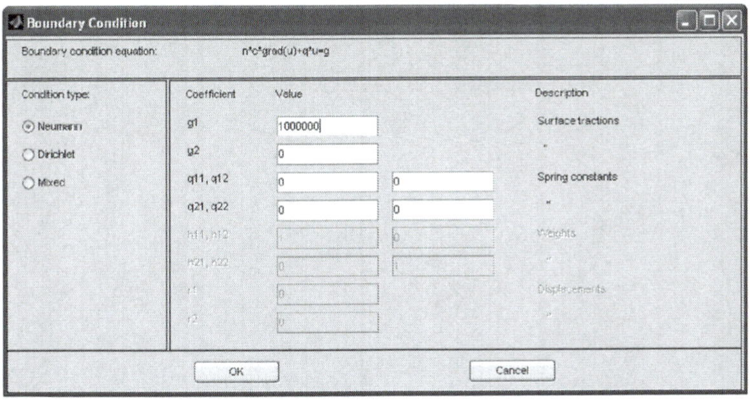

Figure 13.24 **Boundary condition window for input of the Neumann boundary condition. (From MATLAB. With permission.)**

box will pop us as shown in Figure 13.24. Since edge 1 is the Neumann boundary condition (a normal distributed load of 10^6 N/m^2 is applied), select the Neumann boundary type and input the required values. Since g_1 is the surface traction in the x direction and g_2 is the surface traction in the y direction, input 10^6 for g_1 and 0 for g_2 (edge 1 turns to a blue color). Also input 10^6 *nx for g_1 and 10^6 *ny for g_2, where nx and ny are the x and y components of the unit outward normal vector to this edge. The remaining boundaries are free (no normal stress), that is, a Neumann condition with q = 0 and $g_1 = g_2 = 0$. Make sure all the edges are in blue, except edge 3 which is red as shown in Figure 13.23. One can also select mixed boundary conditions, if it is needed.

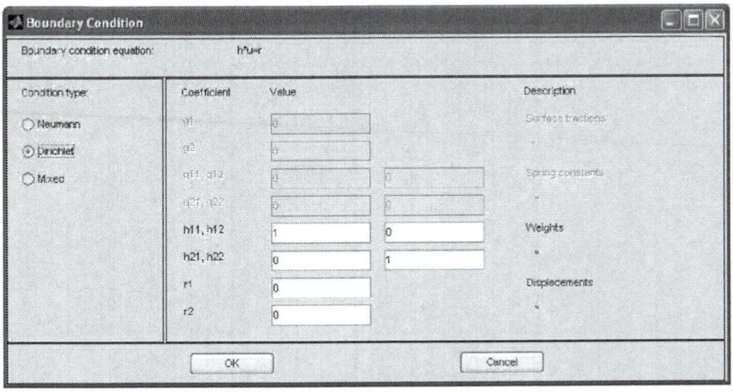

Figure 13.25 **Boundary condition window for input of the Dirichlet boundary condition. (From MATLAB. With permission.)**

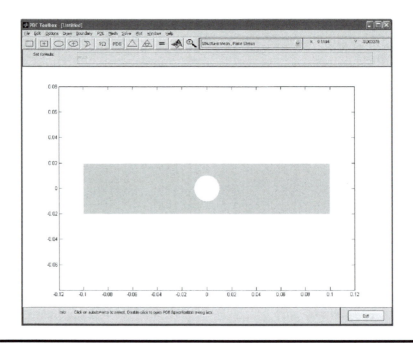

Figure 13.26 **PDF mode to display the solid model. (From MATLAB. With permission.)**

(d) Inputting the physical constants.

Click the "PDE" button and select the "PDE Mode" to turn on the PDE mode, where a solid model (you can turn the grid off now) will be displayed in the window as shown in Figure 13.26. Select the "PDE Specification..."; a dialog box will pop up as shown in Figure 13.27. Input the materials constants (200E9 for E, 0.3 for nu, 0 for kx, 0 for ky, and 1.0 for rho) under the "value heading." The

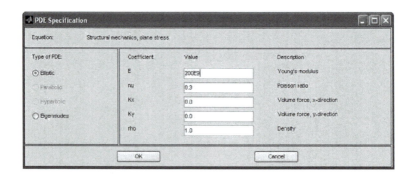

Figure 13.27 **PDF specification for input of the materials constants. (From MATLAB. With permission.)**

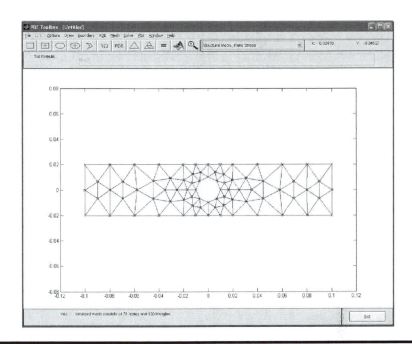

Figure 13.28   Initial mesh with node number. (From MATLAB. With permission.)

body forces, kx and ky, are zero in this example. The density, rho, is not needed in this example. After you input all the data, click the "OK" button.

(e) Generating the mesh.

Click the "Mesh" button in the main menu and select "initialize Mesh" to generate the first mesh as shown in Figure 13.28, where the node number can be displayed by selecting the "Show Node Labels." You can also display element numbers by selecting the "Show Triangle Labels" as shown in Figure 13.29. This mesh does not appear sufficiently fine and further mesh refinement is needed. The mesh can be further refined by selecting "Refine Mesh" or by clicking the button with the four-triangle icon (the Refine button). You can also use the "Jiggle Mesh" option and the mesh can be jiggled to improve the triangle quality. You can also control the jiggling of the mesh, the refinement method, and other mesh generation parameters from the dialog box (see Figure 13.30) that is opened by selecting "Parameters..." from the Mesh menu. You can undo any change to the mesh by selecting the "Undo Mesh Change" from the Mesh menu. The procedure used here was to initialize the mesh, then refine it once, and finally jiggle it once to obtain the mesh shown in Figure 13.31. One can also select the "Display Triangle Quality" from the Mesh menu to show the quality of the triangle as shown in Figure 13.32. A value of 1 indicates the triangle is an equilateral triangle,

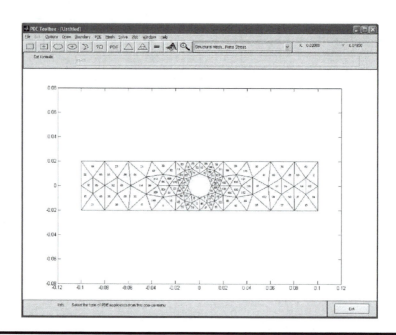

Figure 13.29 **Initial mesh with element number. (From MATLAB. With permission.)**

Figure 13.30 **Mesh generation parameter to control the jiggling and refinement of mesh. (From MATLAB. With permission.)**

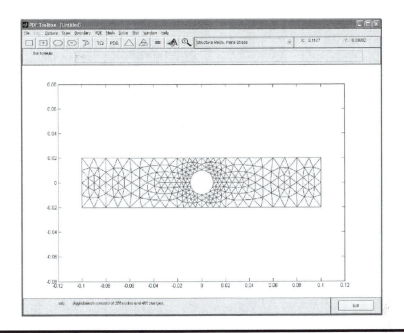

Figure 13.31   Refined mesh. (From MATLAB. With permission.)

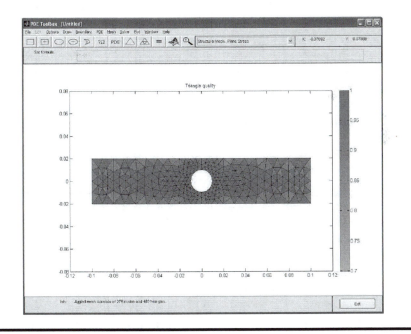

Figure 13.32   Contour plot to display the quality of the triangular element. (From MATLAB. With permission.)

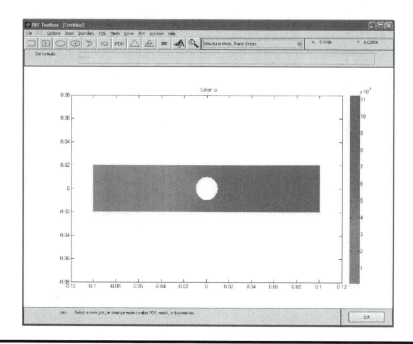

Figure 13.33 **Contour plot for the displacement component u in the x direction. (From MATLAB. With permission.) (See color insert following page 334.)**

which is the best quality. After the solution is obtained with this mesh, the mesh once again can be refined to obtain a second solution. If the first and second solutions are close, the solution is considered convergent to accurate solution and further mesh refinement is not needed. Otherwise, successive refinement of the mesh is needed until a convergent solution is obtained.

2. Obtaining a solution

Click the "Solve" button and select the "Solve PDE" option. The current model will be solved, and a plot of the solution will automatically be created and displayed as shown in Figure 13.33. The solution is the displacement component u in the x direction. One can also select "Parameters..." from the Solve menu causing a dialog box as shown in Figure 13.13 to pop up for entry of PDE adaptive and nonlinear solver.

3. Displaying the results

To control the plotting and visualization of the solution, click on the "Plot" item in the menu bar and select "Parameters..."; a dialog box will pop up as shown in Figure 13.34. The upper part of the dialog box contains four columns: Plot Type, Property, User Entry, and Plot Style, each discussed in the following section [3].

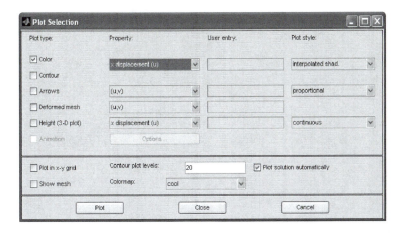

Figure 13.34 **Plot selection window to control the plotting and visualization.** (From MATLAB. With permission.)

- *Plot type* (far left) contains a row of six different plot types (Color, Contour, Arrow, Deformed mesh, 3-D plot, and Animation) that can be used for visualization.
- *Property* contains four pop-up menus containing lists of properties that are available for plotting using the corresponding plot type. From the first pop-up menu one can control the property of visualization using color and/or contour lines. The second and third pop-up menus contain vector valued properties for visualization using arrows and deformed mesh, respectively. From the fourth pop-up menu, finally, one can control which scalar property to visualize using z-height in a 3-D plot. The lists of items are dependent on the current application mode. For the plane stress mode, one can select an item for visualization from the pop-up menus (such as the displacements u and v, the normal strains and stresses in the x and y directions, the shear stress, the von Mises effective stress, and the principal stresses and strains) as shown in Figure 13.35.
- *User entry* contains four edit fields where one can enter their own expression, if the user entry property is selected from the corresponding pop-up menu to the left of the edit fields. If the user entry property is not selected, the corresponding edit field is disabled.
- *Plot style* contains three pop-up menus from which one can control the plot style for the Color, Arrow, and Height plot types, respectively. The available plot styles for color surface plots are Interpolated Shading (the default) and Flat Shading. You can use two different arrow plot styles: Proportional (the default) and Normalized. For Height (3-D plots), the available plot styles are Continuous (the default) and Discontinuous.

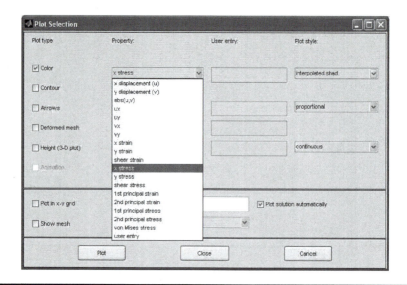

Figure 13.35 **Select the display of normal stress component (σ_x) in the x direction. (From MATLAB. With permission.)**

Now one can select the "Color" from the Plot-type section, "x stress" from the Property section, and all other sections are left as default, and then click the "Plot" button; a contour plot for σ_x will be displayed as shown in Figure 13.36. To find the actual value of σ_x at any point in the model, move the cursor to that point and press the left button on the mouse and the "info" box in the bottom of the window will show the value of σ_x and the triangular element number where the point is located.

Discussion of Results

By moving the cursor to the points on the top and bottom of the hole, it can be determined that $(\sigma_x)_{max}$ is approximately 2.81×10^6 N/m^2, as shown in the "info box" of Figure 13.36 and Figure 13.37a. The average normal stress σ_{avg} at the same point is 2×10^6 N/m^2, as shown in Figure 13.37b. Therefore, the stress concentration factor is $\sigma_x / \sigma_{avg} = 1.405$, which is far away from the stress concentration factor of about 2.37 obtained from strength of material books. Therefore, further mesh refinement is needed to obtain a more accurate solution.

One can also use the adaptive mode, where the mesh is automatically refined until a solution is converged. In areas where the gradient of the solution (the stress) is large, finer meshes are created to increase the accuracy of the solution. To do this, select "Parameters" from the "Solve" menu and check the Adaptive mode box as shown in Figure 13.13. You can use the default options for adaptation, which are the Worst triangles selection method with the Worst triangle fraction set to 0.5. The

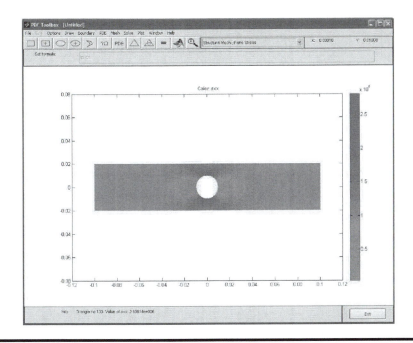

Figure 13.36 **Contour plot of normal stress σ_x. (From MATLAB. With permission.) (See color insert.)**

maximum number of triangles can be changed from 1,000 (default) to 100,000 and the maximum number of refinement can be changed from 10 (default) to 20. Choose the "longest" refinement method. Now solve the plane stress problem again. The actual number of triangles in this solution is 125,984 after 20 iterations. Select the "Show Mesh" option in the Plot Selection dialog box to see how the mesh is refined in areas where the stress is large as shown in Figure 13.38. It clearly shows that finer meshes are created near the circle and corners of the support. The maximum σ_x stress is about 4.28×10^6 N/m^2 on the top and bottom of the circle and the stress concentration is about 2.14, which is close to 2.37 from strength of materials. You can run more iterations to obtain a more accurate solution.

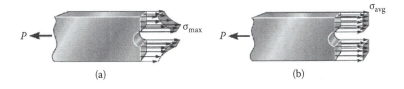

Figure 13.37 **(a) Normal stress distribution from FEA solution. (b) Average normal stress obtained from strength of materials. (From MATLAB. With permission.)**

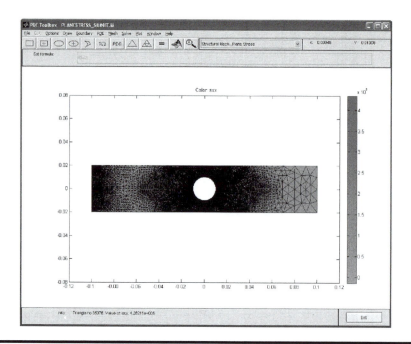

Figure 13.38 Contour plot of σ_x obtained from adaptive solution mode. (From MATLAB. With permission.)

13.5 Structural Mechanics Plane Strain Analysis

For plane strain problems, all the strain components (ε_z, γ_{yz}, and γ_{xz}) are zero. Setting $\varepsilon_z = 0$ in Equation (13.3) gives

$$\varepsilon_z = \frac{1}{E}[\sigma_z - \upsilon(\sigma_x + \sigma_y)] = 0 \Rightarrow\Rightarrow \sigma_z = \upsilon(\sigma_x + \sigma_y) \tag{13.63}$$

Substituting Equation (13.63) into Equations (13.1) and (13.2), gives

$$\varepsilon_x = \frac{1}{E}[\sigma_x - \upsilon(\sigma_y + \upsilon\sigma_x + \upsilon\sigma_y)] = \frac{1+\upsilon}{E}[\sigma_x(1-\upsilon) - \upsilon\sigma_y] \tag{13.64}$$

$$\varepsilon_y = \frac{1}{E}[\sigma_y - \upsilon(\sigma_x + \upsilon\sigma_x + \upsilon\sigma_y)] = \frac{1+\upsilon}{E}[\sigma_y(1-\upsilon) - \upsilon\sigma_x] \tag{13.65}$$

From Equation (13.4),

$$\gamma_{xy} = \frac{\tau_{xy}}{G} \tag{13.66}$$

Equations (13.64) through (13.66) are the stress–strain relations for the plain strain problem that can be expressed in the matrix form as

$$
\begin{Bmatrix} \varepsilon_x \\ \varepsilon_y \\ \gamma_{xy} \end{Bmatrix} = \frac{1+\upsilon}{E} \begin{bmatrix} 1-\upsilon & -\upsilon & 0 \\ -\upsilon & 1-\upsilon & 0 \\ 0 & 0 & 2 \end{bmatrix} \begin{Bmatrix} \sigma_x \\ \sigma_y \\ \tau_{xy} \end{Bmatrix} \tag{13.67}
$$

or

$$
\{\sigma\} = \begin{Bmatrix} \sigma_x \\ \sigma_y \\ \tau_{xy} \end{Bmatrix} = \frac{E}{(1+\upsilon)(1-2\upsilon)} \begin{bmatrix} 1-\upsilon & \upsilon & 0 \\ \upsilon & 1-\upsilon & 0 \\ 0 & 0 & \dfrac{1-2\upsilon}{2} \end{bmatrix} \begin{Bmatrix} \varepsilon_x \\ \varepsilon_y \\ \gamma_{xy} \end{Bmatrix} = [D]\{\varepsilon\} \tag{13.68}
$$

Matrix [D] in Equation (13.68) is the material matrix for the plain strain problem. The material matrix for plane stress and plane strain is different as shown by Equations (13.10) and (13.68). To solve the plain strain problem in MATLAB just click on the "Option" menu and select "Applications/Structural Mechanics, Plane Strain," then Equation (13.68) will be used to solve the plane strain problem. All the other procedures are exactly the same as those discussed in the plane stress analysis problem.

13.6 Model Analysis of 2-D Structures

When a structure is subjected to an impact load, or when loads vary with time, the response of the structure will also vary with time. If the frequency of the load or the excitation applied to the structure is less than about one-third of the structure's lowest natural frequency, the effects of the inertia can be neglected and the problem can still be considered as quasistatic. That is, the static analysis, [K]{Q} = {F}, that was previously discussed is sufficiently accurate since the loads {F} and displacements {Q} vary slowly with time. The general form of the differential equation for dynamic analysis is

$$
[M][\ddot{Q}] + [C][\dot{Q}] + [K][Q] = [F(t)] \tag{13.69}
$$

where
  [M] = the mass matrix of the structure.
  [C] = the damping matrix of the structure.
  [K] = the stiffness matrix of the structure.
  [F(t)] = the loading which is a function of time.
  $[\ddot{Q}]$ = the components of acceleration at each node.
  $[\dot{Q}]$ = the components of velocity at each node.
  [Q] = the components of displacement at each node.

If damping and external forces, {F(t)}, are neglected, Equation (13.69) becomes a free vibration problem, which is

$$[M][\ddot{Q}] + [K][Q] = [0] \tag{13.70}$$

Equation (13.70) can be solved by using eigenvalue analysis to obtain the natural frequencies and their corresponding vibration mode shapes of the structure. The number of natural frequencies in Equation (13.70) is the same as the total number of active degrees of freedom in the finite element model. Each frequency has a corresponding vibration mode shape. Although there are many frequencies (up to infinity) that can be calculated, only a few of the lowest frequencies are of interest for engineering applications.

Example 13.2 Model Analysis of the Plane Stress Problem

For the planes stress or plane strain problem, the natural frequencies and their corresponding vibration mode shapes can be determined. Here the same solid model as in Example 13.1 is used to calculate the several lowest natural frequencies and their corresponding mode shapes. The mass density for this problem is 7850 kg/m³.

To solve this problem, the exact same procedure is used as in Example 13.1, except in this case, select the "Specify Boundary conditions…" from the "Boundary" menu and change the value in g_1 to 0 for edge 1 as shown in Figure 13.24, so that edge 1 becomes traction free. Now only edge 3 is constrained and all other edges are free from loading, which is displayed in Figure 13.23. Next select the "PDE Specification" from the "PDE" menu to open a dialog box. Change the type of PDE from "Elliptic" to "Eigenmodes" and input the mass density of 7850 into the density box as shown in Figure 13.39. For this model, a mesh can also be generated as shown in Figure 13.40.

Now, select the "Solve Parameters…" from the "Solve" menu and a dialog box will pop up as shown in Figure 13.41. One can then input the eigenvalue range for

Figure 13.39 PDF specification for input of the materials constants and density. (From MATLAB. With permission.)

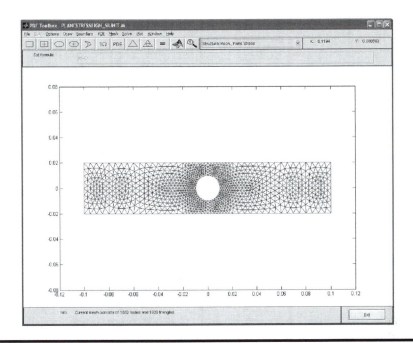

Figure 13.40   Refined mesh. (From MATLAB. With permission.)

this problem (default is from 0 to 100). If eigenvalues in the entered range are not found, keep increasing the range until the eigenvalues are found. When the range from 0 to 5×10^9 as shown in Figure 13.41 is entered, three eigenvalues in this range will be found after it is solved. Select the "Parameters..." in the "Plot" menu, and click the "Deform Mesh" under the plot type, the first vibration mode shape for the lowest eigenvalue ($\lambda_1 = 2.497 \times 10^7$) is shown in Figure 13.42. The second ($\lambda_2 = 7.741 \times 10^8$) and third ($\lambda_3 = 1.413 \times 10^9$) vibration modes are also shown in Figure 13.43 and Figure 13.44, respectively. Since the natural frequency f (Hz) is defined as

$$f = \frac{\sqrt{\lambda}}{2\pi}$$

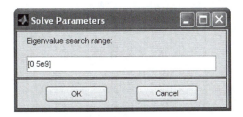

Figure 13.41   Dialog box for input of the eigenvalue range. (From MATLAB. With permission.)

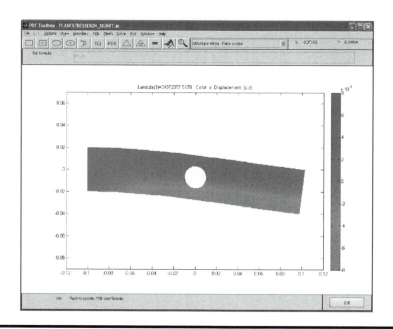

Figure 13.42 The first vibration mode for the lowest natural frequency. (From MATLAB. With permission.) (See color insert.)

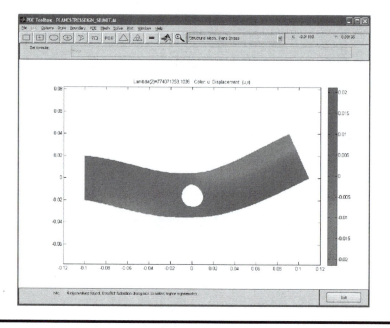

Figure 13.43 The second vibration mode for the second lowest frequency. (From MATLAB. With permission.) (See color insert.)

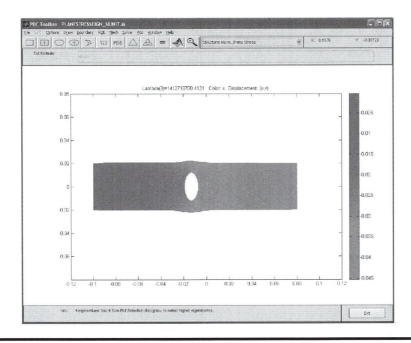

Figure 13.44   The third vibration mode for the third lowest natural frequency. (From MATLAB. With permission.) (See color insert.)

the lowest natural frequency is 795 Hz, the second lowest natural frequency (λ_2) is 4428 Hz, and the third lowest natural frequency (λ_3) is 5982 Hz.

13.7 Finite Element Analysis for Heat Transfer

The governing PDE for heat conduction (for the derivation of Equation 13.71, see Appendices B.1 and B.2) is

$$\frac{\partial(\rho c T)}{\partial t} = \nabla \cdot (k \nabla T) + g \tag{13.71}$$

where
  T = the material temperature.
  ρ = the mass density of the material.
  k = the thermal conductivity of the material.
  c = the specific heat of the material.
  g = the rate of heat generation within the material.

The variables ρ, c, and k are mild functions of temperature and are frequently taken as constants. This is particularly true if one is interested in obtaining a closed-form solution by separation of variables. In that case Equation (13.71) reduces to

$$\frac{1}{a}\frac{\partial T}{\partial t} = \nabla^2 T + g \tag{13.72}$$

where

a = the thermal diffusivity of the material $= \dfrac{k}{\rho c}$.

Example 13.3 Temperature Distribution in a Slab with a Hole

In this example, modeling of a 2-D solid subjected to various boundary conditions is covered. After using any combination of several different boundary conditions, almost any 2-D heat transfer problem can be modeled in MATLAB.

Problem Description

As shown in Figure 13.45, the top is insulated, the right and left sides have constant temperatures, the bottom side has a constant heat flux, and the inside circle

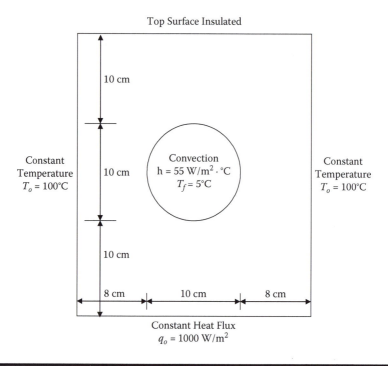

Figure 13.45 Geometry and boundary condition for Example 13.3. (From MATLAB. With permission.)

is subjected to convection. The solid has a thermal conductivity, k, of 25 W/m-C and has a uniform heat generation rate, g, of 30 W/m^3. Determine the temperature distribution in the solid.

Solution Procedure

First click the "Option" item on the main menu bar and select "Application." Then select the "Heat Transfer" option to get into the heat transfer mode. The procedure to set up the drawing area and create a solid model is exactly the same as described in step 1(b) in Example 13.1. To specify the boundary conditions, click on the "Boundary" option in the main menu bar and select the "Boundary Mode" to display the boundary edges as shown in Figure 13.46. As is frequently done in solving a heat transfer problem by separation of variables, MATLAB solves the problem in terms of T' where $T' = T - T_f$. In this equation, T is the actual temperature and T_f is the ambient temperature. As seen in Figure 13.45, $T_f = 5°C$. Once the problem is solved for T', T is obtained by adding 5°C to T' to obtain the actual temperature T. To obtain the solution:

1. Double-click edge 1 causing the dialog box to pop up as shown in Figure 13.47. Select the Dirichlet boundary condition and input h = 1 and r = 95 (subtract 5 from 100) to specify the temperature, T', of 95°C along this edge.

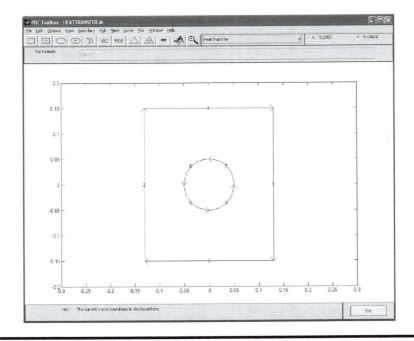

Figure 13.46 **Boundaries of the model; there are eight boundary edges in this model. (From MATLAB. With permission.)**

Figure 13.47 Boundary condition window for input of the Dirichlet boundary condition. (From MATLAB. With permission.)

2. Double-click edge 3 and repeat step 1 to specify the temperature, T', of 95°C along edge 3.
3. Double-click edge 2 and select the Neumann boundary condition and input g = 1000 and q = 0 as shown in Figure 13.48 to specify the heat flux = 1000 W/m² along this edge.
4. Double-click edge 4 and select the Neumann boundary and input g = 0 and q = 0 as shown in Figure 13.49 to specify the insulated boundary condition along this edge.
5. Double-click edge 5 and select the Neumann boundary and input g = 0 and q = 55 as shown in Figure 13.50 to specify the heat transfer coefficient for the convection boundary condition along this edge.
6. Double-click edges 6, 7, and 8 and repeat the process described in step 5 to specify the convection boundary condition along those edges.

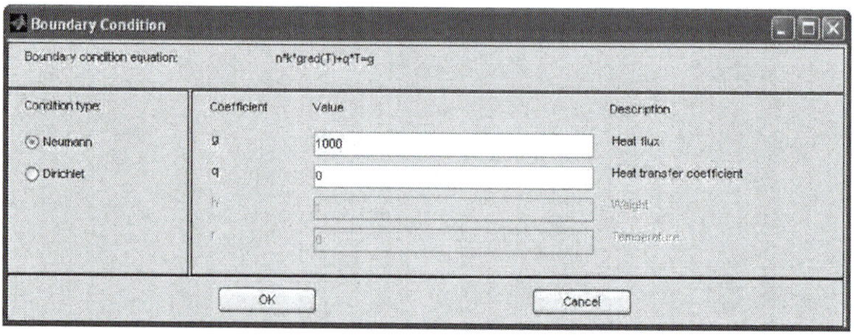

Figure 13.48 Boundary condition window for input of the Neumann boundary condition to specify the heat flux = 1000 W/m². (From MATLAB. With permission.)

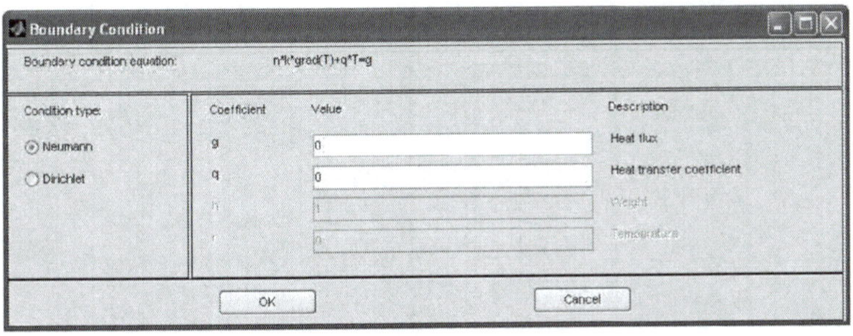

Figure 13.49 **Boundary condition window for input of the Neumann boundary condition to specify the insulated edges. (From MATLAB. With permission.)**

To specify the physical constants, click the "PDE" menu and select the "PDE Specification…" and input k = 25 for thermal conductivity, Q = 30 for heat source, h = 55 for convective heat transfer coefficient, and Text = 0 (subtract 5 from T_f) for ambient temperature as shown in Figure 13.51.

Now click on the "Mesh" option in the main menu and select "Initialize Mesh" to generate the first mesh. Then click on the "Refine Mesh" to obtain finer mesh and, finally, click on the "Jiggle Mesh" option to improve the mesh quality producing the mesh shown in Figure 13.52.

Finally, to solve the model, click on the "Solve" option in the main menu and select "Solve PDE." The finite element model is then solved and the results of the temperature field, T', are automatically displayed as shown in Figure 13.53. To obtain the temperature field, T, 5°C needs to be added to T'. This is accomplished by clicking on the "Plot" option in the main menu and selecting "Parameters…"; a dialog box will pop up. Then select "user entry" in the "Property" column and

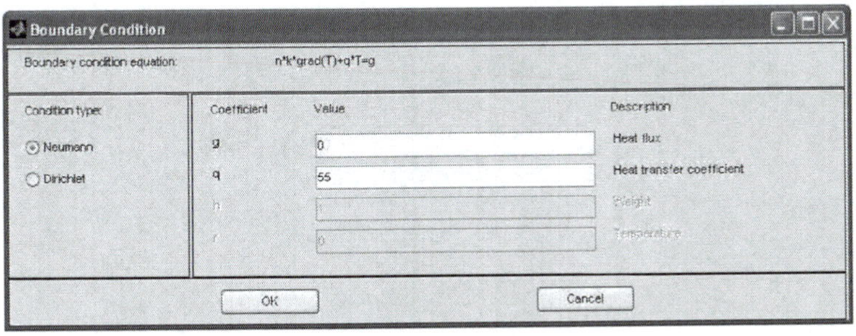

Figure 13.50 **Boundary condition window for input of the Neumann boundary condition to specify the heat transfer coefficient for the convection edge. (From MATLAB. With permission.)**

PDE Specification

Equation: -div(k*grad(T))=Q+h*(Text-T), T=temperature

Type of PDE	Coefficient	Value	Description
⊙ Elliptic	rho	1.0	Density
○ Parabolic	C	1.0	Heat capacity
○ Hyperbolic	k	25	Coeff. of heat conduction
○ Eigenmodes	Q	30	Heat source
	h	55	Convective heat transfer coeff.
	Text	0	External temperature

OK   Cancel

Figure 13.51 PDF specification for input of the physical constants. (From MATLAB. With permission.)

input "u + 5" in the "User entry" column as shown in Figure 13.54. The final solution in terms of T is displayed in Figure 13.55. To display the heat flux flow in the body select the "Arrow" option in the "Plot type" column and the "Heat flux" option in the "property" column in the dialog box shown in Figure 13.54, resulting in the plot shown in Figure 13.56.

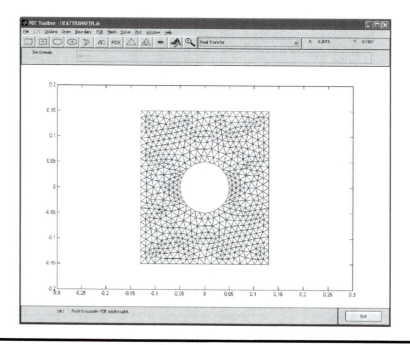

Figure 13.52 Refined mesh. (From MATLAB. With permission.)

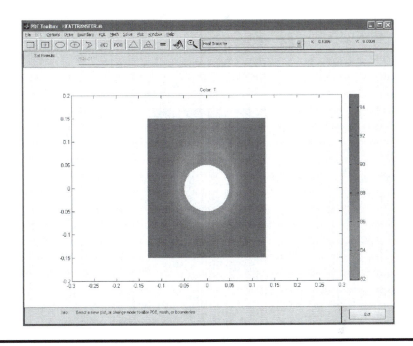

Figure 13.53 Contour plot of the temperature field. (From MATLAB. With permission.) (See color insert.)

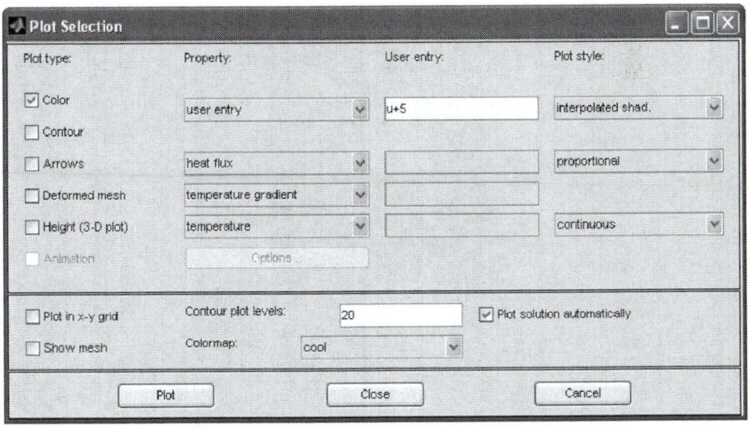

Figure 13.54 Plot selection window to add 5°C to the solutions shown in Figure 13.53. (From MATLAB. With permission.)

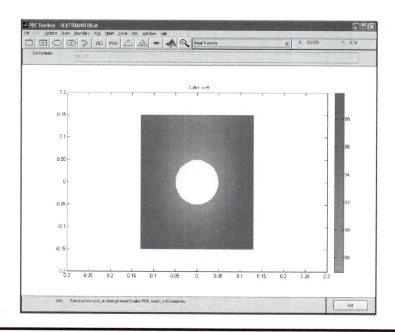

Figure 13.55 Contour plot of the temperature field after adding 5°C to the solutions shown in Figure 13.53. (From MATLAB. With permission.)

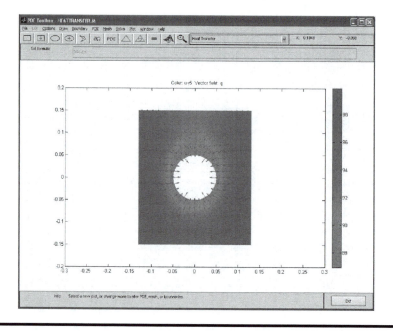

Figure 13.56 Contour plot of the temperature field with heat flux flow. (From MATLAB. With permission.) (See color insert.)

Example 13.4 Second Example of Temperature Distribution in a Slab

A slab as shown in Figure 13.57 is subjected to convection on the top and bottom surfaces, where $T_\infty = 50°C$, $T_W = 500°C$, L = 2 m, w = 0.2 m, k = 59 W/(m·C), and h = 890 W/(m²·C).

Determine the temperature distribution in the slab and compare the results with the solutions obtained from the Gauss–Seidel method.

Solution Procedure

Since this problem is symmetric about the x axis, only half of the slab is modeled.

First click the "Option" menu and go to "Application" and select "Heat Transfer." Then set up the drawing area and create a solid model by following the procedure described in step 1(b) in Example 13.1. To specify the boundary conditions, click on the "Boundary" option in the main menu and select the "Boundary Mode" to display the boundary edges as shown in Figure 13.58. Since MATLAB solves a temperature distribution problem in terms of $T' = T - T_\infty$, the boundary conditions need to be specified in terms of T', thus set $T'_\infty = 0$ and $T'_W = 450°C$. To obtain a solution in terms of T, add 50°C to the solution obtained from MATLAB.

1. Double-click edge 1 and a dialog box will pop up as shown in Figure 13.59, where you can select a Neumann boundary condition and input g = 0 and q = 890 to specify the heat transfer coefficient for the convection boundary condition along this edge.
2. Double-click edge 3 to select the Neumann boundary and input g = 0 and q = 0 as shown in Figure 13.60 to specify the insulated (symmetric) boundary condition along this edge.

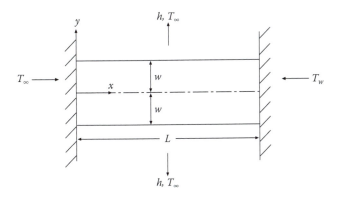

Figure 13.57 Geometry and boundary condition for Example 13.4.

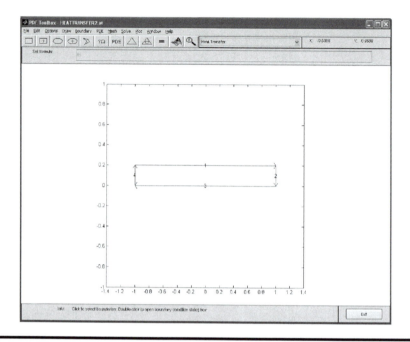

Figure 13.58 **Four boundary edges in this model. (From MATLAB. With permission.)**

3. Double-click edge 4 and a dialog box will pop up as shown in Figure 13.61, where you can select the Dirichlet boundary condition and input h = 1 and r = 0 (subtract 50 from 50) to specify temperature, $T' = 0°C$, along this edge.

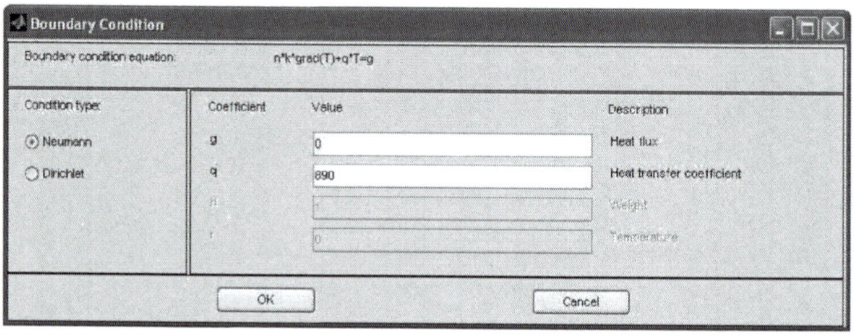

Figure 13.59 **Input the heat transfer coefficient for the convection edge. (From MATLAB. With permission.)**

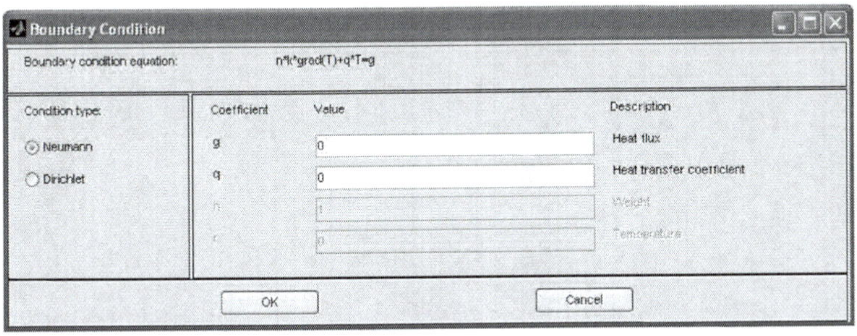

Figure 13.60 Input the insulated (symmetric) boundary condition. (From MATLAB. With permission.)

4. Double-click edge 2 and a dialog box will pop up as shown in Figure 13.62, where you can select the Dirichlet boundary condition and input h = 1 and r = 450 (subtract 50 from 500) to specify temperature, $T' = 450°C$, along this edge.

To specify the physical constants, click the "PDE" option in the main menu and select the "PDE Specification..." and input k = 59 W/m-C for thermal conductivity, Q = 0 (no heat source), h = 890 W/m²-°C for convective heat transfer coefficient, and Text = 0 (subtract 50 from T_∞) for ambient temperature as shown in Figure 13.63.

Now click on the "Mesh" option in the main menu and select "Initialize Mesh" to generate the first mesh. Then click the "Refine Mesh" option to obtain a finer

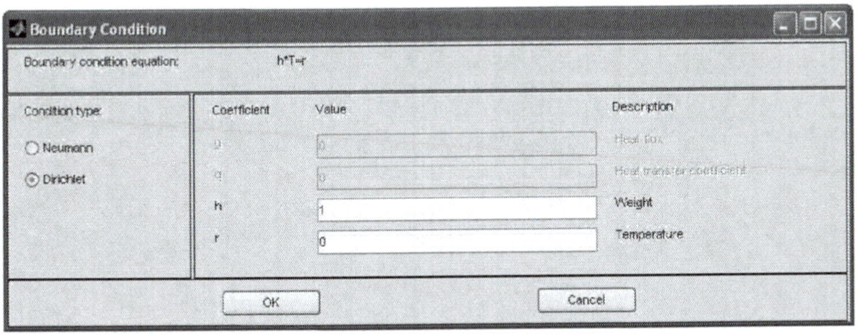

Figure 13.61 Input r = 0 to specify the temperature of 50°C along this edge. (From MATLAB. With permission.)

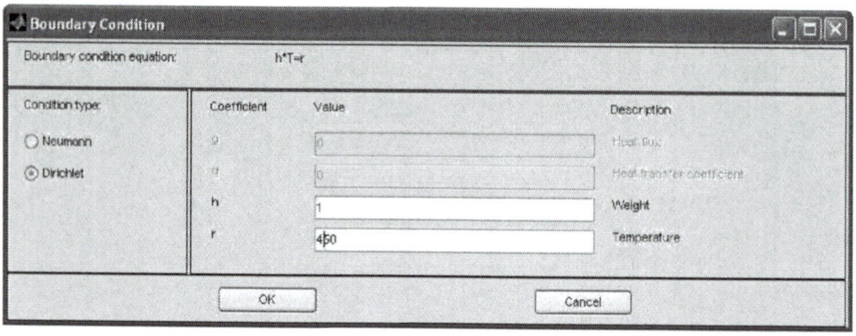

Figure 13.62 Input r = 450 to specify the temperature of 500°C along this edge. (From MATLAB. With permission.)

mesh, and finally click "Jiggle Mesh" to improve the mesh quality, producing the mesh shown in Figure 13.64.

Finally, click the "Solve" option in the main menu and select "Solve PDE." The finite element model is solved and the results of the temperature field are automatically displayed as shown in Figure 13.65. To obtain the final solution in terms of *T*, add 50°C to the solution by clicking on the "Plot" option in the main menu and select "Parameters…"; a dialog box will pop up. Select the "User entry" in the "Property" column and input "u + 50" in the "User entry" column as shown in Figure 13.66. The final solution will display as shown in Figure 13.67. The maximum temperature along edge 2 is about 475°C, which is below the 500°C specified along edge 2. The solution can be improved by refining the mesh.

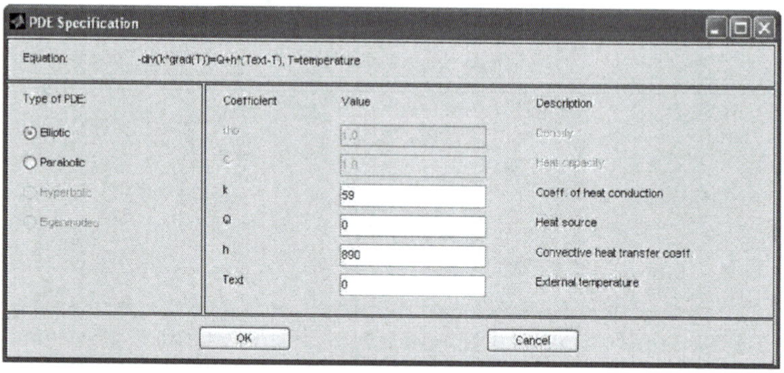

Figure 13.63 PDF specification for input of the physical constants. (From MATLAB. With permission.)

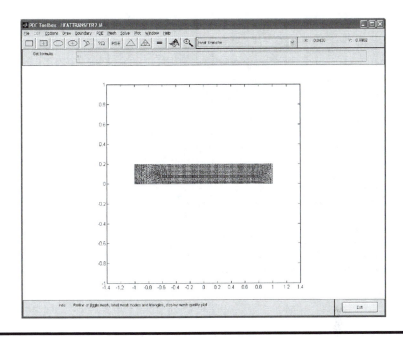

Figure 13.64   Refined mesh. (From MATLAB. With permission.)

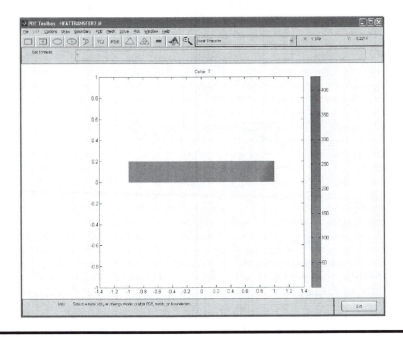

Figure 13.65   Contour plot of the temperature field. (From MATLAB. With permission.) (See color insert.)

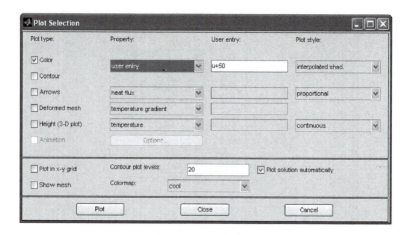

Figure 13.66 Plot selection window to add 5°C to the solutions shown in Figure 13.65. (From MATLAB. With permission.)

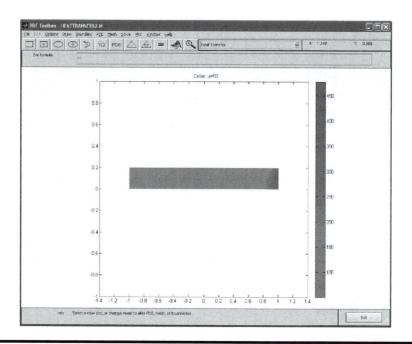

Figure 13.67 Contour plot of the temperature field after adding 50°C to the solutions shown in Figure 13.65. (From MATLAB. With permission.) (See color insert.)

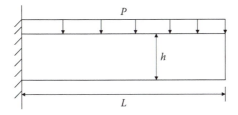

Figure P13.1   Geometry of a cantilevered beam for Project 13.1.

Projects

Project 13.1

A cantilevered beam is subjected to uniform pressure on the top surface as shown in Figure P13.1. E = 200 GPa, v = 0.3, thickness = 0.1 m, P = 10,000 N/m².

(a) If L = 0.25 m and h = 0.1 m, calculate the maximum normal stress, σ_x, at the support and deflection, v, at the center of the free end, and compare the finite element result with the solution from the strength of materials.

(b) If L = 1 m and h = 0.1 m, calculate the maximum normal stress, σ_x, at the support and the deflection, v, at the center of the free end at point A, and compare the finite element result with the solution from the strength of materials.

Project 13.2

A cantilevered beam has a semicircle with a radius of 0.01 m on the top and bottom center as shown in Figure P13.2. There is a stress concentration near the hole. Determine the maximum normal stress near the hole when it is subjected to 10 kN/m of distributed load as shown in the figure. Compare the finite element result with the one obtained from strength of materials using the

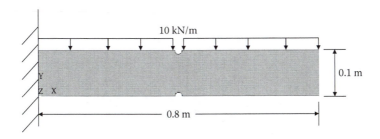

Figure P13.2   Geometry of a cantilevered beam for Project 13.2. (See color insert.)

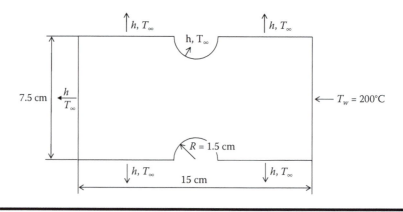

Figure P13.3 Geometry of a cantilevered beam for Project 13.3.

stress concentration factor K (stress concentration factor K can be found in any strength of materials textbook).

Project 13.3

Find the temperature distribution for a plate subject to a temperature of T_w on the right edge and having convection on all other edges as shown in Figure P13.3, where ambient temperature $T_\infty = 20°C$, $T_w = 200°C$, heat transfer coefficient h = 10 W/m².°C, and thermal conductivity k = 386 W/m.°C.

Project 13.4

Find the temperature distribution for a plate subject to a temperature of T_w on the right edge and having convection on all other edges as shown in Figure P13.4,

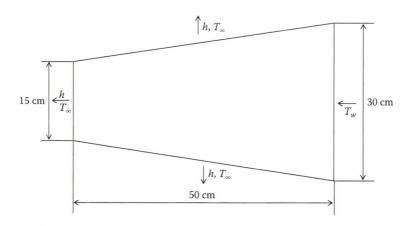

Figure P13.4 Geometry of a cantilevered beam for Project 13.4.

where ambient temperature $T_\infty = 20°C$, $T_w = 200°C$, heat transfer coefficient $h = 10 \ W/m^2 \cdot °C$, and thermal conductivity $k = 386 \ W/m \cdot °C$.

References

1. Timoshenko, S. P. and Goodier, J. N., *Theory of Elasticity*, 3rd ed., McGraw-Hill, New York, 1970.
2. Ugural, A. C. and Fenster, S. K., *Advanced Strength and Applied Elasticity*, 4th ed., Prentice Hall, Upper Saddle River, NJ, 2003.
3. *MATLAB Partial Differential Equation Toolbox User's Guide*, The Mathworks, Inc., Natick, MA, 2008.
4. Reddy, J. N., *Finite Element Method*, 3rd ed., McGraw-Hill, New York, 2006.
5. Cook, R. D., Malkus, D. S., and Pelsha, M. E., *Concepts and Applications of Finite Element Analysis*, 3rd ed., John Wiley & Sons, New York, 1989.
6. Zienkiewicz, O. C. and Taylor, R. L., *The Finite Element Method, Volume 1, Basic Formulation and Linear Problems,* 4th ed., McGraw-Hill International, UK, 1994.

Chapter 14

Control Systems

Oren Masory

14.1 Introduction

A typical block diagram of a control system is shown in Figure 14.1. The plant, which can be a device such as a motor or a process such as heat treatment, is fed with an input, $u(t)$, which affects its output, $y(t)$. Concurrently, the plant is exposed to time-varying unknown loads, $d(t)$, which eventually affect the plant's output, $y(t)$. The purpose of the controller is to vary the input to the plant in such a way that the plant's output will follow the required reference, $r(t)$, in the presence of the unknown loads. In a simple single input single output (SISO) system a sensing device is used to obtain the value of the plant's output. The controller acts on the error, $e(t)$, which is the difference between the reference, $r(t)$, and the measured output by the sensor.

A simple example for such a system is a cruise control system shown in Figure 14.2. In this case, the required speed, $r(t)$, is set by the user; $y(t)$ is the actual vehicle's speed, and in order to determine the error, $e(t)$, the vehicle's speed is obtained from a sensor located inside the transmission. The error is being used by the controller to determine the required amount of fuel, $u(t)$, needed to maintain the required speed under unexpected loads, $d(t)$, exerted by the varying slope of the road and the wind, for example.

Generally there are two types of control systems, regulator and servo. In a regulator system, the reference, $r(t)$, is a constant. A good example for such a system is an air conditioning thermostat that maintains a constant room temperature, in spite of the varying load caused by a change in the number of people in the room, varying exposure of windows to the sun, varying outside temperature, and other

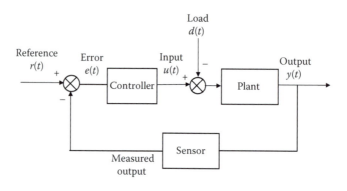

Figure 14.1 A typical structure of control system.

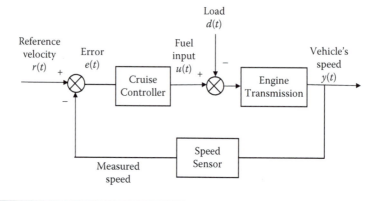

Figure 14.2 Cruise control block diagram.

changes. In a servo system the reference varies with time; a good example for such a system is an antiaircraft gun where the controller should continuously modify the angular position of the gun in order to track the moving target. The design of both systems is very similar and involves two major steps: (1) development of a plant model; and (2) design of a controller that will satisfy one or more requirements related to the dynamic behavior in the plant's output in the presence of changes in the reference and in the loads. The dynamic response of a typical system to unit step input is shown in Figure 14.3. Any or a combination of the parameters, shown in this figure, such as Rise Time or Maximum Overshoot, can be used as design criteria for the controller.

Design criteria are not always related to the system's response in the time domain. In some cases the system's response in the frequency domain is more important because it provides an indication of the capability of the system to track a time-varying reference. For example, a pen plotter is used to draw a circle of radius R and the pen

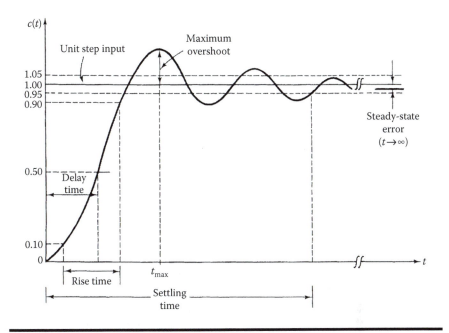

Figure 14.3 Typical system's response to unit step input.

moves along the circle with a speed V as shown in Figure 14.4. The references to the X and Y plotter's axes are given by $V_x = -V sin(\omega t)$ and $V_y = V cos(\omega t)$, where $\omega = V/R$. Thus, the controller designer is more interested in the system frequency response (response to a frequency reference) than its time response.

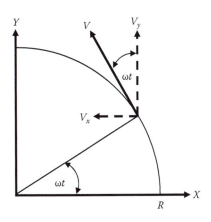

Figure 14.4 Plotting a circle.

MATLAB® provides functions by which the time and frequency response of the system can be determined. In addition, graphical tools that assist in the design of the controller, such as Bode diagrams, Nyquist plots, and root locus plots (see Sections 14.2 and 14.3 for a description of Bode diagrams, Nyquist plots, and Root Locus plots), are available. For the design of most SISO systems these tools are sufficient.

14.2 Representation of Systems in MATLAB

Ignoring the load disturbance, $d(t)$, the relationship between the plant's input, $u(t)$, and its output, $y(t)$, in the case of linear systems can be represented by a linear time-invariant ordinary differential equation of the form

$$a_n \frac{d^n y(t)}{dt^n} + a_{n-1} \frac{d^{n-1} y(t)}{dt^{n-1}} + \cdots + a_0 y(t) = b_m \frac{d^m u(t)}{dt^m} + b_{m-1} \frac{d^{m-1} u(t)}{dt^{m-1}} + \cdots + b_0 u(t)$$

(14.1)

where $n \geq m$ and the coefficients a_j and b_j are constant real numbers.

For a linear controller, the relationship between the controller input, $e(t)$, and its output, $u(t)$, can be described in the same way.

Using MATLAB, these input–output relationships can be represented in three different ways:

1. Transfer function representation (tf)
2. Zero-Pole-Gain representation (zpk)
3. State space representation (ss)

To demonstrate the different representation methods, the coupled tanks system shown in Figure 14.5 will be used as an example.

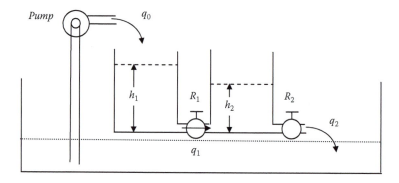

Figure 14.5   Coupled tanks.

The change in the volume of fluid in each tank, v_1 and v_2, is given by

$$\frac{dv_1}{dt} = q_0 - q_1$$

$$\frac{dv_2}{dt} = q_1 - q_2 \tag{14.2}$$

If the cross-section areas of the tanks are A_1 and A_2, respectively, then

$$A_1 \frac{dh_1}{dt} = q_0 - q_1$$

$$A_2 \frac{dh_2}{dt} = q_1 - q_2 \tag{14.3}$$

To simplify the process, we will assume that the flow through the valve is proportional to the pressure drop across the valve (see Figure 14.6). Furthermore, for a slow flow, we can neglect the kinetic energy of the fluid and take the pressure at the entrance and exit of the valve to be what would exist in a static fluid ($p = \gamma h$) (where γ is the specific weight of the fluid and h is the depth below the fluid-free surface). Thus, flow through the valve is given by

$$q \propto p_1 - p_2 \approx \gamma(h_1 - h_2) \tag{14.4}$$

or

$$q = \frac{1}{R}(h_1 - h_2) \tag{14.5}$$

where γ has been absorbed in the proportionality constant $1/R$. Thus, the flows through the valves are given by

$$q_1 = \frac{h_1 - h_2}{R_1}$$

$$q_2 = \frac{h_2}{R_2} \tag{14.6}$$

Substituting Equation (14.6) into Equation (14.3) will produce the dynamic equations for the system:

$$\frac{dh_1}{dt} = \frac{1}{A_1}\left(q_0 - \frac{h_1 - h_2}{R_1}\right)$$

$$\frac{dh_2}{dt} = \frac{1}{A_2}\left(\frac{h_1 - h_2}{R_1} - \frac{h_2}{R_2}\right) \tag{14.7}$$

14.2.1 Transfer Function Representation

Transfer function is the Laplace Transforms of the output divided by the Laplace Transforms of the input, assuming zero initial conditions. This presentation is very simple and intuitive. Referring to Equation (14.1), the transfer function of this equation is given by

$$H(s) = \frac{Y(s)}{U(s)} = \frac{b_m s^m + b_{m-1} s^{m-1} + \cdots + b_0}{a_n s^n + a_{n-1} s^{n-1} + \cdots + a_1 s + a_0} \tag{14.8}$$

where $n \geq m$ and $a_0 \neq 0$.

Applying Laplace Transform to Equation (14.7) yields

$$H_1(s)[s - h_1(0)] = \frac{1}{A_1}\left[Q_0(s) - \frac{H_1(s) - H_2(s)}{R_1} \right]$$

$$H_2(s)[s - h_2(0)] = \frac{1}{A_2}\left[\frac{H_1(s) - H_2(s)}{R_1} - \frac{H_2(s)}{R_2} \right] \tag{14.9}$$

where $H_1(s) = \mathcal{L}\mu(h_1(t))$, $H_2(s) = \mathcal{L}(h_2(t))$, and $Q_0(s) = \mathcal{L}(q_0(t))$. With zero initial conditions, $h_1(0) = h_2(0) = 0$, and some algebraic manipulation of Equation (14.9) can be reduced to

$$H_1(s)\left[s + \frac{1}{A_1 R_1} \right] = \frac{Q_0(s)}{A_1} + \frac{H_2(s)}{A_1 R_1}$$

$$H_2(s)\left[s + \frac{R_1 + R_2}{A_2 R_1 R_2} \right] = \frac{H_1(s)}{A_2 R_1} \tag{14.10}$$

It is clear that the input to the system is q_0. However, from looking at Figure 14.5 it is not so clear what the output is. Realizing that $h_2 < h_1$ and that overflow has to be avoided, it becomes clear that the output of the system is h_1. Solving Equation (14.10) the transfer function of the system, $H_1(s)/Q_0(s)$, is obtained and is given by

$$\frac{H_1(s)}{Q_0(s)} = \frac{A_2 R_1 R_2 s + (R_1 + R_2)}{A_1 A_2 R_1 R_2 s^2 + (A_1 R_1 + A_2 R_2 + A_1 R_2)s + 1} \tag{14.11}$$

In MATLAB, the transfer function is represented by two vectors that contain the coefficients of the numerator and the denominator polynomials of the transfer function. The number of coefficients indicates the degree of the polynomial. The

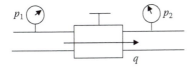

Figure 14.6   Flow through a valve.

following *function* generates the above transfer function using the following values:
$A_1 = 5$ m², $A_2 = 3$ m², $R_1 = 0.8$ s/m², and $R_2 = 0.5$ s/m² [1].

```
function PlantTF=CoupledTanksTF
% Define constants
A1=5;
A2=3;
R1=0.8;
R2=0.5;
% Define the numerator of the transfer function
Num=[A2*R1*R2 (R1+R2)];
% Define the denominator of the transfer function
Den=[A1*A2*R1*R2 (A1*R1+A2*R2+A1*R2) 1];
% Call the tf function
PlantTF=tf(Num,Den);
```

The function was saved under the name *CoupledTanksTF*. Typing the command *CoupledTanksTF* in the command window (make sure that the current directory is the one where this file resides) the following will be displayed:

```
Transfer function:
 1.2 s + 1.3
---------------
6 s^2 + 8 s + 1
```

As an SISO system, MATLAB provides the facility to determine the response of the system for a unit step input. In the command window type *step(CoupledTanksTF)* [2]. The graph shown in Figure 14.7 will be displayed.

Have in mind that even though this response seems logical and acceptable, it is your responsibility to check it out. In this case the explicit transfer function is given by

$$\frac{H_1(s)}{Q_0(s)} = \frac{1.2s + 1.3}{6s^2 + 8s + 1} \tag{14.12}$$

Because at steady state all derivatives are zero, the steady state can be determined by substituting $s = 0$ into the transfer function. Thus for a unit step input ($Q(s) = 1/s$ m³/s) the steady-state value of h_2 can be determined:

$$h_2(\infty) = \frac{1.3}{1} \times 1 = 1.3\text{m} \tag{14.13}$$

This result is the same as the one obtained by the simulation (see Figure 14.7).

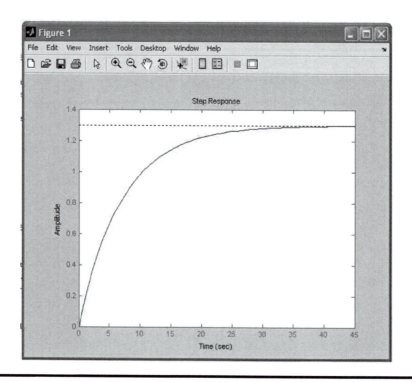

Figure 14.7   System response to a step input. (From MATLAB. With permission.)

When a linear SISO system (refer to Figure 14.1) is excited by an input of the form $u(t) = A\cos(\omega t)$, the output of the system is given by $y(t) = B\cos(\omega t + \phi)$, where A and B are the amplitudes of the input and output, respectively, and ϕ is a phase angle. A Bode plot provides the ratio B/A, usually in decibels [3], as a function of the frequency ω.

The frequency response of the system, which describes the way $h_2(t)$ responds to an input given by $q_0 = \sin(\omega t)$, can be obtained by typing in the command window *bode(CoupledTanksTF)*. The graph shown in Figure 14.8 will display a Bode plot [4].

14.2.2 Zero-Pole-Gain Format of Transfer Function Representation

Another representation of a transfer function is the Zero-Pole-Gain format. Because the transfer function, as the one shown in Equation (14.8), is a rational polynomial function, its numerator and denominator can be factored; that is,

$$H(s) = K\frac{(s-z_1)(s-z_2)\dots(s-z_m)}{(s-p_1)(s-p_2)\dots(s-p_n)} \tag{14.14}$$

The poles, p_j, and the zeros, z_j, can be real, imaginary, or complex numbers and the gain, K, is a real positive number. The zeros, the poles, and the gain of the transfer

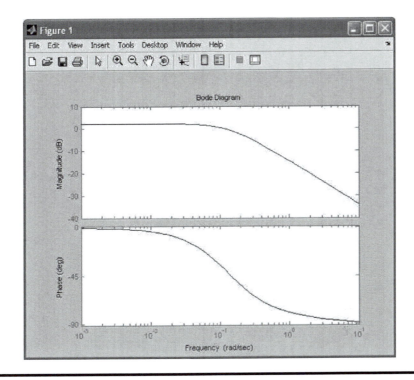

Figure 14.8 **Bode plots for the Coupled Tanks system. (From MATLAB. With permission.)**

function can be obtained by the *pole*, *zero*, and *dcgain* commands. In the command window type [5–7]:

```
>> pole(CoupledTanksTF)
```

The following will be displayed in the command window:

```
Num =
 1.2000 1.3000
Den =
  6 8 1
ans =
 -1.1937
 -0.1396

>> zero(CoupledTanksTF)
Num =
 1.2000    1.3000
Den =
  6 8 1
ans =
 -1.0833
```

```
>> dcgain(CoupledTanksTF)
Num =
  1.2000    1.3000
Den =
   6 8 1
ans =
  1.3000
```

The results show that there are two poles, one at −1.1937 and the other at −0.1396, a zero at −1.0833, and the gain of the system is 1.3 (which is confirmed by the steady-state value of h_2 found by Equation 14.13).

The conversion between transfer function to Zero-Pole-Gain representation is supported by the command *zpk* [8]. The following function performs the conversion and draws the response for a unit step input:

```
function PlantZPK=CoupledTanksZPK
% Running the transfer function object
CoupledTanksTF;
% Finding the poles
PlantPoles=pole(CoupledTanksTF);
% Finding the zeros
PlantZeros=zero(CoupledTanksTF);
% Finding the DC gain
PlantGain=dcgain(CoupledTanksTF);
% Finding the zero-pole-gain presentation
PlantZPK=zpk(PlantZeros,PlantPoles,PlantGain);
% Plotting the step response
step(PlantZPK);
```

Running the program yields the following results and the graph shown in Figure 14.7. Typing in the command window *CoupledTanksZPK* displays the following:

Zero/pole/gain:

1.3 (s+1.083)

(s+1.194) (s+0.1396)

14.2.3 State Space Representation

State space models are first-order coupled differential equations that describe the system in the following general form:

$$\dot{x}(t) = Ax(t) + Bu(t)$$

$$y(t) = Cx(t) + Du(t) \tag{14.15}$$

where x is the system state, u is the input to the system, and y is the system's output. For linear invariant systems the matrices A, B, C, and D are constants. This

State Space presentation will be demonstrated using the Coupled Tanks example. Rewriting the dynamic equation of the tanks (Equation 14.7),

$$\frac{dh_1}{dt} = \frac{1}{A_1}\left(q_0 - \frac{h_1 - h_2}{R_1}\right)$$

$$\frac{dh_2}{dt} = \frac{1}{A_2}\left(\frac{h_1 - h_2}{R_1} - \frac{h_2}{R_2}\right)$$

(14.7)

Define the states, the input, and the output of the system by

$$x_1 = h_1$$

$$x_2 = h_2$$

$$u = q$$

(14.16)

$$y_1 = h_1$$

$$y_2 = h_2$$

The state space representation is directly derived from Equation (14.7):

$$\begin{bmatrix} \dot{x}_1 \\ \dot{x}_2 \end{bmatrix} = \begin{bmatrix} -\dfrac{1}{A_1 R_1} & \dfrac{1}{A_1 R_1} \\ \dfrac{1}{A_2 R_1} & -\dfrac{1}{A_2 R_1} - \dfrac{1}{A_2 R_2} \end{bmatrix} \begin{bmatrix} x_1 \\ x_2 \end{bmatrix} + \begin{bmatrix} \dfrac{1}{A_1} & 0 \\ 0 & 0 \end{bmatrix} \begin{bmatrix} u \\ 0 \end{bmatrix}$$

(14.17)

$$\begin{bmatrix} y_1 \\ y_2 \end{bmatrix} = \begin{bmatrix} 1 & 0 \\ 0 & 1 \end{bmatrix} \begin{bmatrix} x_1 \\ x_2 \end{bmatrix} + \begin{bmatrix} 0 & 0 \\ 0 & 0 \end{bmatrix} \begin{bmatrix} u \\ 0 \end{bmatrix}$$

The following program will define the state space model in MATLAB [9]:

```
function PlantSS=CoupledTanksSS
% Define the constants
A1=5;
A2=3;
R1=0.8;
R2=0.5;
% Define the matrices A, B, C and D
A=[-1/(A1*R1) 1/(A1*R1);1/(A2*R1) -1/(A2*R1)-1/(A2*R2)];
B=[1/A1 0;0 0];
C=[1 0; 0 1];
D=[0 0; 0 0];
% Finding the state space representation
PlantSS=ss(A,B,C,D);
```

Executing the program yields

```
>> CoupledTanksSS
a =
          x1     x2
     x1 -0.25 0.25
     x2 0.4167 -1.083
b =
          u1   u2
     x1 0.2 0
     x2 0   0
c =
          x1 x2
     y1 1 0
     y2 0 1
d =
          u1 u2
     y1 0 0
     y2 0 0
Continuous-time model.
```

In the command window type *step(CoupledTanksSS)*. The time response of h_1 and h_2 will be drawn as shown in Figure 14.9. Note that since the input $u_2 = 0$, the corresponding outputs are zero.

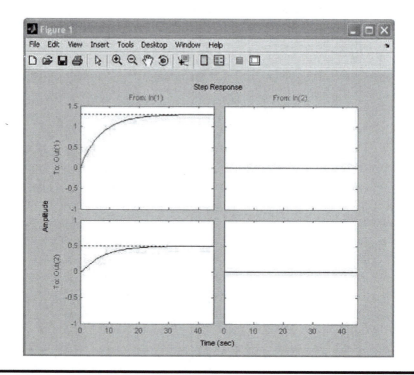

Figure 14.9 System's response to unit step input. (From MATLAB. With permission.)

14.3 Closed-Loop Systems

Up to this point different representations of the plant (see Figure 14.1) were discussed. By nature, the plant itself is an open-loop system that responds to the inputs without any internal corrections or compensations. In a closed-loop control system, the plant's output is measured by a sensor and its signal is fed back and compared with the desired plant's output (reference). The difference between the two, the error, is fed into a controller which in turn produces the input to the plant. Obviously, at this point the closed-loop response of the system is of interest and the problem is how to design a controller that will satisfy a set of requirements related to the dynamic behavior of the plant's output.

In this section several commonly used design tools that are available in MATLAB will be discussed and demonstrated. The design of a simple controller, which regulates the speed of a DC motor, will be used to demonstrate these tools.

14.3.1 DC Motor Modeling

Two equations dominate the dynamic response of a DC motor: the voltage equation, which describes the electromagnetic behavior of the motor, and the torque equation, which describes the electromechanical behavior of the motor. The voltage equation is given by

$$L_m \frac{dI_m}{dt} + R_m I_m + E_m = U \tag{14.18}$$

where
L_m = rotor inductance (Henry).
R_m = rotor resistance (+brushes) (ohm).
E_m = induced voltage (volt).
I_m = current (ampere).
U = control Voltage (volt).

The induced voltage is proportional to the angular velocity of the motor.

$$E_m = K_b \omega_m \tag{14.19}$$

where ω_m is the rotor angular velocity [rad/second] and K_b is the motor constant (Volt/[rad/second]). Thus, the voltage equation is reduced to

$$L_m \frac{dI_m}{dt} + R_m I_m + K_b \omega_m = U \tag{14.20}$$

The torque equation is given by

$$J \frac{d\omega_m}{dt} + B\omega_m + T_e = K_i \phi I_m \tag{14.21}$$

where

J = total inertia reflected to the motor shaft (including the rotor inertia)(kg m²).
B = total damping torque reflected to the motor shaft (N m/s).
T_e = total mechanical torque acting on the motor shaft (N m).
ϕ = magnetic flux generated by the stator.

For a permanent magnet (or constant excitation field) DC motor, ϕ is a constant. Defining the motor constant $K_t = K_i\phi$, the torque, T_m, produced by the motor is given by

$$T_m = K_t I_m \tag{14.22}$$

and therefore Equation (14.21) becomes

$$J\frac{d\omega_m}{dt} + B\omega_m + T_e = K_t I_m \tag{14.23}$$

In most cases $L_m \ll 1$ and $B \ll 1$ and thus the associated term in Equations (14.20) and (14.23) can be neglected. The reduced model becomes

$$R_m I_m + K_b \omega_m = U \tag{14.24}$$

$$J\frac{d\omega_m}{dt} + T_e = K_t I_m \tag{14.25}$$

Combining Equations (14.24) and (14.25) gives

$$J\frac{d\omega}{dt} + T_e = K_t \frac{U - K_b\omega}{R_m} \tag{14.26}$$

In the Laplace domain

$$\Omega(s) = K\frac{U(s) - W(s)}{\tau s + 1} \tag{14.27}$$

where

$\tau = \dfrac{JR_m}{K_t K_b}$ motor time constant (second)

$W = T_e \dfrac{R_m}{K_t}$ load disturbance (volt)

$K = \dfrac{1}{K_b}$ motor gain ([rad/second]/volt)

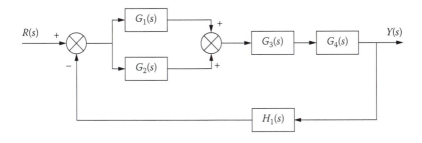

Figure 14.10   A sample block diagram.

14.3.2 Block Diagrams

Up to now our block diagrams were used to describe the connectivity between the different elements of the control system. However, block diagrams can also be used to describe the dynamic behavior of each element by including its transfer function. The block diagram may be used to determine the closed-loop transfer function of the system, which in turn can be used to determine the stability and the time and frequency response of the system.

The diagram shown in Figure 14.10 will be used to demonstrate some of the basic operations and to determine the closed-loop transfer function $Y(s)/R(s)$. To simplify, the solution follows these steps:

1. The parallel connected blocks, $G_1(s)$ and $G_2(s)$, are reduced to $G_1(s) + G_2(s)$.
2. The serially connected blocks, $G_3(s)$ and $G_4(s)$, are reduced to $G_3(s)*G_4(s)$.
3. The two new blocks are connected in serial and can be reduced to $[G_1(s) + G_2(s)]*G_3(s)*G_4(s)$.

These three steps will reduce the block diagram to the one shown in Figure 14.11.

The procedure by which the system's transfer function is determined follows. The error, $E(s)$, using the feedback path of the diagram, is given by

$$E(s) = R(s) - H_1(s)Y(s) \tag{14.28}$$

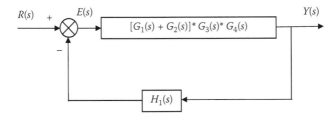

Figure 14.11   Simplified block diagram of the system shown in Figure 14.8.

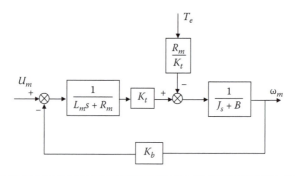

Figure 14.12   Block diagram of a DC motor.

The output, $Y(s)$, using the forward path of the diagram, is given by

$$Y(s) = \{[G_1(s) + G_2(s)] * G_3(s) * G_4(s)\}E(s) \qquad (14.29)$$

Substituting Equation (14.28) into Equation (14.29) yields the closed-loop transfer function

$$\frac{Y(s)}{R(s)} = \frac{[G_1(s) + G_2(s)] * G_3(s) * G_4(s)}{1 + [G_1(s) + G_2(s)] * G_3(s) * G_4(s)H_1(s)} \qquad (14.30)$$

The block diagram that describes the DC model given by Equations (14.20) and (14.23) is shown in Figure 14.12.

14.3.3 MATLAB Tools for Defining the Closed-Loop System

As we have seen in many cases, systems (in particular SISO systems) are described by block diagrams. MATLAB provides tools by which a closed-loop transfer function can easily be determined. As an example, the system shown in Figure 14.10 will be used to demonstrate some of MATLAB's functions. The following transfer functions will be used:

$$G_1(s) = \frac{1}{s+1}$$

$$G_2(s) = \frac{1}{s+2}$$

$$G_3(s) = \frac{s+1}{s^2 + 7s + 1} \qquad (14.31)$$

$$G_4(s) = \frac{1}{s+5}$$

$$H_1(s) = \frac{s}{s+1}$$

The following program simplifies the block diagram shown in Figure 14.10 and determines the closed-loop transfer function following the steps in Section 14.3 [10–12].

```
function PlantTF=Close_loopTF
% Definitions of the blocks of the transfer functions
NumG1=[1]; DenG1=[1 1];
G1TF=tf(NumG1,DenG1);
NumG2=[1]; DenG2=[1 2];
G2TF=tf(NumG2,DenG2);
NumG3=[1 1]; DenG3=[1 7 1];
G3TF=tf(NumG3,DenG3);
NumG4=[1]; DenG4=[1 5];
G4TF=tf(NumG4,DenG4);
NumH1=[1 0]; DenH1=[1 1];
H1TF=tf(NumH1,DenH1);
% Connecting G1 and G2 in parallel
G1G2TF=parallel(G1TF,G2TF);
% Connecting G3 and G4 in serial
G3G4TF=series(G3TF,G4TF);
% Connecting G1G2 and G3G4 in serial
G1G2G3G4TF=series(G1G2TF,G3G4TF);
% Closing the loop with H1;
PlantTF=feedback(G1G2G3G4TF,H1TF);
```

Typing *Close_loopTF* in the command window yields the closed-loop transfer function of the system:

```
Transfer function:
           2 s^3 + 7 s^2 + 8 s + 3
-------------------------------------------------------
s^6 + 16 s^5 + 89 s^4 + 213 s^3 + 229 s^2 + 100 s + 10
```

The block diagram of the DC motor shown in Figure 14.12 will be used as a second example. Neglecting the load, T_e, and using the following constants, the response of the motor to a unit step input, $U_m = 1$, will be determined.

$$K_b = 0.824 \text{ (volt/[rad/second])} \qquad K_t = 7.29 \text{ (lb in./A)}$$
$$R_m = 0.41 \text{ } (\Omega) \qquad J = 0.19 \text{ (lb in. s}^2)$$
$$L_m = 0.005 \text{ (H)} \qquad B = 0.002 \text{ (lb in. second)}$$

```
function PlantTF=DC_Close_LoopTF
Rm=0.41;
Lm=0.005;
Kb=0.824;
Kt=7.29;
B=0.002;
J=0.19;
%Define the electrical block
NumE=[Kt];
DenE=[Lm Rm];
```

```
ElectricalTF=tf(NumE,DenE);
%Define the mechanical block
NumM=[1];
DenM=[J B];
MechanicalTF=tf(NumM,DenM);
%Define the feedback block
NumFB=[Kb];
DenFB=[1];
FeedbackTF=tf(NumFB,DenFB);
%Connect the electrical and mechanical blocks in series
Elec_MechTF=series(MechanicalTF,ElectricalTF);
%Closing the loop
DC_Close_LoopTF=feedback(Elec_MechTF,FeedbackTF)
%Plot the time response to a unit step
step(DC_Close_LoopTF)
```

Executing the program yields the following transfer function, and the time response plot is shown in Figure 14.13.

```
Transfer function:
             7.29
-------------------------------
0.00095 s^2 + 0.07791 s + 6.008
```

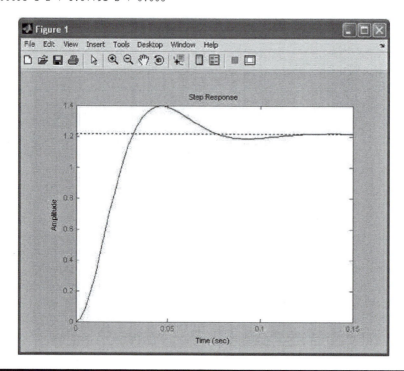

Figure 14.13 **Step response for the model shown in Figure 14.10. (From MATLAB. With permission.)**

14.4 MATLAB Tools for the Performance Analysis of Closed-Loop Systems

Up to this point two MATLAB functions were used to determine the system behavior in time and frequency domains, that is, the *step* and *bode* functions. Two other functions, *rlocus* and *nyquist*, are discussed in the following sections.

14.4.1 Root Locus Plots

A simple SISO closed-loop system is shown in Figure 14.14. Notice that the gain of the forward path, K, was separated from the forward transfer function, $G(s)$.

The closed-loop transfer function is given by

$$\frac{Y(s)}{R(s)} = \frac{KG(s)}{1 + KG(s)H(s)} \tag{14.32}$$

The system characteristic equation, which is the denominator of the closed-loop transfer function, is given by

$$1 + KG(s)H(s) = 0 \tag{14.33}$$

This simple SISO system will be stable as long as the roots of the characteristic equation reside on the left-hand side of the complex plane. The root locus is a graph showing the location of the roots of the characteristic equation as a function of the gain, K. Thus, using the locus, the designer can determine the range of the gain, K, that will ensure the stability of the system. The function *rlocus* uses a slightly different block diagram structure, shown in Figure 14.15, that produces the same characteristic equation.

The following program defines a closed-loop transfer function and calls the *rlocus* function. The root locus plot is shown in Figure 14.16. From the plot it is clear that for a range of gain values some roots are located on the right-hand side of the complex plane. The designer should avoid this range in order to maintain the stability of the system.

```
% Define the Closed-Loop transfer function
N=[ 1  2  4];
D=[1 7 6 5 3];
Closed_LoopTF=tf(N,D);
% Call the rlocus function
rlocus(Closed_LoopTF);
```

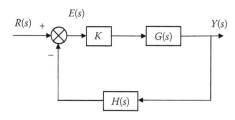

Figure 14.14 A simple closed-loop system.

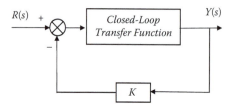

Figure 14.15 System block diagram definition for *rlocus* function.

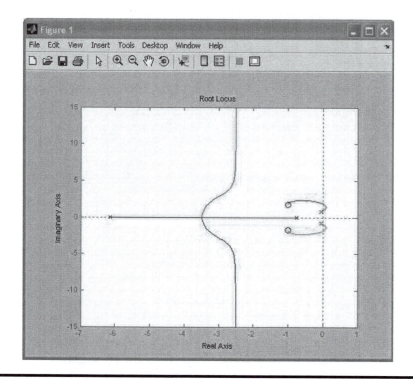

Figure 14.16 Root locus plot of the above closed-loop transfer function. (From MATLAB. With permission.) (See color insert following page 334.)

14.4.2 *Nyquist Plots*

Because the Laplace variable, s, is a complex variable that can be expressed by $s = j\omega$, the term $L(s)$ of a closed-loop system with a characteristic equation given by $1 + L(s) = 0$ can be expressed by its magnitude, M, and its angle ϕ:

$$M = |L(j\omega)|$$
$$\phi = /L(j\omega$$

(14.34)

The Nyquist plot is a M-ϕ plot as ω changes from zero to infinity. The plot is fundamentally used to determine the stability of a system. The Nyquist's stability criteria state that the system is stable if and only if the Nyquist plot of $L(s)$

1. Does not encircle the point $(-1,0)$ when the number of poles of $L(s)$ on the right-hand side of the s plane is zero.
2. Encircles the point $(-1,0)$ counterclockwise n times, where n is the number of poles of $L(s)$ with positive real parts.

As an example, consider the closed-loop system shown in Figure 14.17. The transfer function of the closed-loop system shown in Figure 14.17 is given by

$$\frac{Y(s)}{R(s)} = \frac{\frac{5}{s(s+1)}}{1 + \frac{5}{s(s+1)}\frac{1}{(s+2)}}$$

(14.35)

and thus,

$$L(s) = \frac{5}{s(s+1)(s+2)}$$

(14.36)

The following program will produce the Nyquist plot, shown in Figure 14.18, and the closed-loop system response to a unit step, shown in Figure 14.19.

```
% Define the transfer function L
N_L=[5];
D_L=[1 3 2 0 ];
L_TF=tf(N_L,D_L);
```

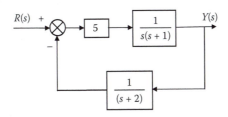

Figure 14.17   Closed-loop system.

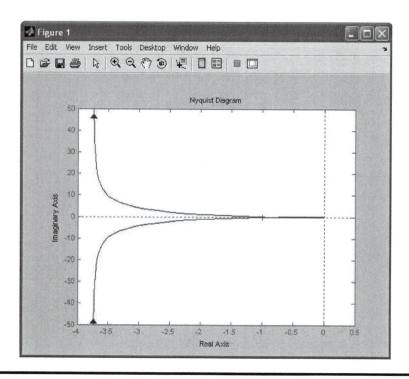

Figure 14.18 **Nyquist plot of *L*(s) defined by Equation (14.36). (From MATLAB. With permission.)**

```
% Call the rlocus function
nyquist(L_TF);
pause;
% Define the closed-loop transfer function L
N_CL=[1 2];
D_CL=[1 3 2 5];
CL_TF=tf(N_CL,D_CL);
step(CL_TF);
```

Note that the plot is symmetrical about the real axis due to the conjugate solution of the roots. Only one of the graphs is needed for the evaluation of the above criteria.

The scales in this plot were selected automatically by MATLAB. The graph that was produced did not provide a good view near the (–1,0) point. To visualize the geometry of the plot in the nationhood of (–1,0), left click on the X scale and change the X limits to –2 and 0.5 and the Y limits to –0.5 and 0.5. As a result, the plot shown in Figure 14.19 was produced. From this graph it is clear that the system is stable because the point (–1,0) is not encircled.

The response to the unit step is shown in Figure 14.20. As shown, the response is very oscillatory, indicating that the system is close to instability. Increasing the gain from 5 to 10 will cause the system to become unstable. The corresponding

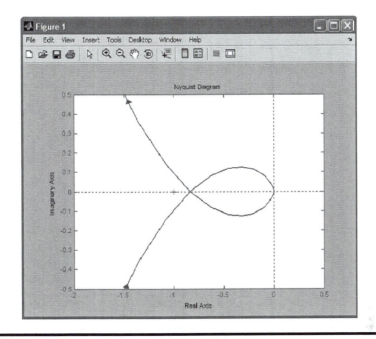

Figure 14.19   Modified Nyquist plot of *L*(*s*). (From MATLAB. With permission.)

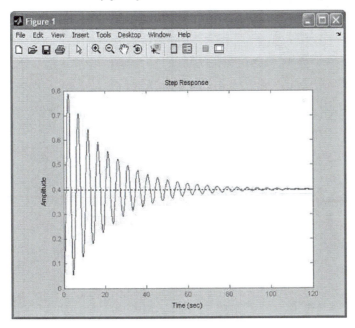

Figure 14.20   Unit step response of the system given in Equation (14.35). (From MATLAB. With permission.)

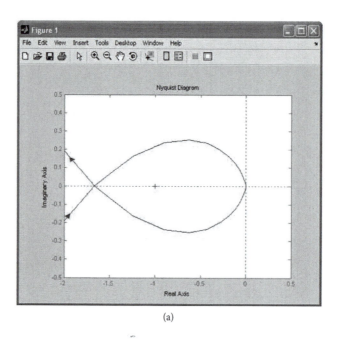

(a)

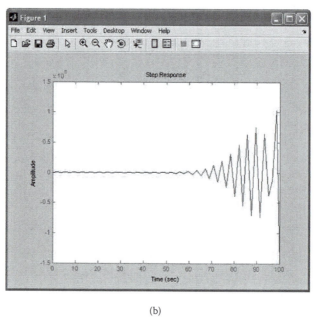

(b)

Figure 14.21 (a) Nyquist plot of $L(s)$ defined by Equation (14.35) with gain = 10. (b) Unit step response of the system given in Equation (14.35) with gain = 10. (From MATLAB. With permission.)

Nyquist plot is shown in Figure 14.21a. This figure shows that the plot encircles the point (–1,0) and therefore the system is not stable. Figure 14.21b shows the response of the closed-loop system to a unit step input and it is clear that the system is not stable.

14.5 MATLAB's SISOtool

MATLAB provides a graphical user interface (GUI) that allows the user to design an SISO system. The GUI provides a few selectable control schemes given by their block diagrams. The controller is programmable within the GUI and its gain can be adjusted "on the fly" using the design tools mentioned above.

14.5.1 Example to Be Used with SISOtool

To demonstrate the capabilities of *SISOtool* the system shown in Figure 14.22 will be used. In this system a ferromagnetic mass, m, is suspended on a spring, k, and a damper, c. An electromagnetic force, f, is applied to the mass, which is proportional to the current, i, flowing through the R-L circuit. The circuit is excited by the voltage, v_{out}, which is proportional to the control signal, v_{in}. The position of the mass is measured by a linear potentiometer of length d and is excited by bipolar reference voltage $\pm v$. The controller is fed with the error, e, and accordingly produces the control signal, v_{in}.

As a first step the system has to be decomposed to subsystem where the input and output of each subsystem are clearly established, as shown in Figure 14.23. Note that the diagram does not include the explicit model of each subsystem.

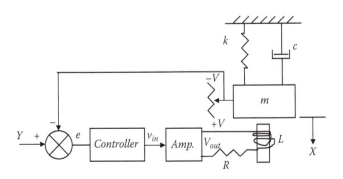

Figure 14.22   Example for *SISOtool*.

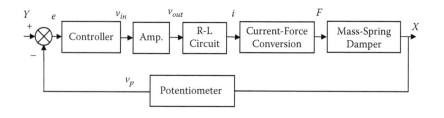

Figure 14.23 **Schematic block diagram of the system shown in Figure 14.22.**

At this point the transfer function of each block has to be determined:

Controller

An I (integrator) controller will be used:

$$\frac{V_{in}(s)}{E(s)} = \frac{K_I}{s} \tag{14.37}$$

where K_I is the controller gain.

Amplifier

The amplifier is a pure gain, K_A:

$$\frac{V_{out}(s)}{V_{in}(s)} = K_A \tag{14.38}$$

R-L Circuit

The current equation for the R-L circuit is given by

$$v_{out} = iR + L\frac{di}{dt} \tag{14.39}$$

where R is the resistance and L is the inductance of the coil. The transfer function $I(s)/V_{out}(s)$ is given by

$$\frac{I(s)}{V_{out}(s)} = \frac{1}{R + Ls} \tag{14.40}$$

Electromagnetic Force

The electromagnetic force, F, is proportional to the current i:

$$\frac{F(s)}{I(s)} = K_F \tag{14.41}$$

where K_F is proportionality constant.

Mass-Spring-Damper

The transfer function $X(s)/F(s)$ is given by

$$\frac{X(s)}{F(s)} = \frac{1}{ms^2 + cs + k} \tag{14.42}$$

Linear Potentiometer

The potentiometer provides a feedback signal that is proportional to the displacement, X, and it depends on its stroke and excitation. It has a gain given by

$$\frac{V_p(s)}{X(s)} = \frac{2V}{d} \tag{14.43}$$

Input Filter

Notice that the reference input, Y, has to be modified in order to match the units at the summation junction by multiplying Y by the feedback device gain.

The complete block diagram of the system is shown in Figure 14.24 and can be simplified to the form shown in Figure 14.25.

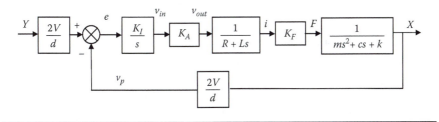

Figure 14.24   Block diagram of the system shown in Figure 14.22.

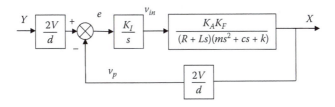

Figure 14.25   Modified block diagram of the system shown in Figure 14.22.

14.5.2 SISOtool Main Features

To start the SISOtool type in the command window *sisotool*. The following window shown in Figure 14.26 will be displayed. There are many options available in this tool and in the following the most important one will be discussed (the reader should open the tool in order to follow these short descriptions):

- *Control Architecture:* Allows the user to select one of the architectures provided, such as change signs at the summation junction, and change name of block and signals.
- *System Data:* Allows the user to specify the transfer function, written ahead of time, for each block.
- *Compensator Editor:* A tool by which the transfer functions of the controller and the input filter can be specified.
- *Graphics Tuning:* Allows the user to select the different plots to be displayed. Initially the Root Locus, and open and closed system Bode diagrams are displayed as shown in Figure 14.27. Using these plots the user can modify the controller interactively.
- *Analysis Plots:* Allow the user to plot the time response to a step input, an impulse input, and other plots needed for analysis.

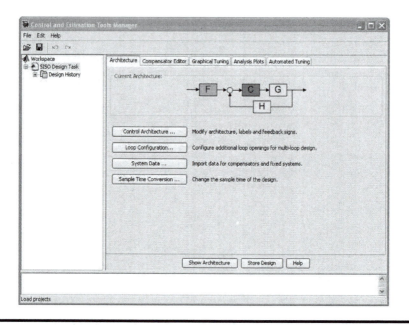

Figure 14.26 The *SISOtool* window. (From MATLAB. With permission.)

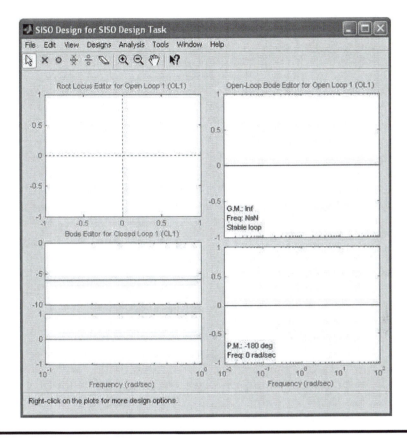

Figure 14.27 Plots initially displayed by *SISOtool*. (From MATLAB. With permission.)

14.5.3 Using SISOtool to Design the Controller for Example at Beginning of This Section

The design of the controller for the above example will be explained step by step:

1. The common call for SISOtool is *sisotool(G,C,H,F)*, where *G*, *C*, *H*, and *F* are transfer functions describing the Plant, Controller, Feedback device, and Input Filter, respectively. These transfer functions are established using the *tf*, *zpk*, or *ss* functions that were explained in Section 14.2.

2. The following values will be used:

$$m = 0.1 \text{ kg} \qquad k = 500 \text{ N/m} \qquad c = 100 \text{ N/(m/s)}$$
$$R = 5 \text{ ohm} \qquad L = 0.01 \text{ H}$$
$$V = 5 \text{ V} \qquad d = 0.1 \text{ m}$$
$$K_A = 5 \qquad K_F = 10 \text{ N/A} \qquad K_I = 1 \text{ V/s}$$

The transfer functions are given by

$$G(s) = \frac{50}{0.001s^3 + 1.5s^2 + 505s + 2500} \tag{14.44}$$

$$C(s) = \frac{1}{s} \tag{14.45}$$

$$H(s) = 50 \tag{14.46}$$

$$F(s) = 50 \tag{14.47}$$

3. The following MATLAB function defines these transfer functions:

```
% The function is in the file GTF.m
function PlantTF=GTF
% Definitions of the Plant's transfer functions
NumG=[50]; DenG=[0.001 1.5 505 2500];
PlantTF=tf(NumG,DenG);

% The function is in the file CTF.m
function ControllerTF=CTF
% Definitions of the Controller's transfer functions
NumC=[1]; DenC=[1 0];
ControllerTF=tf(NumC,DenC);

% The function is in the file HTF.m
function FeedbackTF=HTF
% Definitions of the Controller's transfer functions
NumH=[50]; DenH=[1];
FeedbackTF=tf(NumH,DenH);

% The function is in the file FTF.m
function InputFilterTF=FTF
% Definitions of the Controller's transfer functions
NumF=[50]; DenF=[1];
InputFilterTF=tf(NumF,DenF);
```

4. In the command window type *sisotool(GTF,CTG,HTF,FTF)*, where *GTF, CTG,HTF,FTF* are the transfer function objects established in the program above. Two windows are displayed: The first is the same as in Figure 14.26 and the second, shown in Figure 14.28, is the same as in Figure 14.27 but the plots correspond to the system in this example. Note that the Gain Margin (*GM* = 50.5 db) and the Phase Margin (*PM* = 78.8 degrees) are displayed on the Bode graph and the zeros and poles are displayed on the Root locus plot. The values for the *PM* and *GM* indicate that the system is stable. However, they do not provide any indication regarding the time response of the system.

To see the response of the closed-loop system to a unit step input click on Analysis Plots and fill the form as shown in Figure 14.29. The step response

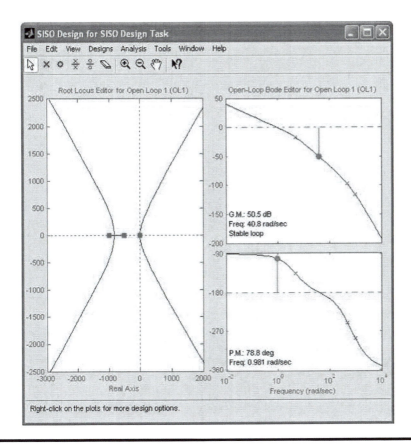

Figure 14.28 Plots initially displayed by *SISOtool* for the example. (From MATLAB. With permission.)

plot, shown in Figure 14.30, will be displayed. As shown, the response is stable. Steady state is reached after approximately 4.5 seconds with a zero steady-state error.

5. The graph shown in Figure 14.28 is a design tool by which the *GM* and *PM* or the open-loop system's gain can be modified.

Using the Bode Graph Tool

To modify *GM* and *PM* put the cursor to the yellow dot on the Bode amplitude diagram (the cursor will change to a hand), right click and drag the point up or down, decreasing and increasing the *GM*, respectively. Simultaneously, the plot of time response to a unit step will change. As the *GM* increases the response becomes

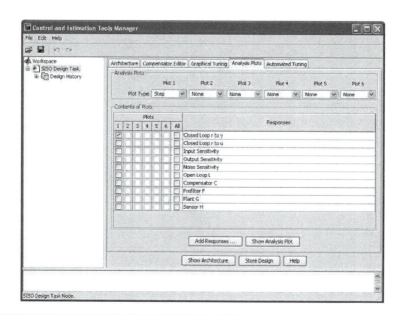

Figure 14.29 Analysis Plots selection. (From MATLAB. With permission.)

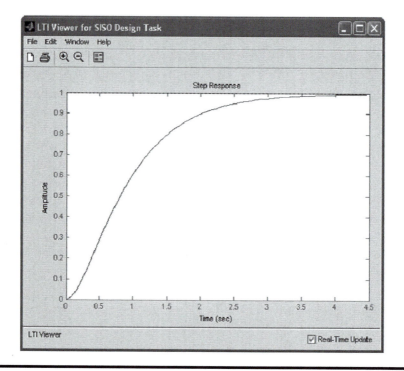

Figure 14.30 Time response to a unit step. (From MATLAB. With permission.)

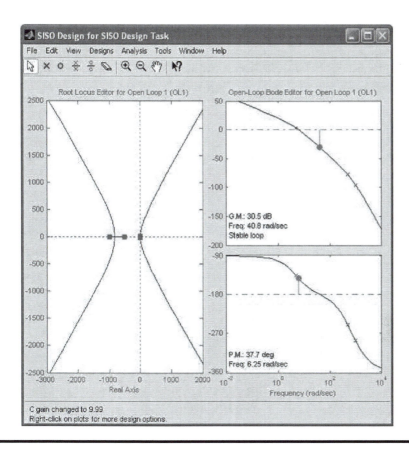

Figure 14.31   Bode diagram of the system after modification of the *GM*. (From MATLAB. With permission.)

faster but oscillatory. Once the *GM* is positive, the system becomes unstable. Figure 14.31 shows the Bode diagram at *GM* = 30.5 db and *PM* = 37.7 degrees and Figure 14.32 shows the corresponding unit step time response of the closed-loop system. The forced change in *GM* corresponds to a change in the open-loop gain of the system. Because only the controller's gain, K_I, can be adjusted in this system, the design tool indicates (at the bottom left of the Bode plot window) the controller gain for these values of *GM* and *PM*—"C gain changes to 9.99."

Using the Root Locus Graph Tool

The default scales of the Root locus plot have to be changed in order to observe the location of the roots close to (0,0). Right double-click on one of the axes to open

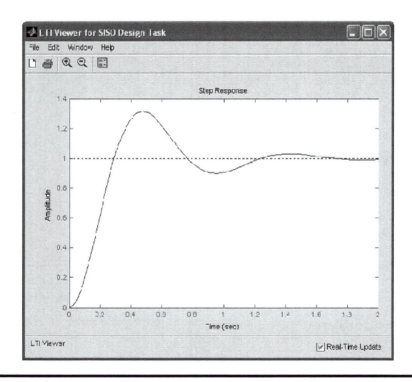

Figure 14.32 Time response to a unit step of the closed-loop system after modification of the *GM*. (From MATLAB. With permission.)

a new window in which the scales can be changed. Change the scales to 10 and −10 on the imaginary axis and to −100 and 100 on the real axis (see Figure 14.33). It is clear that the system is stable because the roots are on the left-hand side of the complex plane. Note that there are two additional zeros that are not shown in Figure 14.33 that are not shown in Figure 14.31. The location of the roots of the characteristic equation can be manipulated by dragging one root along the curve, the same way that the *GM* was manipulated. Notice that

(a) As long as the roots lie on the left-hand side of the complex plane, the system is stable.
(b) If the roots are located on the real axis, the response of the system to a unit step input has no overshoot.
(c) If the roots are complex conjugate the response of the system to a unit step input is oscillatory.

For demonstration, the roots of the characteristic equation were relocated as shown in Figure 14.33. As expected, the system is stable and its response to a unit step is oscillatory, as shown in Figure 14.34.

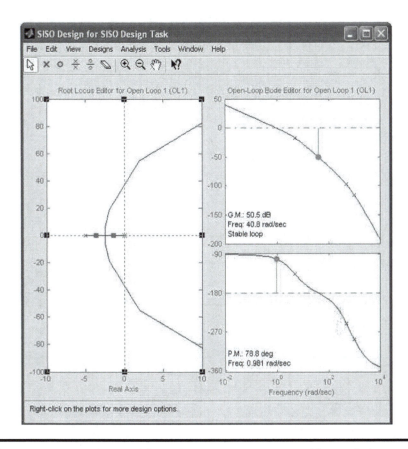

Figure 14.33 Root locus of the system. (From MATLAB. With permission.)

14.6 Application of Simulink in Controls and Dynamic Systems

Simulink® is a friendly, graphical interface by which a control system, or any other dynamic system, is described by a block diagram much the same as was described in Section 14.3.2. As such, the programming is intuitive and easy. Moreover, Simulink supports programming elements, such as saturation, that are very difficult to implement in MATLAB. Some of the main features of Simulink will be demonstrated in the following with several examples. Also see Chapter 7.

14.6.1 Example of Control of the Fluid Level in Coupled Tanks

The goal in this example is to design a controller that will regulate the level of the fluid, h_1, in the coupled tanks system shown in Figure 14.5. To make this example

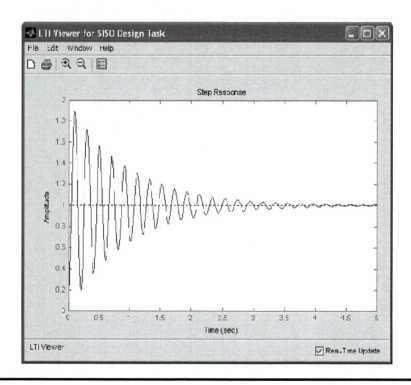

Figure 14.34 **Response of the system in Figure 14.33 to a unit step. (From MATLAB. With permission.)**

more realistic, an additional unknown flow, q_d, will be added as disturbance to the first tank. To this end, the control scheme block, shown as the block diagram in Figure 14.35, is proposed.

The modeling of the whole system will be done block by block, which will make it easier for the implementation in Simulink.

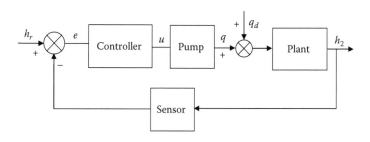

Figure 14.35 **Proposed block diagram for the fluid level control system.**

Modeling the Plant

Recalling Equations (14.19) through (14.22) and adding the additional flow, q_d, yields

$$H_1(s)\left[s+\frac{1}{A_1 R_1}\right]=\frac{Q_0(s)}{A_1}+\frac{Q_d(s)}{A_1}+\frac{H_2(s)}{A_1 R_1}$$

$$H_2(s)\left[s+\frac{R_1+R_2}{A_2 R_1 R_2}\right]=\frac{H_1(s)}{A_2 R_1}$$

(14.48)

Using the same constants—$A_1=5\,m^2, A_2=3\,m^2, R_1=0.8\,s/m^2$, and $R_2=0.5\,s/m^2$—the above transfer functions reduce to

$$H_1(s)=\frac{0.8}{4s+1}Q_0(s)\frac{0.8}{4s+1}Q_d(s)+\frac{1}{4s+1}H_2(s)$$

$$H_2(s)=\frac{0.384}{0.923s+1}H_1(s)$$

(14.49)

The block diagram of the plant, given by Equation (14.49), is shown in Figure 14.36 and its Simulink equivalent is shown in Figure 14.37.

Using the subsystem option the block diagram of the plant is reduced to one block with two inputs, q_0 and q_d, and one output, h_2. The names of the input and output ports, IN1 IN2 and OUT1, should be changed so they will appear in the subsystem block (see top right corner in Figure 14.39).

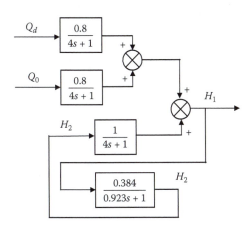

Figure 14.36   Plant's block diagram.

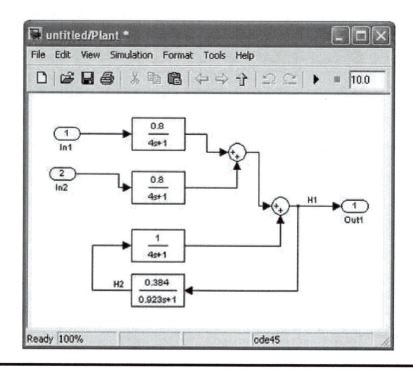

Figure 14.37 Simulink description of the plant. (From MATLAB. With permission.)

Modeling the Pump

A constant displacement pump driven by a DC motor is assumed. The pump has a displacement of $K_d = 0.003$ m³/rev. The reduced model of a DC motor, given by Equation (14.35), will be used and its constants are $K_m = 20$ rpm/V and $\tau = 0.1$ second. The motor is driven by an amplifier with a gain of $K_a = 15$. The block diagram of the Amplifier-Motor-Pump combination is shown in Figure 14.38. Intentionally, the units of the variables were added in order to emphasize that the units at the output of the block should correspond to the units of the input and the units of the gain. The corresponding Simulink diagram is shown in Figure 14.39. The Simulink model for the pump is also reduced to one block.

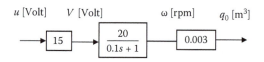

Figure 14.38 The pump unit block diagram.

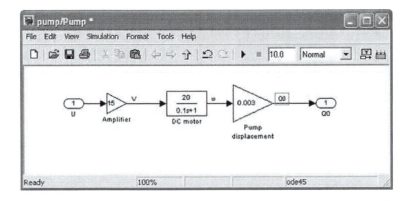

Figure 14.39 Simulink description of the pump unit. (From MATLAB. With permission.)

Modeling the Sensor

The sensor is a device that measures the level of the fluid, h_2, and provides a signal proportional to the measured height. A suitable sensor has an order-of-magnitude faster response than the process variable it is intended to measure. Therefore, its dynamic behavior is usually neglected and it is considered as a pure gain. In this example a sensor with a gain of $K_s = 0.5$ V/m will be used. Note that the same gain is used to filter the input reference, H_r, as explained in Section 14.3.

Modeling the Controller

An integral controller will be added with the transfer function:

$$\frac{U(s)}{E(s)} = \frac{K_I}{s} \tag{14.50}$$

where K_I is a tunable gain.

Reference and Disturbance

Since the controller is a regulator, designed to maintain the fluid level constant in the presence of disturbances, the reference to the system is the required height, H_r. Assuming that the required height is 5 m, the reference is represented as a *step* function of a value 5.

For a start it is assumed that there is no disturbance and therefore $Q_d = 0$.

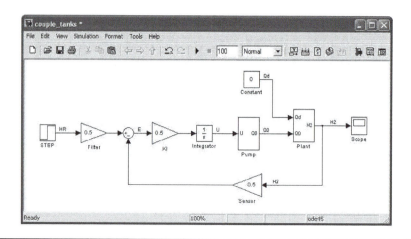

Figure 14.40 Complete Simulink block diagram. (From MATLAB. With permission.)

Executing the Simulation

The complete Simulink block diagram is shown in Figure 14.40, with $K_I = 0.5$. The simulation was executed for the duration of 100 seconds and the results are shown in Figure 14.41.

Controller Gain Tuning

The controller gain, K_I, is the only adjustable variable in this system. Thus, the designer has to select its value according to certain criteria and to the physical constraints of the system, such as the maximum fluid level allowed before overflow and maximum anticipated disturbance flow.

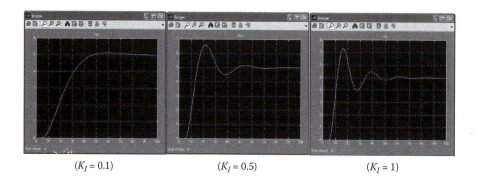

$(K_I = 0.1)$ $(K_I = 0.5)$ $(K_I = 1)$

Figure 14.41 Simulation results—fluid level as function of time. (From MATLAB. With permission.)

Intuitively, it should be clear that as the gain, K_I, is increased, the faster the response will be with the "penalty" of increased overshoot. Figure 14.41 demonstrates this statement by showing the response for K_I= 0.1, 0.5, 1.0. Obviously an overshoot of over 40% (for K_I= 1) is not acceptable because it means that a large portion of the tank capacity cannot be used.

Physical Constraints

Up to this point it was assumed that the system is linear, which is correct for a certain range of operation, for example, when the demand from the pump unit is within its capability. In reality, there are physical constraints that have to be embedded in the simulation:

1. The amplifier output cannot exceed a maximum output voltage of $|V_{max}|$. Therefore a saturation element with the limits of $\pm V_{max}$ has to be added to the output of the amplifier.
2. The pump is unidirectional, which means it can pump water only into the tank but not from the tank. This means that an additional saturation element has to be used at the pump's output, which will limit the output only to a positive value with a maximum value of

$$q_{max} = u_{max} K_a K_m K_d$$

If the controller output signal, u, is limited to ±5V, the maximum flow rate of the pump is given by

$$q_{max} = 5 \times 15 \times 20 \times 0.003 = 4.5 \ m^3/s$$

3. With current technology the controller is implemented by a microcontroller where the control signal, u, is produced by a digital-to-analog converter (DAC). A typical DAC's output varies in the range of ±5V. Thus, another saturation element has to be added to the controller output in order to limit the signal u to that range.

The changes to the pump unit block diagram are shown in Figure 14.42.

The modified simulation was executed for the same controller's gains. Since the pump's flow rate is the major physical constraint, its trace was added to the *scope* using a *multiplexer*. This way, both the fluid level, h_2, as well as the flow rate, q_0, can be observed on the same time scale. The results of the simulations are shown in Figure 14.43, where the yellow line corresponds to the fluid level and the red line to the flow rate, q_0.

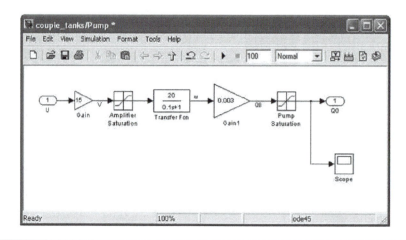

Figure 14.42 **Changes in the pump unit block diagram. (From MATLAB. With permission.)**

As shown, for a low controller gain, $K_I = 0.1$, the flow rate $q_0 < q_{max}$, and therefore saturation does not occur and the system is linear. As the controller's gain increases saturation occurs and the system becomes nonlinear but still stable.

Up to this point in the discussion, no disturbance was introduced ($q_d = 0$). Because the disturbance is not known, but bounded by a known value, a uniform random number bounded by [0,1] is assumed. The *constant* block that produces q_d (see Figure 14.40) will be replaced by a *uniform random number* as shown in Figure 14.44. The disturbance, q_d, is recorded as well. The simulation results are shown in Figure 14.45 and Figure 14.46. As shown, in spite of the fluctuations in the disturbance, the controller maintains the required fluid level for both the linear and nonlinear cases discussed above.

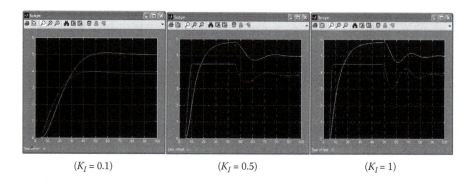

| ($K_I = 0.1$) | ($K_I = 0.5$) | ($K_I = 1$) |

Figure 14.43 **Simulation results of the modified model. (From MATLAB. With permission.) (See color insert.)**

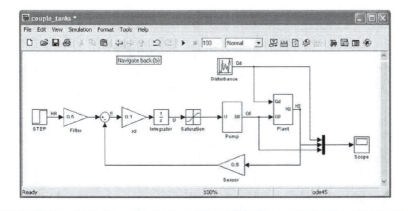

Figure 14.44 Simulation block diagram of the modified model with disturbance. (From MATLAB. With permission.)

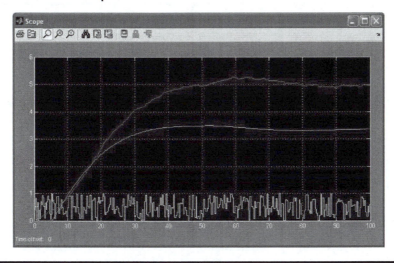

Figure 14.45 Simulation results of the modified model with disturbance ($K_I = 0.1$). (From MATLAB. With permission.) (See color insert.)

14.6.2 Design of a Feed-Forward Loop Using Optimality Criteria

Figure 14.47 illustrates a control scheme that employs a feed-forward loop with a gain K_f in addition to an integral controller.

The system transfer function is given by

$$\frac{X(s)}{Y(s)} = \frac{K_f s + K_I}{\tau s^2 + s + K_I} \tag{14.51}$$

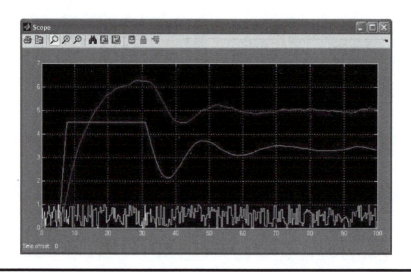

Figure 14.46 Simulation results of the modified model with disturbance ($K_1 = 1.0$). (From MATLAB. With permission.) (See color insert.)

It should be emphasized that the additional feed-forward loop does not change the characteristic equation of the system and therefore its stability (without the feed-forward loop the transfer function is the same where $K_f = 0$). For this linear system the gains K_I and K_f can be selected by different design tools such as root locus and Bode graphs.

However, one has to realize that in a physical system the control signal, u, is bounded and a limiter has to be used. As a result, the system is not linear anymore and the above-mentioned tools cannot be used.

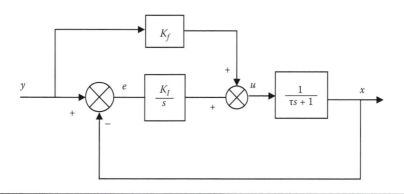

Figure 14.47 A control system with a feed-forward loop.

Performance Indices

Performance index is a measure indicating how well the controller performs. In most cases these indices are applied to regulators (constant reference) where the system reaches a steady state with zero or constant error. These indices can be determined analytically, numerically, or experimentally. Four commonly used indices are given in the following:

1. Integral absolute error (IAE):

$$J = \int_0^\infty |e(t)| \, dt \qquad (14.52)$$

2. Integral of time multiplied by absolute error (ITAE):

$$J = \int_0^\infty t \, |e(t)| \, dt \qquad (14.53)$$

3. Integral of squared error (ISE):

$$J = \int_0^\infty e^2(t) \, dt \qquad (14.54)$$

4. Integral of time multiplied by squared error (ITSE):

$$J = \int_0^\infty te^2(t) \, dt \qquad (14.55)$$

The indices that involve *time* (ITAE and ITSE) put emphasis on the error occurring late in the response because t is small in the early stages of the response. Both indices, IAE and ISE, intend to reduce the errors at the early stages of the response (during the transient) regardless of the error sign, and the ISE index puts higher emphasis on large errors.

Selection of the Gains K_I and K_f

The values of the gains, K_I and K_f, will be determined by minimizing the ITSE performance index numerically. The simulation program, shown in Figure 14.48, will be called by MATLAB using two nested loops where the gains will be varied in a certain range. The gain combination that minimizes the ITSE index is the optimum solution.

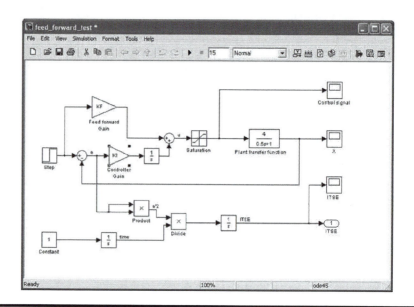

Figure 14.48 Simulink program for the system in Figure 14.47 with ITSE. (From MATLAB. With permission.)

Note the following changes in the simulation program:

1. No numerical values were assigned to the feed-forward and the controller gains. Instead, the names KI and KF were inserted. These names will be used by MATLAB when the simulation is being called.
2. A saturation block was added to limit the maximum/minimum value of the control signal.
3. The value of the performance index will be available to MATLAB using the *out* block.

To test the simulation, the gains were manually set to $K_I = 0.5$ and $K_f = 0.5$ and the input reference was set to a unit step starting at $t = 1$ second. The results of this run are shown in Figure 14.49. The following observations can be made:

1. The control signal, u, is saturated at the early stages of the response because the error, e, is large.
2. While the control signal is saturated, the output, x, remains constant and thus the error is constant and therefore the value of the performance index increases parabolically.
3. As the response reaches steady state with no error the value of the performance index remains constant. This is the value that will be used for determination of optimality.

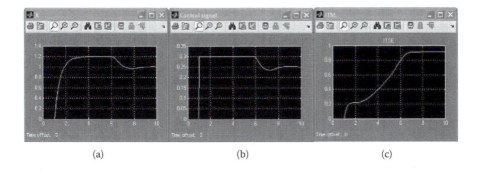

(a)   (b)   (c)

Figure 14.49   Simulation results: (a) X; (b) u; (c) ITSE. (From MATLAB. With permission.)

Multiple Runs of Simulink Model via MATLAB

To call a Simulink program from MATLAB the *sim* function is used [13]. Since the constants in the Simulink model are global variable, they can be changed by MATLAB. The following program changes the values of K_I and K_F using double-nested loops, calls the Simulink model, and prints out the values of K_I and K_F and the corresponding value of the performance index.

```
% File name: feed_forward.m
% Calls simulink model feed_forward_test
% retrieve the vector y which is the outport 1 containing the value of
% the performance index
for KI=0.1:0.1:0.3;
    for KF=0.0:0.1:0.5;
        [t,x,y]=sim('feed_forward_test');
% Extracting the maximum value of the performance index
        ITSE_max=max(y);
        fprintf('KF=%10.2f KI=%10.2f ITSE=%10.3f
      \n',KF,KI,ITSE_max);
    end
end
```

The results of the program are shown below:

```
KF=     0.10    KI=     0.10    ITSE=     1.237
KF=     0.15    KI=     0.10    ITSE=     0.675
KF=     0.20    KI=     0.10    ITSE=     0.387
.

.

.
KF=     0.40    KI=     0.30    ITSE=     0.875
KF      0.50    KI=     0.30    ITSE=     1.353
```

The results were rearranged in Table 14.1. As shown, the minimum value of the performance index occurs when $K_I = 0.2$ and $K_F = 0.2$. These values can be adopted as the optimal solution or the same process can be used to refine these values.

Table 14.1   ITSE Values for Different Combinations of K_I and K_F

		K_I		
		0.1	0.2	0.3
	0.0	3.186	1.391	0.965
	0.1	1.237	0.666	0.537
K_F	0.2	0.387	**0.366**	0.369
	0.3	0.696	0.551	0.509
	0.4	2.009	1.163	0.875
	0.5	2.070	1.961	1.352

In this case, K_I will be changed from 0.1 to 0.3 in increments of 0.05 and K_F will be changed from 0.1 to 0.3 with the same increments. The results of this run are shown in Table 14.2.

The optimal value of the ITSE index, as shown in Table 14.2, is 0.365 and the corresponding optimal values of the gains are $K_I = 0.25$ and $K_F = 0.2$. Setting these values in the Simulink model and executing the program will yield the results shown in Figure 14.50.

A comparison of the simulation's results shown in Figures 14.49 and 14.50 yields the following:

1. The value of the performance index dropped from 0.9 to 0.36.
2. The maximum overshoot of the output, X, dropped from 20% to 15%.
3. The time during which the controller is in saturation dropped from 5 seconds to about 0.5 seconds, which means that the actuation element of the system is not stressed to perform at its limit, thereby extending its useful life.

Table 14.2   ITSE Values for Refined Values of K_I and K_F

		K_I				
		0.1	0.15	0.2	0.25	0.3
	0.1	1.237	0.825	0.666	0.585	0.537
	0.15	0.675	0.519	0.462	0.435	0.419
K_F	0.2	0.387	0.369	0.366	**0.365**	0.369
	0.25	0.375	0.376	0.384	0.391	0.397
	0.3	0.696	0.596	0.551	0.525	0.509

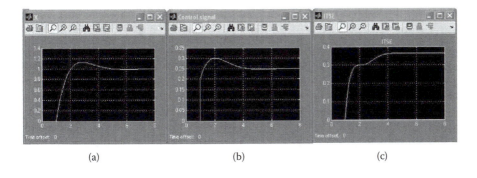

(a)	(b)	(c)

Figure 14.50 Optimum response: (a) X; (b) u; (c) ITSE. (From MATLAB. With permission.)

14.6.3 Active Suspension

Figure 14.51 illustrates a simple model of suspension of one wheel of a vehicle. The mass, m, represents one quarter of the vehicle mass and k and c are the damper and the spring constants. The variable x_v represents the displacement of the vehicle relative to the road and the variable x_r represents the displacement of the axial relative to the road. The vehicle is traveling at the speed V over a bump, which is input to the suspension, with a given geometry.

The governing equation of the above model is given by

$$m\ddot{x}_v + c(\dot{x}_v - \dot{x}_r) + k(x_v - x_r) = 0 \tag{14.56}$$

The input bump, x_r, and its derivative, $\dot{x}_r$, depend on the vehicle's speed and the geometry of the bump. In this case this input can be generated as the summation of four *ramp* inputs whose slopes and starting times are detailed in Table 14.3.

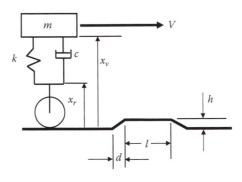

Figure 14.51 A model of the suspension of a single wheel.

Table 14.3   Generation of x_r

	Start Time	Slope
Ramp 1	1	h/(d/V)
Ramp 2	1 + d/V	–h/(d/V)
Ramp 3	1 + d/V + l/V	–h/(d/V)
Ramp 4	1 + 2d/V + l/V	h/(d/V)

The Simulink implementation bump is shown in Figure 14.52 for $V = 45$ kmh, $h = 0.1$ m, $d = 0.25$ m, and $l = 1.5$ m. The Simulink implementation of the model is shown in Figure 14.53. The results for the vehicle displacement and vertical acceleration are shown in Figure 14.54. These results were obtained for a fixed suspension with the following parameters: $m = 225$ kg, $c = 250$ N/m/s, and $k = 5000$ N/m.

The results shown in Figure 14.54 indicate

1. The vehicle response is very oscillatory, which means that the vehicle bounces many times after passing the bump with maximum amplitude of 0.055 m.
2. The occupants of the vehicle experience peak acceleration of 8 m/s², which in not comfortable.

To improve the performance of the suspension, a controller, which will modify the damping coefficient by changing the size of the orifice in the damper, is added to the system. The controller is a proportional derivative (PD) controller with a transfer function given by

$$\frac{U(s)}{E(s)} = K_p + K_d s \qquad (14.57)$$

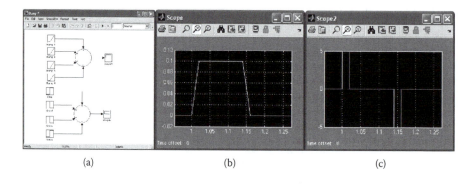

(a)   (b)   (c)

Figure 14.52   Bump simulation: (a) Simulink model; (b) x_r; (c) $\dot{x}_r$. (From MATLAB. With permission.)

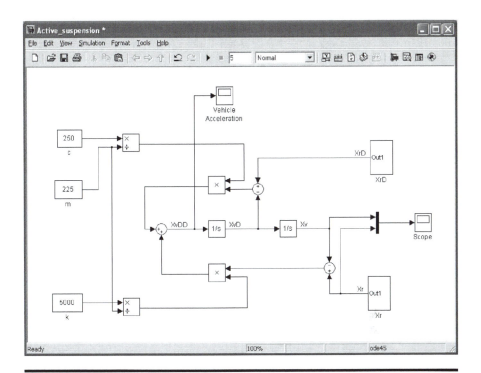

Figure 14.53 Simulink implementation of the suspension model. (From MATLAB. With permission.)

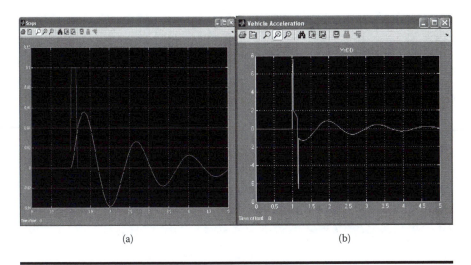

(a)   (b)

Figure 14.54 Simulation results: (a) displacement; (b) acceleration. (From MATLAB. With permission.)

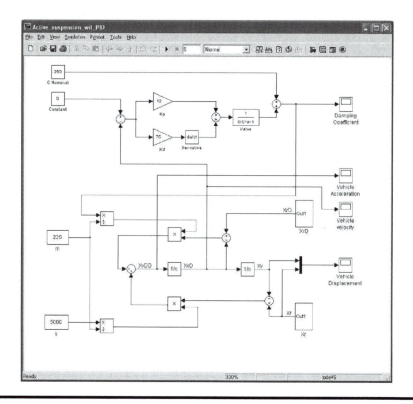

Figure 14.55 Controller implementation. (From MATLAB. With permission.)

The controller drives a small actuator with a time constant $\tau = 0.01$ seconds and gain of one. The initial value of the damper's coefficient is 250 N/m/s.

Figure 14.55 illustrates the implementation of the control scheme in Simulink and Figure 14.56 shows the displacement of the car where the controller's gains were $K_p = 10$ and $K_p = 75$. As seen, the maximum magnitude of the displacement was reduced to 0.03 m compared with 0.055 m for the fixed suspension case. However, the comfort of the occupants is compromised as shown in Figure 14.57; the maximum value of the vehicle's acceleration reaches 3 g.

The reader should realize that an actual active suspension system is by far more complicated and outside the context of this book. The above discussion is just an example to demonstrate that parameters of a system can be controlled as well.

14.6.4 Sampled Data Control System

Most controllers today are implemented by software using microprocessors or microcontrollers. A typical block diagram of a sampled data control system is shown in Figure 14.58. This system functions as follows:

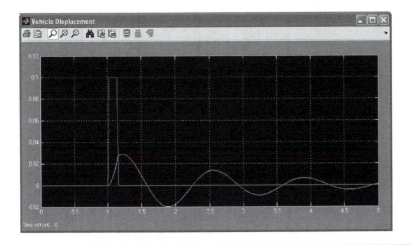

Figure 14.56 **Vehicle's displacement. (From MATLAB. With permission.) (See color insert.)**

1. The reference $r(nT)$ is generated by a program where T is the sampling time and n is an index starting at zero (when t = 0) and increments every T seconds. Thus, nT represents time and as $T \rightarrow 0$, $nT \rightarrow t$.
2. The feedback signal from the sensor is sampled every T seconds (symbolized by the switch that closes at the frequency $1/T$). If the feedback signal is analog, an analog-to-digital converter (ADC) is being used to obtain the numerical value of the signal.
3. Once the reference, $r(nT)$, and the the numerical value of the feedback signal are known, the error, $e(nT)$, can be calculated.

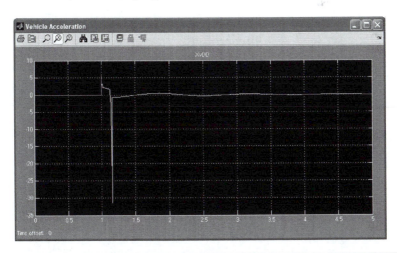

Figure 14.57 **Vehicle's acceleration. (From MATLAB. With permission.)**

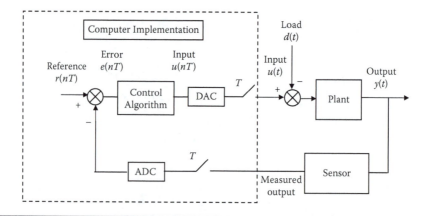

Figure 14.58   Block diagram of a sampled data control system.

4. Using the error signal, a control algorithm is used in order to determine the control signal $u(nT)$.
5. If the control signal needs to be an analog one, the computed value of the control signal is converted to an analog signal using a DAC.
6. The signal produced by the DAC, $u(t)$, is kept constant during the period T and will change in the next iteration.

The design of a sampled data control system requires additional considerations:

1. "Conversion" of a continuous controller to an algorithm. In this case the controller is designed as a continuous one using a variety of available tested tools, then it is converted to an algorithm by different numerical methods for integration and differentiation (replacing $1/s$ and s in the Laplace Transform).
2. Selecting appropriate sampling time T. A simple "rule of thumb" is to select a sampling frequency, $1/T$, three to five times higher than the highest frequency of interest.
3. Selecting appropriate ADC and DAC. Both converters come in different voltage ranges of operation: –5V to +5V, 0V to +5V, and others. In case of ADC the analog signal is converted to a binary number proportional to the input voltage. On the other hand, a DAC is fed with a binary number and it produces a voltage proportional to the value of that number. Both converters are available in different lengths of the binary number (number of bits), which determines the resolution of the converter. For example, a 12-bit DAC with the range of 5V to +5V has a resolution of 0.00244V. This means that an increase in its input number by 1 will increase the output voltage by 0.00244V.
4. Selecting a suitable processor. The processor should be able to calculate the reference and the control algorithm and to acquire the feedback signal (and other tasks) within one sampling period T. Thus, a processor with appropriate computational capabilities has to be chosen.

14.6.5 Implementation of ADC and DAC in Simulink

Implementation of ADC

Figure 14.59 demonstrates the implementation of ADC in Simulink. The implementation consists of three elements:

1. A saturation element that binds the input analog signal to the operating range of the converter (e.g., –5V to +5V).
2. A zero-order hold (ZOH) that samples the signal at the required frequency, $1/T$, and keeps it constant during the duration of the sampling time T.
3. An ADC that converts the analog signal to the corresponding binary number.

The model shown in Figure 14.59 was executed where the input analog signal is $V = 12\sin(2\pi t)$. The results where $T = 0.1$ second and a –10V to +10V 8-bit converter was used are shown in Figure 14.60. The following comments should be made:

1. As can be seen in Figure 14.60, part of the information contained in the signal was lost (10V < V < 12V and –12V < V < –10V) due to the fact that the operating range of the converter (–10V to +10V) did not cover the full range of the signal. Another converter has to be used, or the signal has to be attenuated to the converter operating range.
2. As can be seen in Figure 14.60, the selected sampling frequency is too low because the sampled signal did not resemble the original analog one and a lot of information was lost. A higher sampling rate is needed.

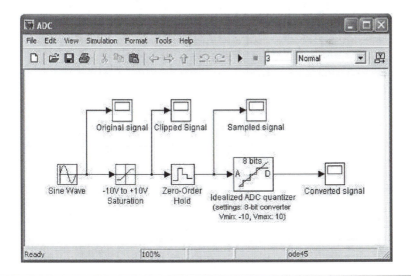

Figure 14.59 Analog-to-digital converter implementation. (From MATLAB. With permission.)

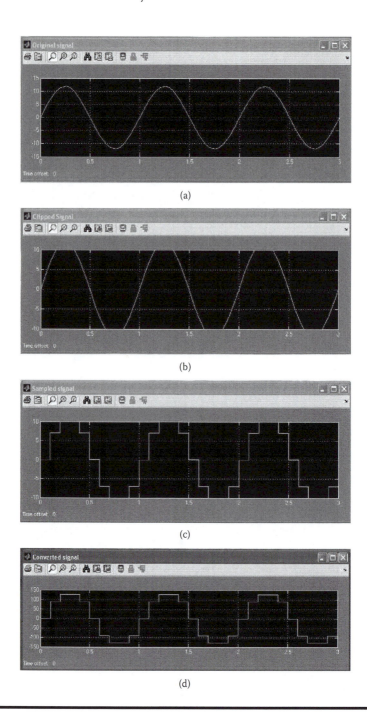

Figure 14.60   (a) The original analog signal; (b) the clipped analog signal; (c) the sampled and held signal; (d) the conversion results. (From MATLAB. With permission.)

3. With an 8-bit binary number, numbers can present in the range 0 to 255. In this case a bipolar converter was selected; the range 0 to 255 was split such that the number 127 represents 10V and −127 represents −10V. This is shown in Figure 14.60d, which illustrates the final results of the conversion.

Implementation of DAC

Figure 14.61 demonstrates the implementation of DAC in Simulink. The implementation consists of five elements:

1. A saturation element that binds the input analog signal to the operating range of the converter (e.g., −5V to +5V).
2. A gain that converts the value of the signal to a real number representing the corresponding input number to be converted.
3. A rounding function that rounds the real number to the closest integer, which is the actual input to the converter.
4. A ZOH that samples the signal at the required frequency, $1/T$, and keeps it constant during the duration of the sampling time T.
5. A gain that converts the integer number to voltage. This gain is determined by the operating range of the ADC and the number of bits of the integer number input.

The model shown in Figure 14.61 was executed where the input analog signal was $V = 6sin(2\pi t)$. The results were $T = 0.05$ second and a −5V to +5V 12-bit converter was used. The results are shown in Figure 14.62. The following comments should be made:

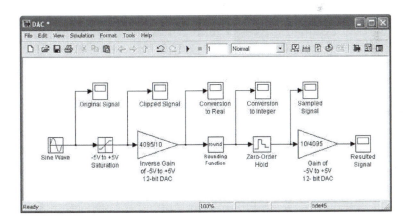

Figure 14.61 **Digital-to-analog converter implementation. (From MATLAB. With permission.)**

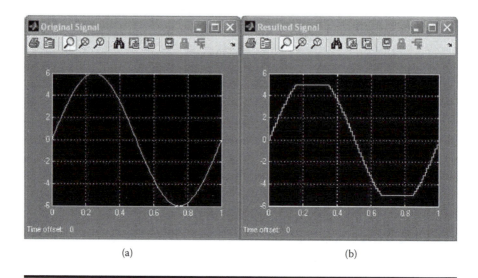

Figure 14.62 **(a) The original analog signal; (b) the conversion results. (From MATLAB. With permission.)**

1. Part of the information contained in the signal was lost (5V < V < 6V and −6V < V < −5V) due to the fact that the operating range of the converter (−5V to +5V) does not cover the full range of the signal. Another converter has to be used, or the signal has to be attenuated to the converter operating range.
2. The selected sampling frequency seems to be sufficient. However, for a real application it should be determined by the dynamics of the application (see Figure 14.62b).
3. A 12-bit binary number with the range of 0 to 4095 provides a resolution of 0.00244V for this operating range.

Position Control of a Hydraulic Piston

To demonstrate the capabilities of Simulink in the design of sampled data control systems the example shown in Figure 14.63 will be used.

The model of each physical component has to be determined before the implementation of the controller:

Servo valve: The servo valve provides a flow rate proportional to its voltage, V, excitation, and its time response, which can be described by a first-order lag with a time constant τ:

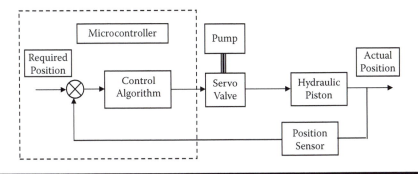

Figure 14.63 Block diagram of a position sampled data control system.

$$\frac{Q(s)}{V(s)} = \frac{K_v}{\tau s + 1} \tag{14.58}$$

where K_v is the gain given in cubic meters per second per volt.

Piston: The piston speed is given by q/A, where A is the cross-section of the piston. In this example a double-acting double-rod piston is assumed and therefore the piston cross-area is the same in forward or backward motion. Since the hydraulic liquid is incompressible, the displacement of the piston, p_p, is given by

$$\frac{P_p(s)}{Q(s)} = \frac{1}{As} \tag{14.59}$$

Position sensor: The position sensor is a linear device with a voltage output, v_p, proportional to the displacement (e.g., linear potentiometer). Thus, its gain is defined by

$$\frac{V_s(s)}{P_p(s)} = K_s \tag{14.60}$$

To proceed with this example the following values will be used:

1. Valve gain: $K_v = 0.00001 \frac{m^3}{s \times volt}$
2. Input voltage to the valve: $V = -10V$ to $+10V$
3. Valve time constant: $\tau = 0.02s$
4. Piston's cross-section area: $A = 0.0005\ m^2$
5. Piston's maximum stroke: $L = 0.5\ m$
6. Sensor gain: $K_s = 20 \frac{V}{m}$
7. Sensor's output: $V_s = 0$ to $10V$

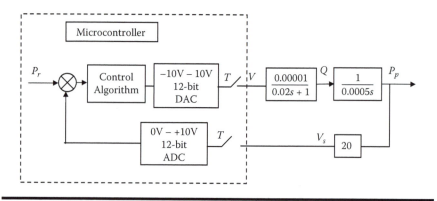

Figure 14.64 **Block diagram of a position sampled data control system.**

Selection of Converters

ADC Converter: An ADC with a range of 0 to 10 volts will be sufficient since the maximum stroke of the piston is 0.5 m and with a sensor gain of 20V/m the maximum output will be 10V. The required resolution on the ADC depends on the needed position accuracy. If an 8-bit ADC is to be used, every increment in the ADC output (binary number) represents a change in position of 0.002 m (0.5/255). If a 12-bit ADC is used, this position increment is 0.00012 m. In this case the latter is selected.

DAC Converter: A DAC with a range of −10 to 10 volts will match the input range of the valve. However, one has to realize that the signal produced by the DAC cannot supply the power needed to drive the valve. A power amplifier, in this case with a gain of one, is needed. The resolution of the DAC should match the resolution of the ADC, otherwise the full capability of the ADC is not used. Thus, a 12-bit DAC is selected.

The block diagram shown in Figure 14.64 summarizes the discussion up to this point.

Simulink Implementation

The implementation, in Simulink, of the system shown in Figure 14.64, where the controller is just a proportional controller, is shown in Figure 14.65. Executing the program with a controller gain of 0.005 yields the results shown in Figure 14.66. As shown, the piston reaches the required position after 7 seconds, during which the position error drops to zero. Also, it should be noticed that the valve is in saturation for a short time.

Increasing the controller's gain to 0.05 yields the results shown in Figure 14.67. In this case the response is faster, but with a small overshoot, and the valve is saturated for a longer time.

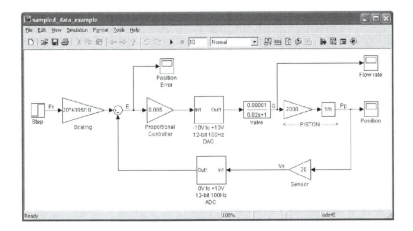

Figure 14.65 Simulink implementation of the position control system that is shown in Figure 14.63. (From MATLAB. With permission.)

In this example, the speed of the piston is not controlled and therefore the error is very large for a long time causing the saturation of the valve even for a small controller gain. A careful examination of the response reveals that the piston's position increases linearly because the valve is saturated and the flow to the piston is constant, equal to the maximum flow rate.

In the case where the velocity of the piston has to be controlled, the reference has to be changed to a ramp where the slope corresponds to the required speed. In the following example the required velocity was set to 0.1 m/s, resulting in the final position of 0.4 m. To generate the desired position ramp (Figure 14.68c) two ramp inputs (shown in Figure 14.68a,b) were added together to produce the position ramp shown in Figure 14.68c.

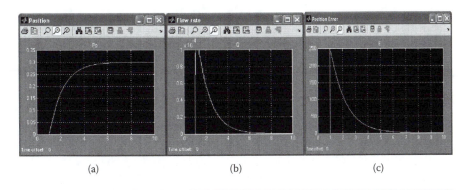

(a)   (b)   (c)

Figure 14.66 Simulation results (for a controller gain of 0.005 and *T* = 0.01 second): (a) piston position; (b) flow rate through the valve; (c) position error. (From MATLAB. With permission.)

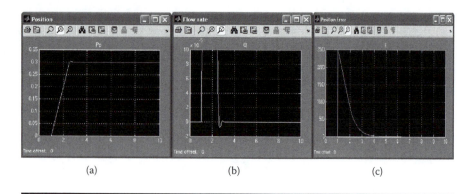

(a)   (b)   (c)

Figure 14.67   Simulation results (for a controller gain of 0.01 and T = 0.01 second): (a) piston position; (b) flow rate through the valve; (c) position error. (From MATLAB. With permission.)

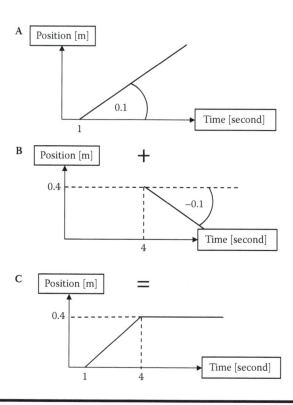

Figure 14.68   Reference generation for speed of 0.1 m/s and final position of 0.4 m.

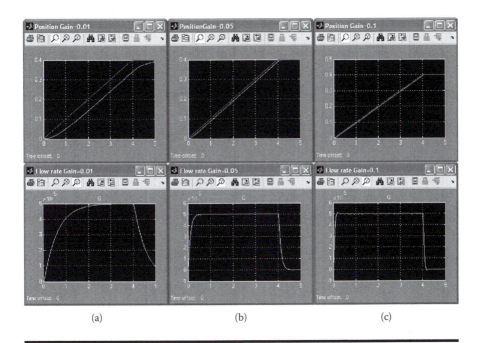

(a)   (b)   (c)

Figure 14.69   Simulation results for different controller gains (*T* = 0.01 second): (a) gain = 0.01; (b) gain = 0.05; (c) gain = 0.1. (From MATLAB. With permission.) (See color insert.)

The simulation results for different gains are shown in Figure 14.69. As shown, the response is very sluggish for a small gain because the valve's response is slow. It takes about 3 seconds to reach the required steady-state flow for the given velocity. As a result, the actual position of the piston lags the reference by a large distance. For applications where only a single piston is used, the mentioned lag is not important. However, in a multipiston system where their motions are coordinated in order to follow a prescribed trajectory, these lags might cause deviations from the trajectory.

14.7   Simulink's Data Acquisition Toolbox

Simulink provides a Data Acquisition Toolbox that makes it possible to program a data acquisition system installed on your computer. If the hardware and the supporting software are installed correctly, Simulink will identify your system automatically.

Selecting the Data Acquisition Toolbox from the Simulink main menu (see Figure 14.70) will show the six supported functions that will be explained in the following sections.

Figure 14.70 Data Acquisition functions. (From MATLAB. With permission.)

14.7.1 Analog Input

Double-clicking on the analog input block will open the window shown in Figure 14.71. As shown, the data acquisition (DAQ) device and the PMD-1208FS board were recognized by Simulink and automatically set up the available channels, the mode of operation (single ended or differential), and range of operation (−20V to +20V), and allowed the user to input the sampling rate. This block is usually used to sample a continuous signal at a certain rate. The acquired data can be processed in real time or stored for off-line processing.

Analog Input (Single)

The difference between this block (see Figure 14.72) and the previous one is that in this case a single sample is being acquired. Thus, this block is usually used as part

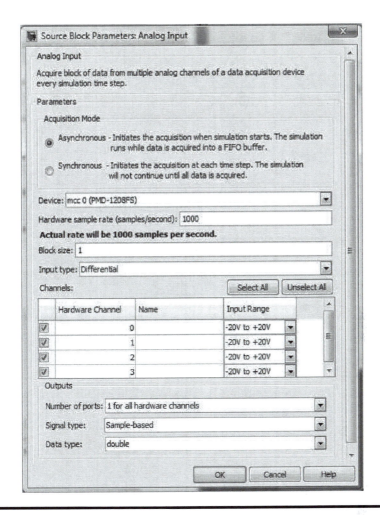

Figure 14.71   Analog Input block parameters. (From MATLAB. With permission.)

of a sampled data control system or a monitoring system where a sensor (or muliple sensors) is being sampled at a relatively low rate.

14.7.2 Analog Output

Analog Output (Single)

The difference between this block (see Figure 14.72) and the previous one is that in this case a single analog output will be genertated for each iteration of the program.

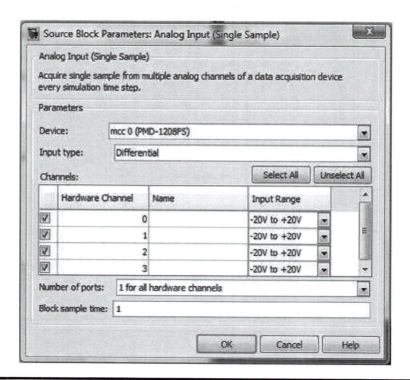

Figure 14.72   Analog Input—Single Sample block's parameters. (From MATLAB. With permission.)

Thus, this block is usually used as part of a sampled data control system where the control signal is generated once for each iteration of the controller.

14.7.3 Digital Input

The PMD-1208FS board has eight discrete Input/Output lines that can be configured as input or output lines. When lines are configured as input, the user can detect the state of on/off sensors such as limit switches or proximity switches. Note that an interface might be needed between the sensor and the input line in order to make sure that the signal is at Transistor-Transistor Logic (TTL) level. The parameters on the Digital Input block are shown in Figure 14.75.

14.7.4 Digital Output

Similarly, the PMD-1208FS discrete lines can be configured as output lines. In this case, the user can use each line to turn on or turn off a discrete device

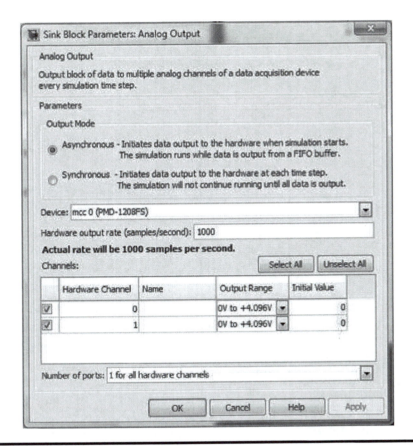

Figure 14.73   Analog Output block. (From MATLAB. With permission.)

such as a solenoid or an electric motor. Note that an electronic interface should be used between the board output and the control device in order to provide the required power needed to drive the device (the board provides a control signal with no power). The parameters on the Digital Output block are shown in Figure 14.76.

Example—Sampling an Analog Signal from an Accelerometer

In this example the signal from an accelerometer, CroosBow CXL02LF3, will be sampled at the rate of 1000 Hz. As shown in Figure 14.77, a simple Simulink program, in which the sampled signal is fed to a scope, was written. Executing the program yields the signal trace shown in Figure 14.78.

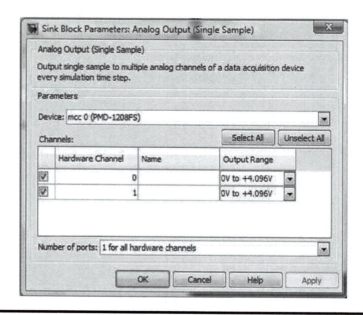

Figure 14.74 Analog Output—Single Output block's parameter. (From MATLAB. With permission.)

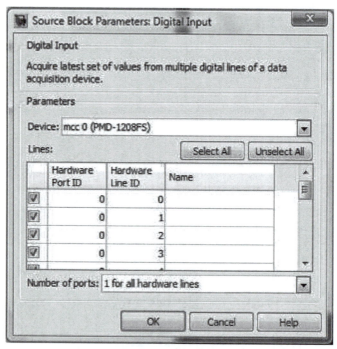

Figure 14.75 Digital Output block's parameter. (From MATLAB. With permission.)

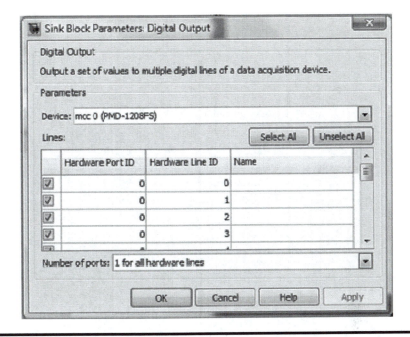

Figure 14.76 Digital Output block's parameter. (From MATLAB. With permission.)

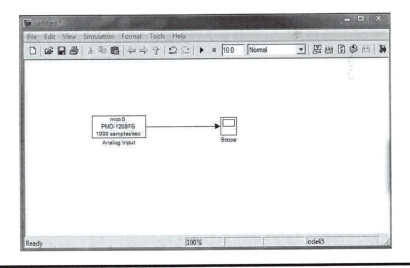

Figure 14.77 Simulink program to acquire data from a sensor. (From MATLAB. With permission.)

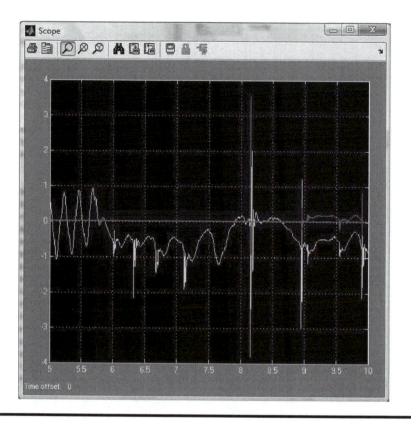

Figure 14.78 Results of the program shown in Figure 14.77. (From MATLAB. With permission.) (See color insert.)

Projects

Project 14.1

For the system shown in Figure P14.1,

   a. Find the closed-loop transfer function.
   b. For $K = 2$ draw the Bode plots and determine the output of the system for an input given by $x(t) = 2\sin(3t) + 3\sin(15t)$.
   c. Draw the Nyquist plot and determine the range of K for which the system is stable.
   d. For $K = 4$ simulate the response of the system to a unit step input.
   e. Determine the maximum value of K so that the overshoot of the output for a unit step input will not exceed 15%.

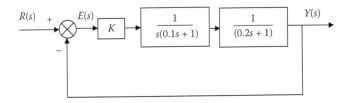

Figure P14.1   Block diagram for Project 14.1.

Project 14.2

The open-loop transfer function of a mechanism that rotates an antenna, used to track a satellite, is given by

$$\frac{\Theta(s)}{U(s)} = \frac{500}{s(s+20)}$$

A phase-lag controller with the following transfer function is used to control the system:

$$\frac{U(s)}{E(s)} = \frac{as+1}{bs+1} \qquad a < b$$

Determine the values of the controller constants, a and b, so that:

a. The maximum overshoot for unit step will be less than 5%.
b. The maximum rise time will be less than 0.2 second.

Project 14.3

The transfer functions of the system shown in Figure P14.3 are

$$G_1(s) = \frac{s+1}{s^2+s+2}$$

$$G_2(s) = \frac{1}{s+5}$$

$$G_3(s) = \frac{K}{s+1}$$

$$G_4(s) = \frac{1}{s+25}$$

$$G_5(s) = \frac{s+1}{s+10}$$

$$H_1(s) = 3$$

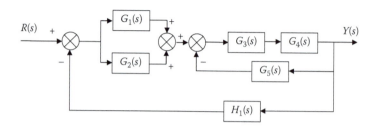

$R(s)$ $+$

$G_1(s)$

$G_2(s)$

$G_3(s)$ $G_4(s)$

$G_5(s)$

$H_1(s)$

$Y(s)$

Figure P14.3   Block diagram for Project 14.3.

a. Determine the transfer function $Y(s)/R(s)$ for $K = 1$.
b. Simulate the system and find its response to a unit step for $K = 1$.
c. Draw the Bode plot for the system (for $K = 1$).
d. Draw the Root Locus of the system and determine the values of K for which the system is stable.

Project 14.4

For the system shown in Figure P14.4 assume that the steady-state level of h_1 is greater than H. Use Simulink to model the system using the following parameters:

$A_1 = 5 \text{ m}^2$, $A_2 = 3 \text{ m}^2$, $R_1 = 0.8 \text{ s/m}^2$, $R_2 = 0.5 \text{ s/m}^2$, $R_3 = 0.25 \text{ s/m}^2$, $H = 5 \text{ m}$

a. Determine, by simulation, what is the steady-state value of h_1 if q_0 is a step function of 4 m³/s.

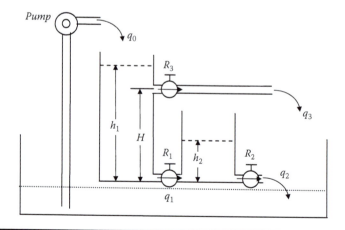

Figure P14.4   Schematic of the tanks for Project 14.4.

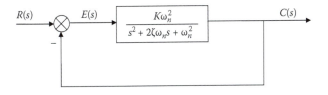

Figure P14.5a   Block diagram of the closed-loop DC drive.

b. The loop is closed with a level sensor with a gain of 1 V/m. The pump with its drive mechanism has the following transfer function:

$$\frac{Q(s)}{U(s)} = \frac{20}{0.25s+1} \left[\frac{m^3}{V} \right]$$

where $u_{max} < 5V$.

Design a PI controller that will minimize the ITSE criteria.

Project 14.5

Axes of an XY plotter are driven by two small DC motors that are controlled separately. The block diagram of one axis is shown in Figure P14.5a, where r is the velocity reference and c is the actual velocity. The nominal values of the system's parameters are $K = 25$, $\zeta = 0.7$, and $\omega_n = 25$ rads/second. As part of a drawing, it was needed to draw a 125-mm-long straight line at 30° as shown in Figure P14.5b. The speed along the trajectory has to be 100 mm/s for proper dispensing of ink.

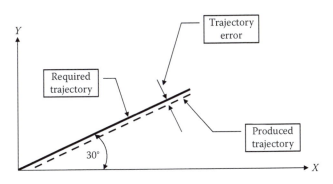

Figure P14.5b   Plotter trajectory.

(a) Simulate the system and draw the produced trajectory (in XY plane) and trajectory error as function of time.

(b) In the physical implementation of the system, the X and Y controllers are not identical. If there is a 5% difference between the gains of the X-axis and Y-axis controllers (i.e., $K_x = 1.05 \ K_y$), how will the trajectory error be affected?

(c) Determine how a 10% difference in ω_n (i.e., $\omega_{nx} = 0.90 \ \omega_{ny}$) will affect the trajectory error.

Project 14.6

A 250-mm-long arm is attached to the shaft of a DC motor as shown in Figure P14.6. A 1-kg mass is attached to the end of the arm.
The motor constants are

Motor inductance — $L = 0.002 \ [H]$
Motor resistance — $R = 4 \ [\Omega]$
Motor damping — $B = 0.002 \ [Nm/rps]$
Motor constant — $K_b = 0.66 \ [V/rps \ ] = 0.105 \ [V/(rad/s)]$
Motor torque constant — $K_t = 0.5 \ [Nm \ /A]$
Motor + arm inertia — $J_R = 0.015 \ [kg \ m^2]$

(a) Simulate the response of the motor (velocity) for a step input of 1V for the following cases:
   1. The arm rotates in the vertical plane.
   2. The arm rotates in the horizontal plane.

(b) Design an analog P controller to regulate the speed of the motor. An amplifier with a gain of 15 is used to drive the motor and an angular velocity sensor with a gain of 10 [volt/krpm] is used as a feedback device. Evaluate the response of the motor for the above two cases where the required angular velocity is 240 [rpm].

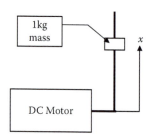

Figure P14.6 Schematic of the motor with the arm.

(c) Design an I controller to regulate the speed of the motor and evaluate the response of the motor for loaded and unloaded conditions where the required angular velocity is 240 [rpm].

(d) Compare the results of items (b) and (c).

Endnotes

1. The *tf* function call format is *TF_object= tf(Num,Den)*, where *Num* and *Den* are vectors containing the coefficients of the numerator and the denominator of the transfer function in descending order of *s*. The result, *TF_object*, is a transfer function object.

2. The *step* function call format is *step(TF_object)*, where *TF_object* is a transfer function object. The function draws the response to unit step applied at t = 0.

3. The value of a scalar *A* in decibel is given by 20log(A).

4. The *bode* function call format is *bode(TF_object)*, where *TF_object* is a transfer function object. The result is a Bode plot.

5. The *pole* function call format is *P=pole(TF_object)*, where *TF_object* is a transfer function object. The result, *P*, is a column vector containing the values of the poles.

6. The *zero* function call format is *Z=zero(TF_object)*, where *TF_object* is a transfer function object. The result, *Z*, is a column vector containing the values of the zeros.

7. The *dcgain* function call format is *K=dcgain(TransferFunction)*, where *TF_object* is a transfer function object. The result, *K*, is the DC gain of the transfer function.

8. The *zpk* function call format is *TF_object=zpk(Zeros,Poles,Gain)*, where *Zeros* and *Poles* are vectors containing the zeros and poles of the transfer function, respectively, and *Gain* is a real number corresponding to the transfer function gain.

9. The *ss* function call format is *TF_object=ss(A,B,C,D)*, where *A, B, C,* and *D* are matrices defining the control system and the output, *TF_Object*, is a transfer function object.

10. The *parallel* function call format is *TF_Object=parallel(TF_object1,TF_Object2)*, where *TF_object1* and *TF_Object2* are transfer function objects. *TF_Object* is a transfer function object generated by parallel connection of *TF_object1* and *TF_Object2*.

11. The *series* function call format is *TF_Object=series1(TF_object1,TF_Object2)*, where *TF_object1* and *TF_Object2* are transfer function objects. *TF_Object* is a transfer function object generated by serial connection of *TF_object1* and *TF_Object2*.

12. The *feedback* function call format is Close_Loop*TF_Object=feedback1(Forward, Feedback)*, where *Forward* is a transfer function object of the forward path and *Feedback* is a transfer function object describing the feedback path. *Close_Loop TF_Object* is a transfer function object of the closed loop.

13. The *sim* function call format is *[t,x,y]=sil('Simulink file name',time,OPTIONS)*, where *t* is the time vector of the simulation, *x* is the state vector (or array) containing the output of all integrators, and *y* is a vector (or array) containing the output of all outports. The variable *time* can specify final time of the simulation or both start and final time of the simulation; OPTIONS is a vector of options defined by the command *Simset*.

Appendix A

A.1 Derivation of Beam Deflection Equation

A horizontal beam subjected to a load will deflect, as shown in Figure A.1a(1) and A.1a(2). The internal moment, M, about the z axis at any section is determined by

$$M = \iint_A \sigma_x(\bar{y})\,\bar{y}\,dA \tag{A.1}$$

where $\sigma_x(\bar{y}) =$ the normal stress at position $\bar{y}$ at the cut section and $\bar{y}$ is measured from the neutral axis of the section. The neutral axis is a section on the beam that does not elongate or shorten during the bending process.

Assuming that $\sigma_x(\bar{y})$ and $\bar{y}$ are positive, M would be in the clockwise direction (see Figure A.1b). For deflection analysis, the moment, M_d, about the z axis is considered positive if the moment at a section is counterclockwise. Therefore,

$$M_d = -\iint_A \sigma_x(\bar{y})\,\bar{y}\,dA \tag{A.2}$$

The beam deflection with a positive M_d will cause an element with a negative $\bar{y}$ to elongate and an element with a positive $\bar{y}$ to shorten. Consider the section between x and $x + dx$ before bending. Element GH and element NS, where NS lies on the neutral axis [see Figure A.1c(1)], are the same length, dx. After bending the element GH becomes $G'H'$ and element NS becomes $N'S'$ [see Figure A.1c(2)]. But length $N'S' =$ length $NS =$ length GH. The section will be curved with a radius of curvature ρ, which varies with position x [see Figure A.1c(2)]. Elongation of element

$$GH = G'H' - GH = (\rho - \bar{y})\Delta\theta - NS = (\rho - \bar{y})\Delta\theta - \rho\Delta\theta$$

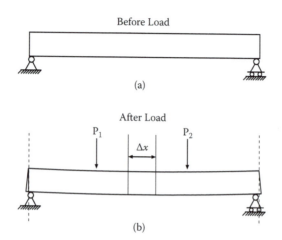

Figure A.1a (a) Beam before being loaded; (b) beam after being loaded.

Then the strain, $\varepsilon_x(\bar{y})$, is given by

$$\varepsilon_x(\bar{y}) = \frac{(\rho - \bar{y})\Delta\theta - \rho\Delta\theta}{\rho\Delta\theta} = -\frac{\bar{y}}{\rho} \qquad (A.3)$$

Hooke's law relates strain to stress as follows:

$$\varepsilon_x = \frac{1}{E}[\sigma_x - v(\sigma_y + \sigma_z)] \qquad (A.4)$$

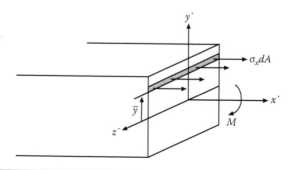

Figure A.1b Stress at beam section.

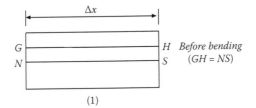

(1)

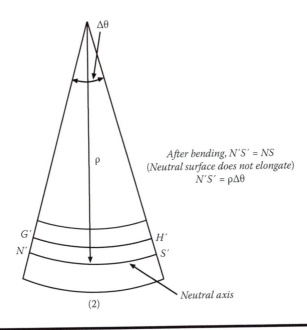

(2)

Figure A.1c   Normal stress causing a section bending moment. (1) Beam element GH before beam bending; (2) beam element GH after beam bending.

where E is Young's Modulus. For the beam problem σ_y and σ_z are negligible, if not zero. Thus,

$$\sigma_x(\overline{y}) = -E\frac{\overline{y}}{\rho} \tag{A.5}$$

(see Figure A.1d). Substituting Equation (A.5) into Equation (A.2) gives

$$M_d = \frac{E}{\rho}\iint_A \overline{y}^2 dA = \frac{EI}{\rho} \tag{A.6}$$

where I is the moment of inertia of cross-sectional area, A.

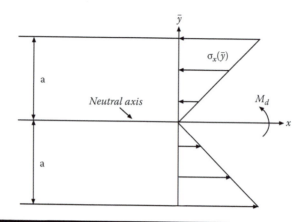

Figure A.1d   Section stress distribution.

The curvature, K, of the deflection curve is given by

$$K = \frac{1}{\rho} = \frac{\dfrac{d^2 y}{dx^2}}{\left[1+\left(\dfrac{dy}{dx}\right)^2\right]^{3/2}}$$

(A.7)

The slope of the deflection curve $\dfrac{dy}{dx} \ll 1$ and therefore

$$\frac{1}{\rho} \approx \frac{d^2 y}{dx^2}$$

(A.8)

Substituting Equation (A.8) into Equation (A.6) gives

$$\frac{d^2 y}{dx^2} = \frac{M_d}{E I}$$

(A.9)

Appendix B

B.1 Derivation of the Heat Transfer Equation in a Solid

- Heat transfer is thermal energy in transit.
- The heat conduction equation provides the means for determining:
    1. The temperature distribution in a solid
    2. The time it takes to transfer a specified amount of heat
    3. The amount of heat transferred in a specified period of time

Heat flow can be represented by the heat flux vector, $\vec{q}$, which is defined as follows:

Select $\Delta s_\perp$ to be perpendicular to the heat flow direction, then $|\vec{q}|$ = the rate that heat flows through $\Delta s_\perp$ per unit surface area and $\vec{q}$ points in the direction of heat flow (see Figure B.1a).

If ΔQ is the rate that heat passes through $\Delta s_\perp$, then

$$|\vec{q}| = \operatorname*{Lim}_{\Delta s_\perp \to 0} \frac{\Delta Q}{\Delta s_\perp}$$

If the surface area, Δs, is not perpendicular to the direction of heat flow, then the rate that heat flows through Δs, say, ΔQ, is given by

$$\Delta Q = \vec{q} \cdot \hat{n} \Delta s = \vec{q} \cdot \Delta \vec{s}$$

where $\hat{n}$ = a unit vector perpendicular to surface Δs (see Figure B.1b).

From Figure B.1a, it can be seen that

$$\Delta s_\perp = \Delta s \cos \alpha = \Delta s\, \hat{n} \cdot \hat{e}_q = \Delta \vec{s} \cdot \hat{e}_q$$

where $\hat{e}_q$ is a unit vector pointing in the direction of heat flow.

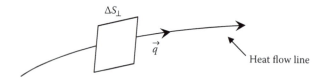

Figure B.1a Heat flux vector, $\vec{q}$.

The heat flow, ΔQ, through Δs is the same as the heat flow through $\Delta s_{\perp}$ (see Figure B.1c). Thus,

$$\Delta Q \approx |\vec{q}|\Delta s_{\perp} = |\vec{q}|\,\hat{e}_q \cdot \Delta \vec{s} = \vec{q} \cdot \Delta \vec{s}$$

For a finite surface area,

$$Q = \iint_s \vec{q} \cdot d\vec{s}$$

There is a large class of materials that obey *Fourier's* heat conduction law, which is

$$\vec{q} = -k\nabla T$$

where k is the thermal conductivity of the material and ∇T is the gradient of the temperature. In Cartesian coordinates

$$\nabla T = \frac{\partial T}{\partial x}\hat{i} + \frac{\partial T}{\partial y}\hat{j} + \frac{\partial T}{\partial z}\hat{k}$$

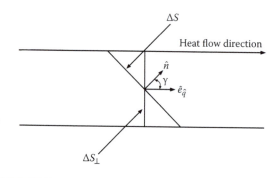

Figure B.1b Heat flow through an arbitrary oriented surface.

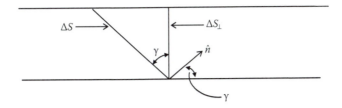

Figure B.1c Relation between ΔS and $\Delta S_\perp$.

where $\hat{i}, \hat{j}, \hat{k}$ are unit vectors in the x, y, z directions, respectively. The significance of the gradient is that its magnitude is the maximum rate of change of the variable with respect to distance and it points in that direction. So Fourier's heat conduction law states that heat will flow down the steepest temperature hill available at a particular point.

B.2 The Heat Conduction Equation for Stationary Solids

The heat conduction equation is based on the First Law of Thermodynamics and Fourier's Heat Conduction Law. First the First Law of Thermodynamics will be discussed.

B.2.1 The First Law of Thermodynamics

For an arbitrary system (region) within the continuum,

> The rate of increase in the total energy in the system = the rate that heat is added to the system plus the rate that heat is generated within the system plus the rate of work done on the system.

We only need to consider energy forms within the system that change during the process. If any of the following phenomena occur within the material—

Electric current
Chemical reactions
Nuclear reactions

then changes in these energy forms need to be accounted for. When this occurs, the work term related to these energy forms are not zero and additional constitutive laws are needed. Since all these processes result in a conversion of some energy form to internal (thermal) energy, the process is accounted for by including a heat generation term in the First Law of Thermodynamics. Under these conditions, the work term for stationary solids in the First Law of Thermodynamics is zero.

We now apply our statement of the First Law to the infinitesimal volume shown in Figure B.2a. *Note*: Evaluate Surface terms at the centroid of the surface and evaluate volume terms at the centroid of the volume. The heat flux vector, $\vec{q}$, can be decomposed into its components; i.e., $\vec{q} = q_1\hat{i} + q_2\hat{j} + q_3\hat{k}$.

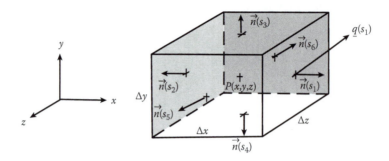

Figure B.2a   Infinitesimal volume in Cartesian coordinates.

The areas, unit normal vectors, and coordinates of the area centroids of the infinitesimal volume shown in Figure B.2a are given in Table B.2a.

Applying the First Law to the infinitesimal system shown gives

$$\left[\frac{\partial}{\partial t}(\rho u)\right]_{(x,y,z,t)} \Delta x \Delta y \Delta z = -\sum_{m=1}^{6} \vec{q}\,(S_m,t)\cdot\vec{n}\,(S_m)\,\Delta S_m + g(x,y,z,t)\,\Delta x \Delta y \Delta z \quad (B.1)$$

Table B.2A   Centroid Coordinates and Normal Unit Vectors of Surfaces S_1 through S_6

Surface	Area	$\vec{n}$	Coord. of Centroid
S_1	$\Delta y \Delta z$	$\hat{i}$	$\left(x+\dfrac{\Delta x}{2},y,z\right)$
S_2	$\Delta y \Delta z$	$-\hat{i}$	$\left(x-\dfrac{\Delta x}{2},y,z\right)$
S_3	$\Delta x \Delta z$	$\hat{j}$	$\left(x,y+\dfrac{\Delta y}{2},z\right)$
S_4	$\Delta x \Delta z$	$-\hat{j}$	$\left(x,y-\dfrac{\Delta y}{2},z\right)$
S_5	$\Delta x \Delta y$	$\hat{k}$	$\left(x,y,z+\dfrac{\Delta z}{2}\right)$
S_6	$\Delta x \Delta y$	$-\hat{k}$	$\left(x,y,z-\dfrac{\Delta z}{2}\right)$

where
> u = internal energy per unit mass.
> g = rate of heat generation per unit volume by internal heat sources.

The minus sign in front of the summation sign in Equation (B.1) results from the fact that $\vec{q} \cdot \vec{n}$ is positive if heat is flowing out of the control volume. The summation term is evaluated as follows:

$$\vec{q}(S_1,t) \cdot \vec{n}(S_1) \, \Delta S_1 = \vec{q}\left(x + \frac{\Delta x}{2}, y, z, t\right) \cdot \hat{i} \, \Delta y \Delta z = q_1\left(x + \frac{\Delta x}{2}, y, z, t\right) \Delta y \Delta z \quad \text{(B.2)}$$

$$\vec{q}(S_2,t) \cdot \vec{n}(S_2) \, \Delta S_2 = \vec{q}\left(x - \frac{\Delta x}{2}, y, z, t\right) \cdot (-\hat{i}) \Delta y \Delta z = - q_1\left(x - \frac{\Delta x}{2}, y, z, t\right) \Delta y \Delta z$$
$$\text{(B.3)}$$

Similarly,

$$\vec{q}(S_3,t) \cdot \vec{n}(S_3) \, \Delta S_3 = q_2\left(x, y + \frac{\Delta y}{2}, z, t\right) \Delta x \Delta z \quad \text{(B.4)}$$

$$\vec{q}(S_4,t) \cdot \vec{n}(S_4) \, \Delta S_4 = - q_2\left(x, y - \frac{\Delta y}{2}, z, t\right) \Delta x \Delta z \quad \text{(B.5)}$$

$$\vec{q}(S_5,t) \cdot \vec{n}(S_5) \, \Delta S_5 = q_3\left(x, y, z + \frac{\Delta z}{2}, t\right) \Delta x \Delta y \quad \text{(B.6)}$$

$$\vec{q}(S_6,t) \cdot \vec{n}(S_6) \, \Delta S_6 = - q_3\left(x, y, z - \frac{\Delta z}{2}, t\right) \Delta x \Delta y \quad \text{(B.7)}$$

Applying Equations (B.2) through (B.7) to Equation (B.1) and dividing by $\Delta x \, \Delta y \, \Delta z$ gives

$$\left[\frac{\partial}{\partial t}(\rho u)\right]_{(x,y,z,t)} = -\left\{ \frac{q_1\left(x + \frac{\Delta x}{2}, y, z, t\right) - q_1\left(x - \frac{\Delta x}{2}, y, z, t\right)}{\Delta x} \right.$$
$$+ \frac{q_2\left(x, y + \frac{\Delta y}{2}, z, t\right) - q_2\left(x, y - \frac{\Delta y}{2}, z, t\right)}{\Delta y}$$
$$\left. + \frac{q_3\left(x, y, z + \frac{\Delta z}{2}, t\right) - q_3\left(x, y, z - \frac{\Delta z}{2}, t\right)}{\Delta z} \right\} + g(x, y, z, t)$$
$$\text{(B.8)}$$

Taking limits as $\Delta x \to 0$, $\Delta y \to 0$, and $\Delta z \to 0$ of both sides of Equation (B.8) gives

$$\frac{\partial}{\partial t}(\rho u) = -\left[\frac{\partial q_1}{\partial x} + \frac{\partial q_2}{\partial y} + \frac{\partial q_3}{\partial z}\right] + g = -\nabla \cdot \vec{q} + g \qquad (B.9)$$

B.2.2 Some Vector Calculus

$$\vec{A} = A_1 \hat{i} + A_2 \hat{j} + A_3 \hat{k}$$

$$\nabla \cdot \vec{A} = \frac{\partial A_1}{\partial x} + \frac{\partial A_2}{\partial y} + \frac{\partial A_3}{\partial z}$$

$$\nabla T = \frac{\partial T}{\partial x}\hat{i} + \frac{\partial T}{\partial y}\hat{j} + \frac{\partial T}{\partial z}\hat{k}$$

Applying Fourier's heat conduction law:

$$\vec{q} = -k\nabla T \qquad (B.10)$$

$$\nabla \cdot \vec{q} = -\nabla \cdot (k\nabla T) \qquad (B.11)$$

$$\nabla \cdot \vec{q} = -\left\{\frac{\partial}{\partial x}\left(k\frac{\partial T}{\partial x}\right) + \frac{\partial}{\partial y}\left(k\frac{\partial T}{\partial y}\right) + \frac{\partial}{\partial z}\left(k\frac{\partial T}{\partial z}\right)\right\} \qquad (B.12)$$

Also, $u = c_v T$, where c_v is the specific heat at constant volume. For solids and liquids, $c_v = c_p = c$.

Because k, ρ, and c are mild functions of temperature, they frequently are taken as constants (especially if analytical techniques are used to solve the problem). The above equation becomes

$$\frac{1}{a}\frac{\partial T}{\partial t} = \left\{\frac{\partial^2 T}{\partial x^2} + \frac{\partial^2 T}{\partial y^2} + \frac{\partial^2 T}{\partial z^2}\right\} + g \qquad (B.13)$$

where

$$a = \text{the thermal diffusivity of the material} = \frac{k}{\rho c}$$

Appendix C

C.1 Bessel's Equation

The solution to Bessel's equation is obtained by applying the power series method. Bessel's equation is

$$x^2 \frac{d^2 y}{dx^2} + x \frac{dy}{dx} + (x^2 - n^2)y = 0 \tag{C.1}$$

Equation (C.1) is the same as Equation (10.84), with n replacing v. Assume that

$$y = \sum_{i=0}^{\infty} a_i x^{i+k} \tag{C.2}$$

$$y' = \sum_{i=0}^{\infty} (i+k) a_i x^{i+k-1}$$

$$y'' = \sum_{i=0}^{\infty} (i+k-1)(i+k) a_i x^{i+k-2}$$

Substituting these expressions into Equation (C.1) gives

$$\sum_{i=0}^{\infty} (i+k-1)(i+k) a_i x^{i+k} + \sum (i+k) a_i x^{i+k} + \sum_{i=0}^{\infty} a_i x^{i+k+2} - \sum n^2 a_i x^{i+k} = 0$$

The coefficients of like powers of x must be zero. When $i = 0$,

$$[(k-1)k + k - n^2]a_0 + a_{-2} = 0$$

429

But a_{-2} does not exist, so $k^2 - n^2 = 0$. Therefore, one solution is for $k = n$. When $i = 1$,

$$[k(k+1) + k + 1 - n^2]a_1 + a_{-1} = [n^2 + 2n + 1]a_1 + a_{-1} = 0$$

Since a_{-1} does not exist and $n^2 + 2n + 1 \neq 0$, therefore $a_1 = 0$. This results in $a_3 = 0$, $a_5 = 0$, $a_7 = 0$, and so on. When $i = 2$,

$$[(n+1)(n+2) + (n+2) - n^2]a_2 + a_0 = 0$$

or

$$a_2 = -\frac{a_0}{(n+2)^2 - n^2} = -\frac{a_0}{2(2n+2)}$$

When $i = 4$,

$$[(n+3)(n+4) + n + 4 - n^2]a_4 + a_2 = 0$$

or

$$a_4 = -\frac{a_2}{(n+4)^2 - n^2} = -\frac{a_2}{4(2n+4)} = \frac{a_0}{4 \cdot 2(2n+4)(2n+2)}$$

When $i = 6$,

$$a_6 = -\frac{a_0}{6 \cdot 4 \cdot 2(2n+6)(2n+4)(2n+2)}$$

etc. Substituting these coefficients into Equation (C.1) gives

$$y = x^n a_0$$

$$\times \left(1 - \frac{x^2}{2(2n+2)} + \frac{x^4}{4 \cdot 2(2n+4)(2n+2)} - \frac{x^6}{6 \cdot 4 \cdot 2(2n+6)(2n+4)(2n+2)} + - \cdots \right)$$

or

$$y = x^n a_0 \left(1 - \frac{(x/2)^2}{1(n+1)} + \frac{(x/2)^4}{2 \cdot 1(n+2)(n+1)} - \frac{(x/2)^6}{3 \cdot 2 \cdot 1(n+3)(n+2)(n+1)} + - \cdots \right)$$

To obtain the J_n expression given by Equation (C.4), set

$$a_0 = \frac{1}{2^n\, n!}$$

then

$$y = \left(\frac{(x/2)^n}{n!} - \frac{(x/2)^{n+2}}{1!(n+1)!} + \frac{(x/2)^{n+4}}{2!(n+2)!} - \frac{(x/2)^{n+6}}{3!(n+3)!} + - \cdots \right) \qquad (C.3)$$

It is left for an exercise to show that y as expressed by Equation (C.3) $= J_n$ as given by Equation (C.4) in the next section.

C.2 Recussion Formulas for $J_n(x)$

The conventional expression for $J_n(x)$ is

$$J_n(x) = \sum_{i=0}^{\infty} \frac{(-1)^i}{i!(i+n)!} \left(\frac{x}{2} \right)^{n+2i} \qquad (C.4)$$

$$J_{n+1}(x) = \sum_{i=0}^{\infty} \frac{(-1)^i}{i!(i+n+1)!} \left(\frac{x}{2} \right)^{n+1+2i} \qquad (C.5)$$

$$J_n'(x) = \sum_{i=0}^{\infty} \frac{(-1)^i (n+2i)}{i!(i+n)!} \left(\frac{1}{2} \right) \left(\frac{x}{2} \right)^{n+2i-1} \qquad (C.6)$$

Now

$$x\, J_n'(x) = \sum_{i=0}^{\infty} \frac{(-1)^i\, n}{i!(i+n)!} \left(\frac{x}{2} \right) \left(\frac{x}{2} \right)^{n+2i-1}$$

$$+ \sum_{i=0}^{\infty} \frac{(-1)^i\, (2i)}{i!(i+n)!} \left(\frac{x}{2} \right) \left(\frac{x}{2} \right)^{n+2i-1} \qquad (C.7)$$

The first term on the right-hand side is just $n\, J_n(x)$ and in the second term on the right-hand side the twos in the $2i$ and in the $x/2$ cancel. Also, the first term in that summation is zero. For the second term on the right side, let $k = i - 1$, then

$$x\, J_n'(x) = n\, J_n(x) + x \sum_{k=0}^{\infty} \frac{(-1)^{k+1} (k+1)}{(k+1)!(k+1+n)!} \left(\frac{x}{2} \right)^{n+2(k+1)-1} \qquad (C.8)$$

or

$$x J_n'(x) = n J_n(x) - x \sum_{k=0}^{\infty} \frac{(-1)^k}{(k)!(k+1+n)!} \left(\frac{x}{2}\right)^{n+1+2k} \tag{C.9}$$

Comparing the second term on the right-hand side of Equation (C.9) with Equation (C.5), it can be seen that

$$x J_n'(x) = n J_n(x) - x J_{n+1}(x) \tag{C.10}$$

This is the first of two recursion relations involving the Bessel functions of interest in this Appendix. Returning to Equation (C.6), write $n + 2i = 2(n + i) - n$, then Equation (C.6) can be written as

$$J_n'(x) = \sum_{i=0}^{\infty} \frac{(-1)^i \, 2(n+i)}{i!(i+n)!} \left(\frac{1}{2}\right) \left(\frac{x}{2}\right)^{n-1+2i}$$

$$+ \sum_{i=0}^{\infty} \frac{(-1)^i (-n)}{i!(i+n)!} \left(\frac{1}{2}\right) \left(\frac{x}{2}\right)^{n-1+2i}$$

$$x J_n'(x) = x \sum_{i=0}^{\infty} \frac{(-1)^i}{i!(i+n-1)!} \left(\frac{x}{2}\right)^{n-1+2i}$$

$$- n \sum_{i=0}^{\infty} \frac{(-1)^i}{i!(i+n)!} \left(\frac{x}{2}\right) \left(\frac{x}{2}\right)^{n+2i-1} \tag{C.11}$$

noting that

$$\left(\frac{x}{2}\right) \left(\frac{x}{2}\right)^{n+2i-1} = \left(\frac{x}{2}\right)^{n+2i}$$

Comparing the first term on the right-hand side of Equation (C.11) with Equation (C.4) and the second term on the right-hand side of Equation (C.11) with Equation (C.4), it can be seen that

$$x J_n'(x) = x J_{n-1}(x) - n J_n(x) \tag{C.12}$$

Equation (C.12) is a second recursion relation of interest. Another relation of interest in this section is

$$\frac{d}{dx}[x^n J_n(x)] = x^n J_{n-1}(x) \qquad (C.13)$$

Proof:

$$\frac{d}{dx}[x^n J_n(x)] = x^n J_n'(x) + n x^{n-1} J_n(x) = x^{n-1}[x J_n'(x) + n J_n(x)] \qquad (C.14)$$

Substituting Equation (C.12) into Equation (C.14) gives

$$\frac{d}{dx}[x^n J_n(x)] = x^n J_{n-1}(x)$$

C.3 Orthogonality of the J_0 Functions

The proof of the orthogonality of the J_0 functions for the problem described in Section 10.2 follows. We wish to show that

$$\int_0^R r J_0(\lambda_m r) J_0(\lambda_n r) dr = \delta_{mn} \frac{\lambda_m^2 R^2 + \left(\frac{hR}{k}\right)^2}{2\lambda_m^2} [J_0(\lambda_m R)]^2$$

Let $\phi_n = J_0(\lambda_n r)$ and $\phi_m = J_0(\lambda_m r)$. Both ϕ_n and ϕ_m satisfy the Bessel's differential equation; that is,

$$r^2 \frac{d^2 \phi_n}{dr^2} + r \frac{d\phi_n}{dr} + \lambda_n^2 r^2 \phi_n = 0 \qquad (C.15)$$

and

$$r^2 \frac{d^2 \phi_m}{dr^2} + r \frac{d\phi_m}{dr} + \lambda_m^2 r^2 \phi_m = 0 \qquad (C.16)$$

Noting that

$$r \frac{d}{dr}\left(r \frac{d\phi}{dr}\right) = r^2 \frac{d^2 \phi}{dr^2} + r \frac{d\phi}{dr} \qquad (C.17)$$

Multiply Equation (C.15) by ϕ_m and Equation (C.16) by ϕ_n, giving

$$\phi_m r \frac{d}{dr}\left(r\frac{d\phi_n}{dr}\right) + \lambda_n^2 r^2 \phi_m \phi_n = 0 \qquad (C.18)$$

$$\phi_n r \frac{d}{dr}\left(r\frac{d\phi_m}{dr}\right) + \lambda_m^2 r^2 \phi_m \phi_n = 0 \qquad (C.19)$$

Subtract Equation (C.19) from Equation (C.18), giving

$$\left(\lambda_m^2 - \lambda_n^2\right)r^2\,\phi_m\,\phi_n = r\phi_m\frac{d}{dr}\left(r\frac{d\phi_n}{dr}\right) - r\phi_n\frac{d}{dr}\left(r\frac{d\phi_m}{dr}\right) \qquad (C.20)$$

Divide Equation (C.20) by r, multiply by dr, and integrate from zero to R, giving

$$\left(\lambda_m^2 - \lambda_n^2\right)\int_0^R r\,\phi_m\,\phi_n\,dr = \int_0^R\left[\phi_m\frac{d}{dr}\left(r\frac{d\phi_n}{dr}\right) - \phi_n\frac{d}{dr}\left(r\frac{d\phi_m}{dr}\right)\right]dr \quad (C.21)$$

The integral on the right-hand side of Equation (C.21) can be made into an exact differential by bringing the ϕ terms inside the derivative expressions. Note that

$$\frac{d}{dr}\left(r\phi_m\frac{d\phi_n}{dr} - r\phi_n\frac{d\phi_m}{dr}\right) = \phi_m\frac{d}{dr}\left(r\frac{d\phi_n}{dr}\right) + r\frac{d\phi_n}{dr}\frac{d\phi_m}{dr}$$

$$-\phi_n\frac{d}{dr}\left(r\frac{d\phi_m}{dr}\right) - r\frac{d\phi_n}{dr}\frac{d\phi_m}{dr}$$

$$= \phi_m\frac{d}{dr}\left(r\frac{d\phi_n}{dr}\right) - \phi_n\frac{d}{dr}\left(r\frac{d\phi_m}{dr}\right)$$

As a result Equation (C.21) becomes

$$\left(\lambda_m^2 - \lambda_n^2\right)\int_0^R r\,\phi_m\,\phi_n\,dr = \int_0^R\frac{d}{dr}\left(r\phi_m\frac{d\phi_n}{dr} - r\phi_n\frac{d\phi_m}{dr}\right)dr$$

$$= \left[r\phi_m\frac{d\phi_n}{dr} - r\phi_n\frac{d\phi_m}{dr}\right]_0^R$$

$$= R\phi_m(R)\frac{d\phi_n}{dr}(R) - R\phi_n(R)\frac{d\phi_m}{dr}(R)$$

or

$$\left(\lambda_m^2 - \lambda_n^2\right) \int_0^R r J_0(\lambda_m r) J_0(\lambda_n r)\, dr = R[J_0(\lambda_m R) J_0'(\lambda_n R) - J_0(\lambda_n R) J_0'(\lambda_m R)] \quad (C.22)$$

But the boundary condition (as shown in Equation 10.88 in Chapter 10),

$$\left[\frac{d}{dr}(J_0(\lambda r)) + \frac{h}{k} J_0(\lambda r)\right]_{r=R} = 0$$

is valid for each $J_0(\lambda_n r)$. Therefore,

$$J_0'(\lambda_n R) = -\frac{h}{k} J_0(\lambda_n R) \quad (C.23)$$

and

$$J_0'(\lambda_m R) = -\frac{h}{k} J_0(\lambda_m R) \quad (C.24)$$

Substituting Equations (C.23) and (C.24) into Equation (C.22) gives

$$\left(\lambda_m^2 - \lambda_n^2\right) \int_0^R r J_0(\lambda_m r) J_0(\lambda_n r)\, dr = R\left[-J_0(\lambda_m R)\frac{h}{k} J_0(\lambda_n R) + J_0(\lambda_n R)\frac{h}{k} J_0(\lambda_m R)\right]$$

$$= 0$$

If $m \neq n$, then

$$\int_0^R r J_0(\lambda_m r) J_0(\lambda_n r)\, dr = 0 \quad (C.25)$$

If $m = n$, then $(\lambda_m^2 - \lambda_n^2) = 0$ and the integral $\int_0^R r[J_0(\lambda_n r)]^2\, dr$ needs to be evaluated. From Equations (C.15) and (C.17)

$$r\frac{d}{dr}\left(r\frac{d}{dr} J_0(\lambda_n r)\right) + \lambda_n^2 r^2 J_0(\lambda_n r) = 0 \quad (C.26)$$

Multiply Equation (C.26) by $2\frac{d}{dr} J_0(\lambda_n r)$, giving

$$2r\frac{d}{dr} J_0(\lambda_n r)\frac{d}{dr}\left(r\frac{d}{dr} J_0(\lambda_n r)\right) = -\lambda_n^2 r^2 2 J_0(\lambda_n r)\frac{d}{dr} J_0(\lambda_n r) \quad (C.27)$$

Let $u = r \dfrac{d}{dr} J_0(\lambda_n r)$ and $v = J_0(\lambda_0 r)$

Substituting these values into Equation (C.27) gives

$$2u \frac{du}{dr} = -\lambda_n^2 r^2 2v \frac{dv}{dr} \tag{C.28}$$

Equation (C.28) can be rewritten as

$$\frac{du^2}{dr} = -\lambda_n^2 r^2 \frac{dv^2}{dr}$$

or

$$\frac{d}{dr}\left[r \frac{d}{dr} J_0(\lambda_n r) \right]^2 = -\lambda_n^2 r^2 \frac{d}{dr} [J_0(\lambda_n r)]^2 \tag{C.29}$$

Multiply both sides of Equation (C.29) by dr and integrate from zero to R, giving

$$\int_0^R \frac{d}{dr}\left[r \frac{d}{dr} J_0(\lambda_n r) \right]^2 dr = -\int_0^R \left\{ \lambda_n^2 r^2 \frac{d}{dr}[J_0(\lambda_n r)]^2 \right\} dr$$

or

$$\int_0^R \left\{ \lambda_n^2 r^2 \frac{d}{dr}[J_0(\lambda_n r)]^2 \right\} dr = [R J_0'(\lambda_n R)]^2 \tag{C.30}$$

By Equation (10.88)

$$J_0'(\lambda_n R) = -\frac{h}{k} J_0(\lambda_n R)$$

Therefore,

$$\int_0^R \left\{ \lambda_n^2 r^2 \frac{d}{dr}[J_0(\lambda_n r)]^2 \right\} dr = \left[\frac{h}{k} R J_0(\lambda_n R) \right]^2 \tag{C.31}$$

Now integrate by parts.

$$\int u\, dv = uv - \int v\, du$$

Let

$$u = \lambda_n^2 r^2 \quad \text{and} \quad dv = d[J_0(\lambda_n r)]^2$$

Then

$$du = 2\lambda_n^2 r\, dr \quad \text{and} \quad v = [J_0(\lambda_n r)]^2$$

Using these relationships, the integral in Equation (C.30) becomes

$$\int_0^R \left\{ \lambda_n^2 r^2 \frac{d}{dr} [J_0(\lambda_n r)]^2 \right\} dr = \left[\lambda_n^2 r^2 (J_0(\lambda_n r)^2 \right]_0^R - 2\lambda_n^2 \int_0^R r[J_0(\lambda_n r)]^2 dr \qquad \text{(C.32)}$$

Combining Equations (C.31) and (C.32) gives

$$\int_0^R r[J_0(\lambda_n r)]^2 dr = \frac{1}{2\lambda_n^2} \left\{ \lambda_n^2 R^2 [J_0(\lambda_n R)]^2 + \left[\frac{h}{k} R J_0(\lambda_n R) \right]^2 \right\} \qquad \text{(C.33)}$$

Therefore,

$$\int_0^R r[J_0(\lambda_n r)]^2 dr = \frac{\lambda_n^2 R^2 + \left(\dfrac{hR}{k} \right)^2}{2\lambda_n^2} [J_0(\lambda_n R]^2 \qquad \text{(C.34)}$$

and

$$\int_0^R r J_0(\lambda_m r) J_0(\lambda_n r)\, dr = \delta_{mn} \frac{\lambda_m^2 R^2 + \left(\dfrac{hR}{k} \right)^2}{2\lambda_m^2} [J_0(\lambda_m R]^2 \qquad \text{(C.35)}$$

Finally, we need to show that

$$\int_0^R r J_0(\lambda_m r)\, dr = \frac{R}{\lambda_m} J_1(\lambda_m R) \qquad \text{(C.36)}$$

Let $x = \lambda_m r$, then $dx = \lambda_m dr$. When $r = 0$, $x = 0$, and when $r = R$, $x = \lambda_m R$.

$$\int_0^R r J_0(\lambda_m r)\, dr = \int_0^{\lambda_m R} \frac{x}{\lambda_m} J_0(x) \frac{dx}{\lambda_m} = \frac{1}{\lambda_m^2} \int_0^{\lambda_m R} x J_0(x)\, dx \qquad \text{(C.37)}$$

By Equation (C.10), with $n = 1$,

$$x J_0(x) = \frac{d}{dx}[x J_1(x)] \tag{C.38}$$

Substituting Equation (C.38) into Equation (C.37) gives

$$\int_0^R r J_0(\lambda_m r) dr = \frac{1}{\lambda_m^2} \int_0^{\lambda_m R} \frac{d}{dx}[x J_1(x)] dx = \frac{1}{\lambda_m^2}[x J_1(x)]_0^{\lambda_m R} \tag{C.39}$$

Thus,

$$\int_0^R r J_0(\lambda_m r) dr = \frac{1}{\lambda_m^2}[\lambda_m R J_1(\lambda_m R)] = \frac{R}{\lambda_m} J_1(\lambda_m R)$$

C.4 Vector Calculus in Cylindrical Coordinates

Scalar $\Phi(r, \vartheta, z)$

$$\nabla \Phi = \frac{\partial \Phi}{\partial r} \hat{e}_r + \frac{1}{r}\frac{\partial \Phi}{\partial r} \hat{e}_\vartheta + \frac{\partial \Phi}{\partial z} \hat{e}_z \tag{C.40}$$

$$\nabla^2 \Phi = \frac{1}{r}\frac{\partial}{\partial r}\left(r\frac{\partial \Phi}{\partial r}\right) + \frac{1}{r^2}\frac{\partial^2 \Phi}{\partial \vartheta^2} + \frac{\partial^2 z}{\partial z^2} \tag{C.41}$$

Vector $\vec{A}(r, \vartheta, z) = A_r \hat{e}_r + A_\vartheta \hat{e}_\vartheta + A_z \hat{e}_z$ (C.42)

$$\nabla \cdot \vec{A} = \frac{1}{r}\left[\frac{\partial}{\partial r}(r A_r) + \frac{\partial A_\vartheta}{\partial \vartheta} + \frac{\partial}{\partial z}(r A_z)\right] \tag{C.43}$$

References

1. Kreyszig, E., *Advanced Engineering Mathematics*, 8th ed., John Wiley & Sons, New York, 1999.
2. http://planetmath.org/encyclopedia/BesselsEquation.html.

Index

439

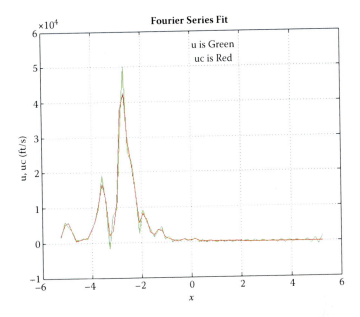

Figure 8.4   Fourier series fit of the data.

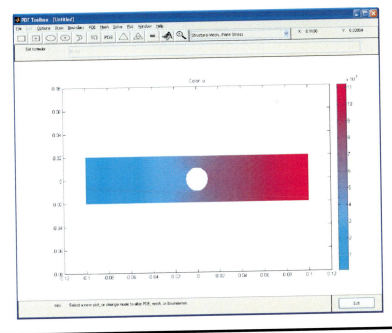

Figure 13.33   Contour plot for the displacement component u in the x direction. (From MATLAB. With permission.)

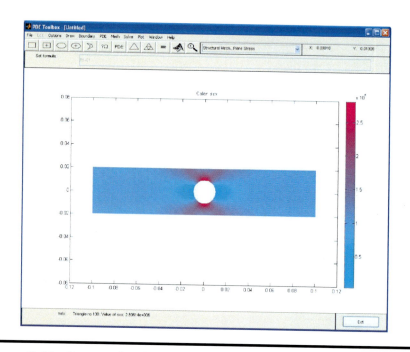

Figure 13.36   Contour plot of normal stress σ_x. (From MATLAB. With permission.)

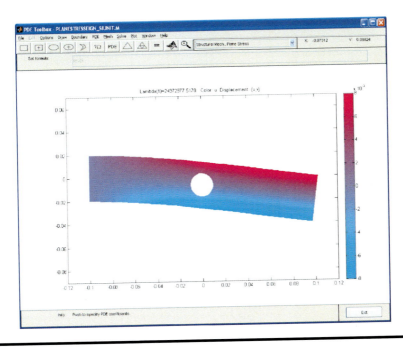

Figure 13.42 The first vibration mode for the lowest natural frequency. (From MATLAB. With permission.)

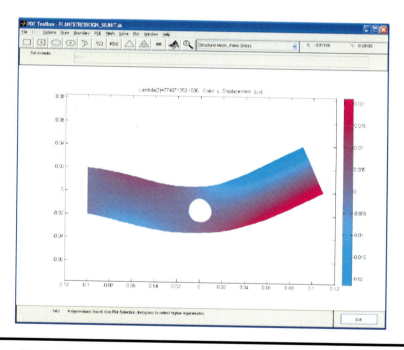

Figure 13.43   The second vibration mode for the second lowest natural frequency. (From MATLAB. With permission.)

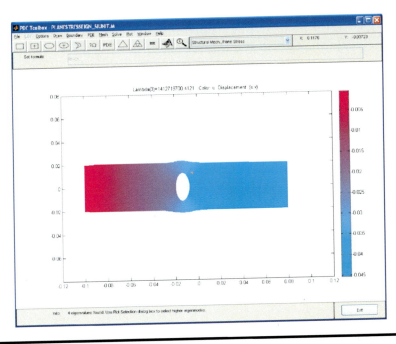

Figure 13.44 The third vibration mode for the third lowest natural frequency. (From MATLAB. With permission.)

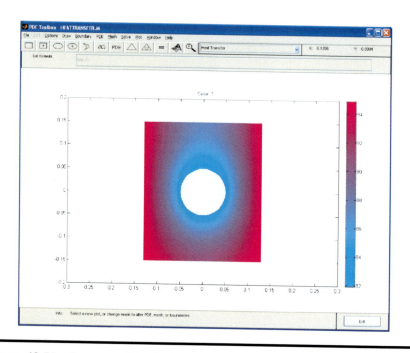

Figure 13.53 Contour plot of the temperature field. (From MATLAB. With permission.)

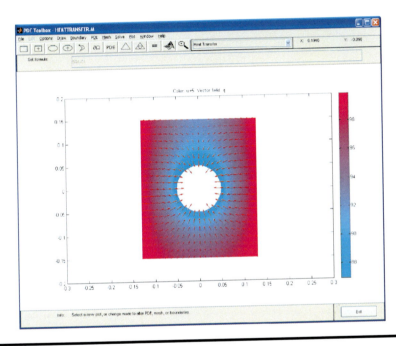

Figure 13.56 Contour plot of the temperature field with heat flux flow. (From MATLAB. With permission.)

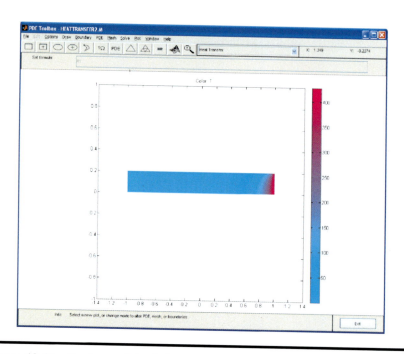

Figure 13.65   Contour plot of the temperature field. (From MATLAB. With permission.)

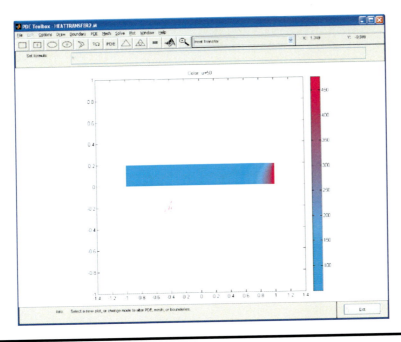

Figure 13.67 Contour plot of the temperature field after adding 50°C to the solutions shown in Figure 13.65. (From MATLAB. With permission.)

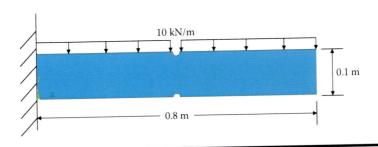

Figure P13.2 Geometry of a cantilevered beam for Project 13.2.

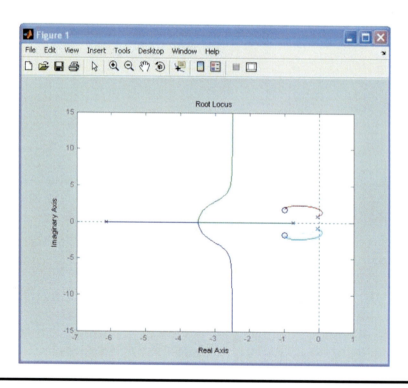

Figure 14.16   Root locus plot of the above closed-loop transfer function. (From MATLAB. With permission.)

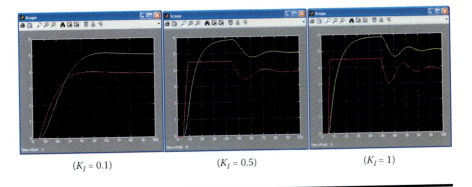

$(K_I = 0.1)$　　　　　　$(K_I = 0.5)$　　　　　　$(K_I = 1)$

Figure 14.43 Simulation results of the modified model. (From MATLAB. With permission.)

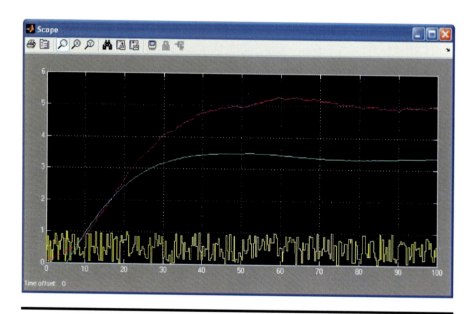

Figure 14.45   Simulation results of the modified model with disturbance ($K_I = 0.1$).
(From MATLAB. With permission.)

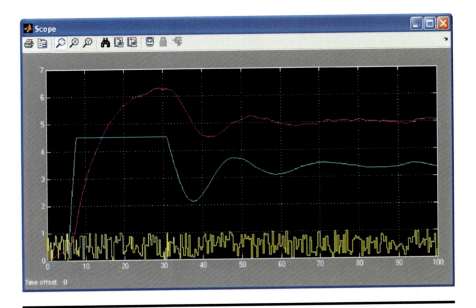

Figure 14.46 Simulation results of the modified model with disturbance ($K_I = 1.0$). (From MATLAB. With permission.)

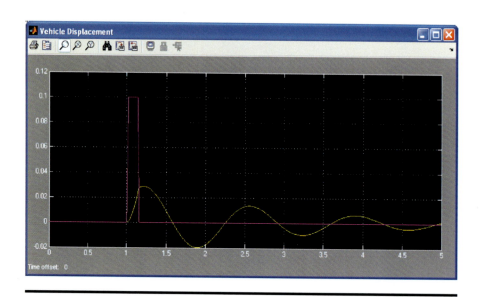

Figure 14.56   Vehicle's displacement. (From MATLAB. With permission.)

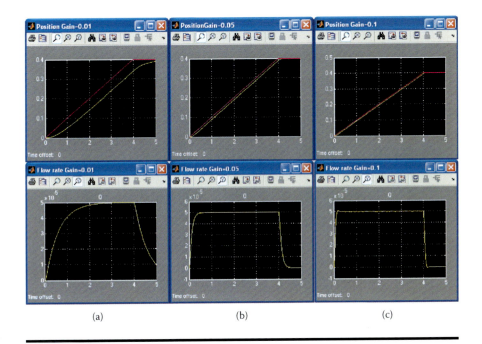

Figure 14.69   Simulation results for different controller gains (T = 0.01 second):
(a) gain = 0.01; (b) gain = 0.05; (c) gain = 0.1. (From MATLAB. With permission.)

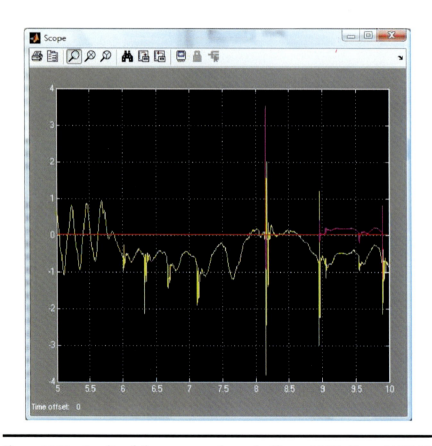

Figure 14.78 Results of the program shown in Figure 14.77. (From MATLAB. With permission.)